MRI ATLAS OF NORMAL ANATOMY

SERIES IN RADIOLOGY

1. J.O. Op den Orth: *The Standard Biphasic-contrast Examination of the Stomach and Duodenum.* Method, Results and Radiological Atlas. 1979 ISBN 90-247-2159-8

2. J.L. Sellink and R.E. Miller: *Radiology of the Small Bowel.* Modern Enteroclysis Technique and Atlas. 1982 ISBN 90-247-2460-0

3. R.E. Miller and J. Skucas: *The Radiological Examination of the Colon.* Practical Diagnosis. 1983 ISBN 90-247-2666-2

4. S. Forgacs: *Bones and Joints in Diabetes Mellitus.* 1982 ISBN 90-247-2395-7

5. Gy. Nemeth and H. Kuttig (eds.): *Isodose Atlas for Use in Radiotherapy.* 1981 ISBN 90-247-2476-7

6. J. Chermet: *Atlas of Phlebography of the Lower Limbs.* Including the Iliac Veins. 1982 ISBN 90-247-2525-9

7. B.K. Janevski: *Angiography of the Upper Extremity.* 1982 ISBN 90-247-2684-0

8. M.A.M. Feldberg: *Computed Tomography of the Retroperitoneum.* An Anatomical and Pathological Atlas with Emphasis on the Fascial Places. 1983 ISBN 0-89838-573-3

9. L.E.H. Lampmann, S.A. Duursma and J.H.J. Ruys: *CT Densitometry in Osteoporosis.* The Impact on Management of the Patient. 1984 ISBN 0-89838-633-0

10. J.J. Broerse and T.J. Macvittie: *Response of Different Species to Total Body Irradiation.* 1984 ISBN 0-89838-678-0

11. C.L'Hermine: *Radiology of Liver Circulation.* 1985 ISBN 0-89838-715-9

12. G. Maatman: *High-resolution Computed Tomography of the Paranasal Sinuses, Pharynx and Related Regions.* Impact of CT Identification on Diagnosis and Patient Management. 1986 ISBN 0-89838-802-3

13. C. Piets, A.L. Baert, G.L. Nijs and G. Wilms: *Computer Tomographic Imaging and Anatomic Correlation of the Human Brain.* A Comparative Atlas of Thin CT-scan Sections and Correlated Neuro-anatomic Preparations. 1987 ISBN 0-89838-811-2

14. J. Valk: *MRI of the Brain, Head, Neck and Spine.* A Teaching Atlas of Clinical Applications. 1987 ISBN 0-89838-957-7

15. J.L. Sellink: *X-Ray Differential Diagnosis in Small Bowel Disease.* A Practical Approach. 1988 ISBN 0-89838-351-X

16. Th. H.M. Falke (ed): *Essentials of Clinical MRI.* 1988 ISBN 0-89838-353-6

17. B.D. Fornage: *Endosonography.* 1989 ISBN 0-7923-0047-5

18. R. Chisin (ed.): *MRI/CT and Pathology in Head and Neck Tumours.* A Correlative Study. 1989 ISBN 0-7923-0227-3

19. G. Gozzetti, A. Mazziotti, L. Bolondi and L. Barbara (eds.): *Intraoperative Ultrasonography in Hepatobiliary and Pancreatic Surgery.* A Practical Guide. With Contributions by Y. Chapuis, J.-F. Gigot and P.-J. Kestens. 1989 ISBN 0-7923-0261-3

20. A.M.A. De Schepper and H.R.M. DeGryse: *Magnetic Resonance Imaging of Bone and Soft Tissue Tumours and Their Mimics.* A Clinical Atlas. With Contributions by F. De Belder, L. van den Houwe, F. Ramon, P. Parizel and N. Buyssens. 1989 ISBN 0-7923-0343-1

21. J.O. Barentsz, F.M.J. Debruyne and S.H.J. Ruijs: *Magnetic Resonance Imaging of Carcinoma of the Urinary Bladder.* With a Foreword by H. Hricak and R. Hohenfellner. 1990 ISBN 0-7923-0838-7

22. C. Depré, J.A. Melin, W.Wijns, R. Demeure, F. Hammer and J. Pringot: *Atlas of Cardiac MR Imaging with Anatomical Correlations.* Foreword by Alexander R. Margulis. 1991 ISBN 0-7923-0941-3

23. J.A. Castelijns, G.B. Snow and J. Valk: *MR Imaging of Laryngeal Cancer.* With Contributions by G.J. Gerritsen and W.N. Hanafee. 1991 ISBN 0-7923-1101-9

24. R.H. Mohiaddin and D.B. Longmore: *MRI Atlas of Normal Anatomy.* 1992 ISBN 0-7923-8974-3

MRI ATLAS OF NORMAL ANATOMY

Raad H. Mohiaddin, MD, MSc

Senior Research Fellow and Head of the Clinical Science Division,
Magnetic Resonance Unit,
Royal Brompton National Heart and Lung Hospital and the National Heart
and Lung Institute, London, United Kingdom

Donald B. Longmore, FRCS, FRCR

Director of the Magnetic Resonance Unit,
Royal Brompton National Heart and Lung Hospital and the National Heart
and Lung Institute, London, United Kingdom

SPRINGER-SCIENCE+BUSINESS MEDIA, B.V.

Library of Congress Cataloging-in-Publication Data

Mohiaddin, Raad H.
  MRI atlas of normal anatomy / Raad H. Mohiaddin, Donald B. Longmore
    p. cm.
  ISBN 978-94-010-5329-7   ISBN 978-94-011-2990-9 (eBook)
  DOI 10.1007/978-94-011-2990-9
    1. Human anatomy—Atlases. 2. Magnetic resonance imaging—Atlases. I. Longmore, Donald. II.
Title.
    [DNLM: 1. Anatomy—atlases. 2. Magnetic Resonance Imaging—atlases. W1 SE719 v.64 / QS 17
M697m]
  QM25.M55   1992
  611′.0022′2—dc20
  DNLM/DLC
  for Library of Congress   92-13450
                                                                                       CIP

CONTENTS

INTRODUCTION

Magnetic resonance (MR) is a safe, non-invasive technique which can be used to produce high resolution thin tomographic slices in any chosen plane or true three-dimensional blocks of information. MR is usually used to image hydrogen nuclei (proton magnetic resonance imaging), which are the most abundant nuclei in the body; however, it is possible to image other nuclei such as sodium. The illustrations in this book are therefore soft-tissue images based on the water content rather than the familiar X-ray shadowgram of mainly hard tissues. The contrast between different tissues is excellent and the parameters displayed on the images vary according to the biochemical environment. Several image-acquisition techniques are used routinely, most of them requiring up to 256 acquisitions to produce an image. Each acquisition contributes to the whole image rather than building it up line by line. The image acquisitions can be repeated to average the signal and to reduce image degradation due to random radio noise arising from the patient and Johnson noise from the aparatus. The quality of images is therefore a trade-off between time and resolution. In organs which move with the heartbeat or respiration, it is necessary to gate the acquisitions to the ECG or to respiration or both thus, adding further to the acquisition time. All existing techniques are sensitive to movement. In order to make MR more cost-effective and to eliminate movement artefact, several methods have been developed which can speed up the acquisition of data and these will eventually become routine. These include Echo Planar Imaging (EPI), which acquires all the data needed to make an image in one excitation taking a few milliseconds, and various sub-second imaging techniques which can be used to produce a movie of the heart in 15 seconds (well within one breath hold). Other non-imaging techniques exist, including MR spectroscopy, and the Superconducting Quantum Interference Device (SQUID) can be used to obtain information non-invasively about biochemistry.

Despite the relatively high cost of contemporary equipment, magnetic resonance has become the method of choice for studying the central nervous system, the vertebral column and many joints. The main manufacturers are still amortising the high costs of developing the first generation of machines often working at high field strengths which introduce signal to movement sensitivity and chemical shift artefacts. It has not, however, come into routine use in the abdomen, where respiratory motion and bowel movements frequently degrade the image. It has also not yet gained general acceptance in the cardiovascular system, although cardiac gating can often overcome the problems of cardiac movement to produce excellent cardiac images. The real strength of MR, however, lies in its ability to produce functional information, particularly the measurement of velocity and acceleration of blood in the cardiac chambers and peripheral vessels.

Cardiovascular imaging, however, is slow because each sequence of events in the cardiac cycle must be studied at the same interval after the ECG trigger from the R wave. Many of the cardiac images in this book are 10 mm slices which took between 4 and 8 minutes to acquire. These images have been produced over a period of six years and it is now possible to acquire 16 or more images of the heart within 15 second, allowing a complete cardiac movie to be made during one breath hold. Individual images produced in this way appear to be of much higher resolution than those in this book because the eye can integrate moving images and reject noise. Even such rapid images of the heart are degraded by inter-beat variations caused by ectopic beats, atrial fibrillation or sinus arrhythmia.

Magnetic resonance is not, however, a universal panacea for imaging in the body. Up to 4%

of the normal population are claustrophobic, though most sick patients will tolerate an examination. Implanted pacemakers are an absolute contraindication and although all metallic objects and implants are a theoretical contraindication, in practice metallic objects in the chest, such as sternal wires, metallic clips and prosthetic valves, are not ferromagnetic, and although in some sequences they induce an area of signal loss around them, MR examinations are still satisfactory.

The work in this book was done with a low- to mid-field machine operating at 0.26 and 0.5 T. It is generally thought that for still parts of the body and the CNS, a higher field strength is desirable. At higher fields, however, signal to movement artefacts become a serious problem and it is likely that machines will be developed for specific purposes rather than for the production of images of all parts of the body with a general-purpose machine such as was used for the production of this book. Specialized units will be able to show images of their regions of interest which will frequently appear better than those in this production. However, it is important to remember that image quality is subjective and that the chemical shift artefact which puts black lines round the edges of structures in higher-field machines may obscure detail and the contrast obtained using some sequences at higher fields may not be as good.

The purpose of this book is to provide the student and the radiologist with a reference which can be used to identify the major structures in the body, bearing in mind that, in each region, a more detailed high-resolution study can usually be obtained.

DIRECTIONS FOR USE

Imaging parameters: A Picker International Vista MR 2055 scanner operating at 0.5 Tesla was used with a surface receiver coil and a spin echo sequence (echo time 40 ms). The slice thickness was between 5 mm and 10 mm. The field of view was 30 cm with a resolution of 256 pixels in the frequency encoding direction and 256 pixels in the phase encoding direction. Thoracic and cardiac images were acquired using an electrocardiographic gating system.

Image presentation: Because images were acquired in multiple places in the same subject, we have provided a small orientation picture showing the imaging plane of each section. Each MR anatomical section is also accompanied by a line drawing featuring the essential characteristic elements of this section.

Nomenclature: The choice of nomenclature is always difficult. To make the terms as accessible as possible, all labels have been reviewed for their appropriateness to the English-speaking clinician and strict adherence to Nomina Anatomica has been abandoned in favour of practical terminology.

PREFACE

The production of this atlas has been a joint effort and the following people are effectively co-authors:

Dr Makoto Amanuma MD, Department of Radiology, Saitama Medical School, Iruma, Japan

Dr Carla Manzara MD, Department of Cardiology, University of Rome

Dr David Firmin PhD, Magnetic Resonance Unit, Royal Brompton National Heart and Lung Hospital and The National Heart and Lung Institute, London, UK

Karl Lotey MSc, Royal Brompton National Heart and Lung Hospital, London, UK

Sharmin Obeyesekara (3rd year medical student), Royal Free Hospital, London, UK

John MacMillan (3rd year medical student), Royal Free Hospital, London, UK

We would also like to thank Miss Victoria Harding for being the subject of most of the studies.

Raad H. Mohiaddin
Donald B. Longmore

HEAD

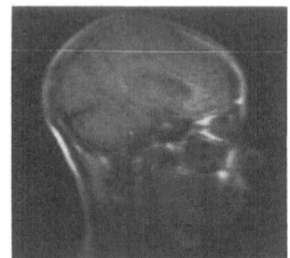

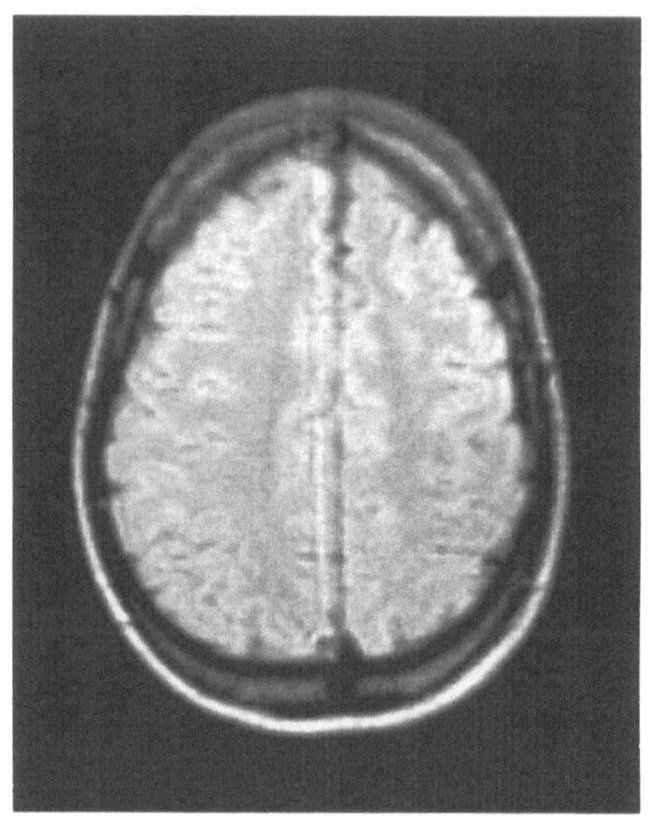

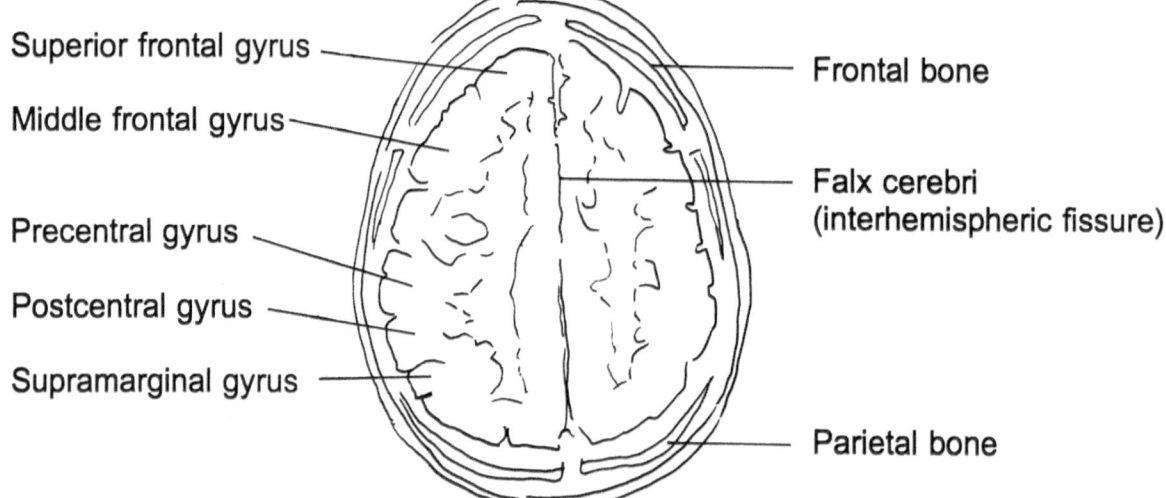

Superior frontal gyrus ——————————— Frontal bone

Middle frontal gyrus ——————

Precentral gyrus ——————————— Falx cerebri (interhemispheric fissure)

Postcentral gyrus ——————

Supramarginal gyrus ——————

Parietal bone

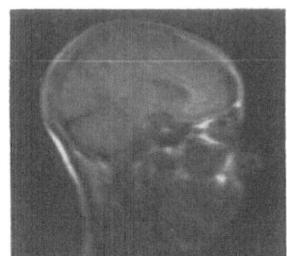

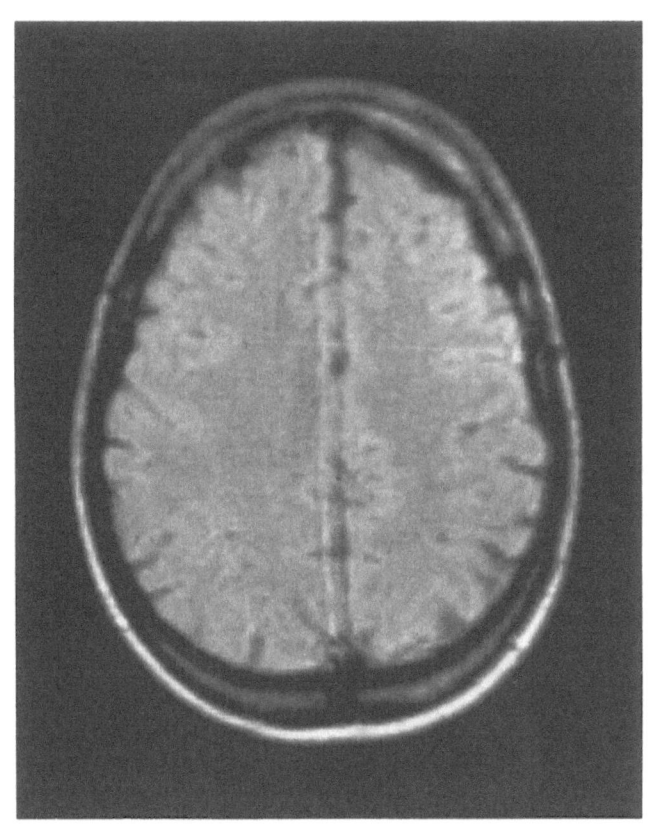

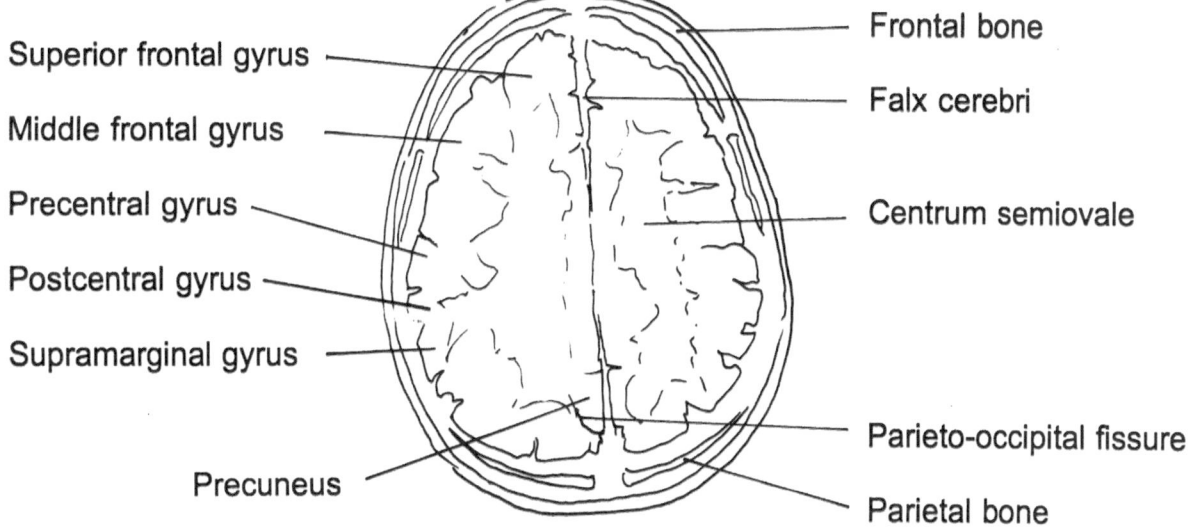

Superior frontal gyrus

Middle frontal gyrus

Precentral gyrus

Postcentral gyrus

Supramarginal gyrus

Precuneus

Frontal bone

Falx cerebri

Centrum semiovale

Parieto-occipital fissure

Parietal bone

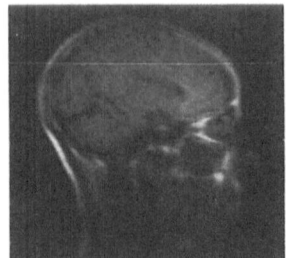

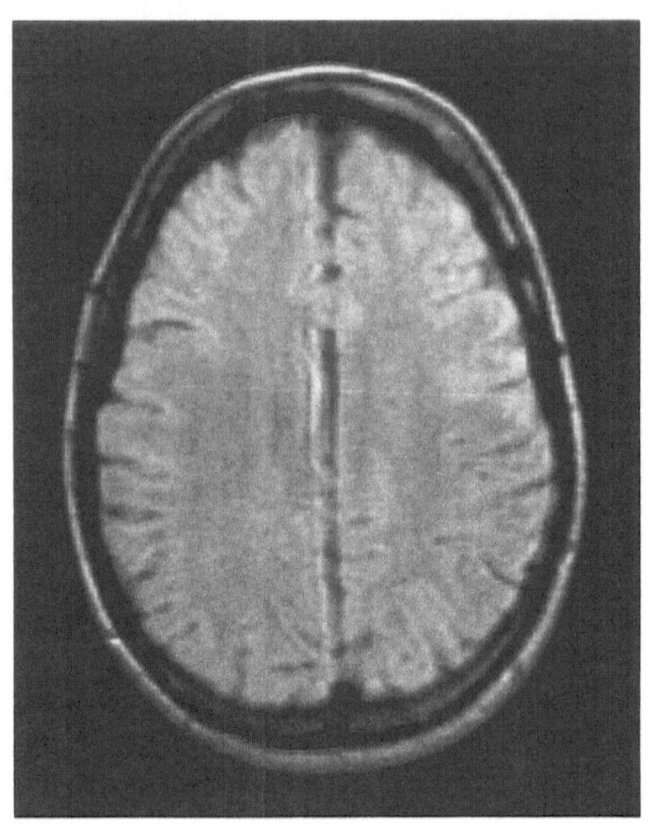

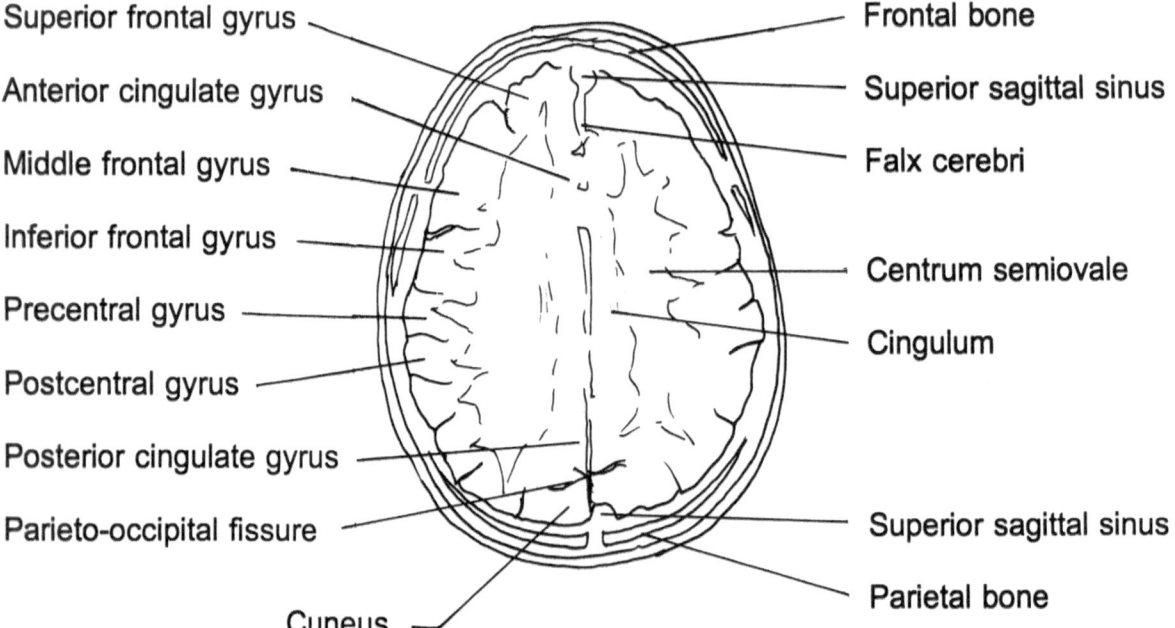

Superior frontal gyrus	Frontal bone
Anterior cingulate gyrus	Superior sagittal sinus
Middle frontal gyrus	Falx cerebri
Inferior frontal gyrus	
Precentral gyrus	Centrum semiovale
Postcentral gyrus	Cingulum
Posterior cingulate gyrus	
Parieto-occipital fissure	Superior sagittal sinus
	Parietal bone
Cuneus	

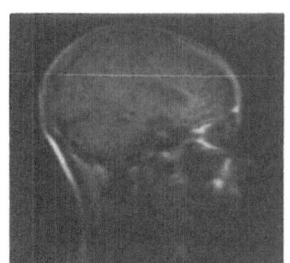

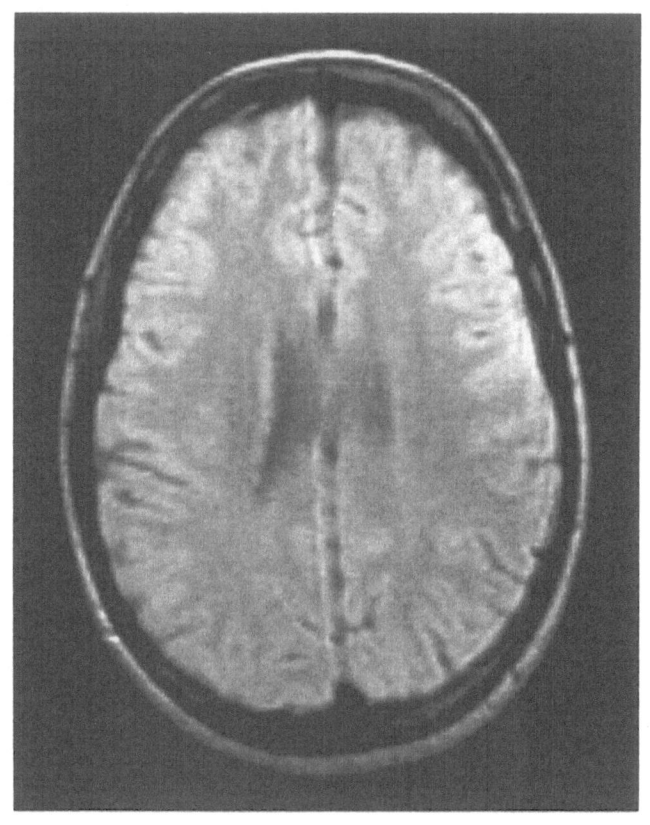

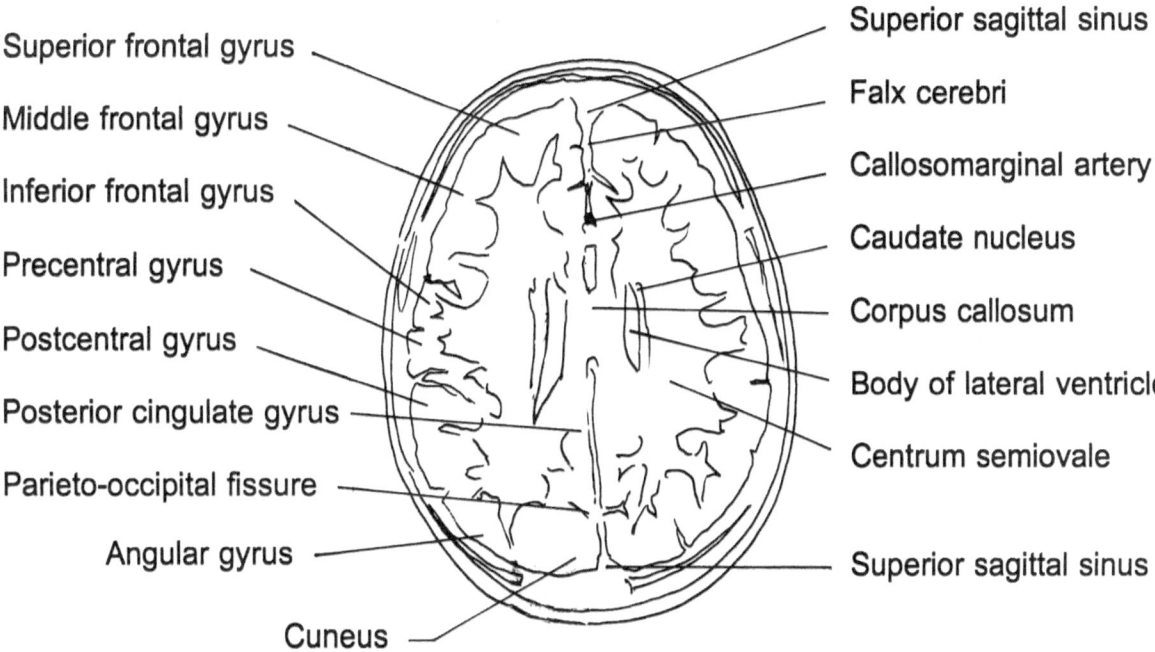

Superior frontal gyrus

Middle frontal gyrus

Inferior frontal gyrus

Precentral gyrus

Postcentral gyrus

Posterior cingulate gyrus

Parieto-occipital fissure

Angular gyrus

Cuneus

Superior sagittal sinus

Falx cerebri

Callosomarginal artery

Caudate nucleus

Corpus callosum

Body of lateral ventricle

Centrum semiovale

Superior sagittal sinus

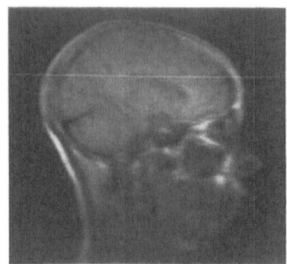

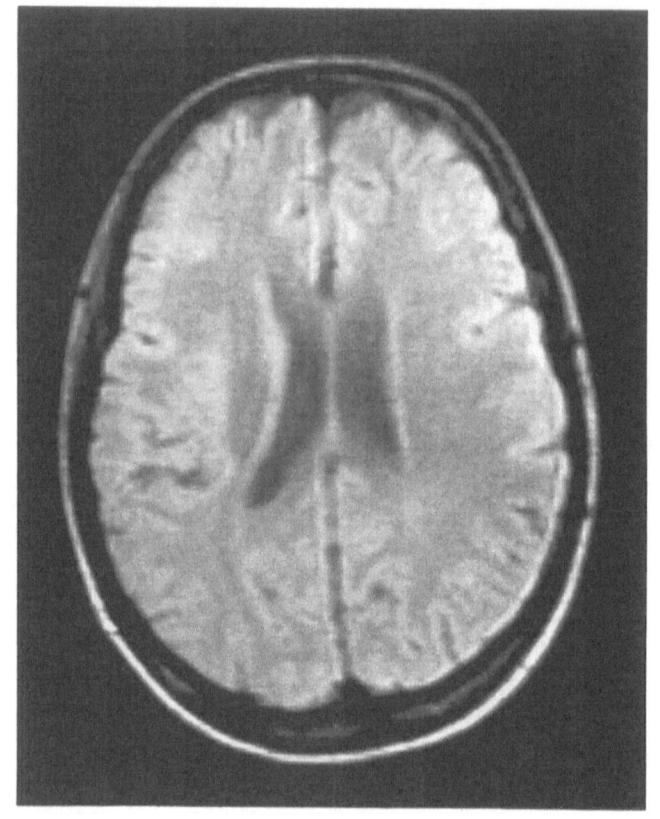

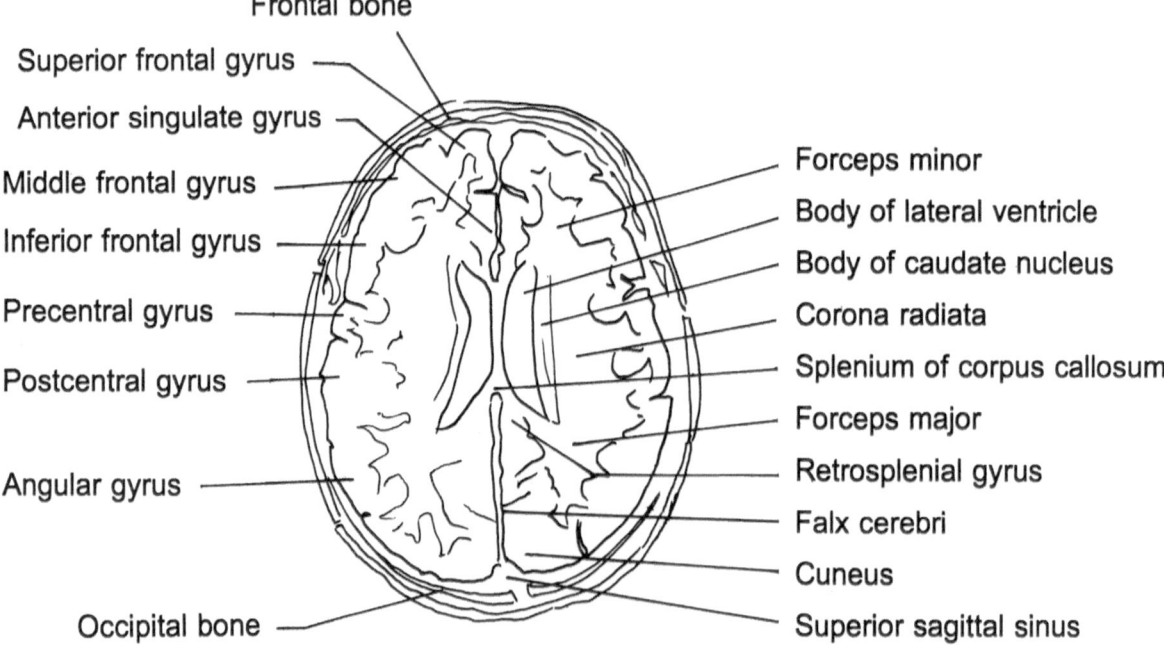

Frontal bone

Superior frontal gyrus

Anterior singulate gyrus

Middle frontal gyrus

Inferior frontal gyrus

Precentral gyrus

Postcentral gyrus

Angular gyrus

Occipital bone

Forceps minor

Body of lateral ventricle

Body of caudate nucleus

Corona radiata

Splenium of corpus callosum

Forceps major

Retrosplenial gyrus

Falx cerebri

Cuneus

Superior sagittal sinus

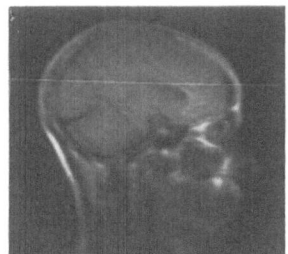

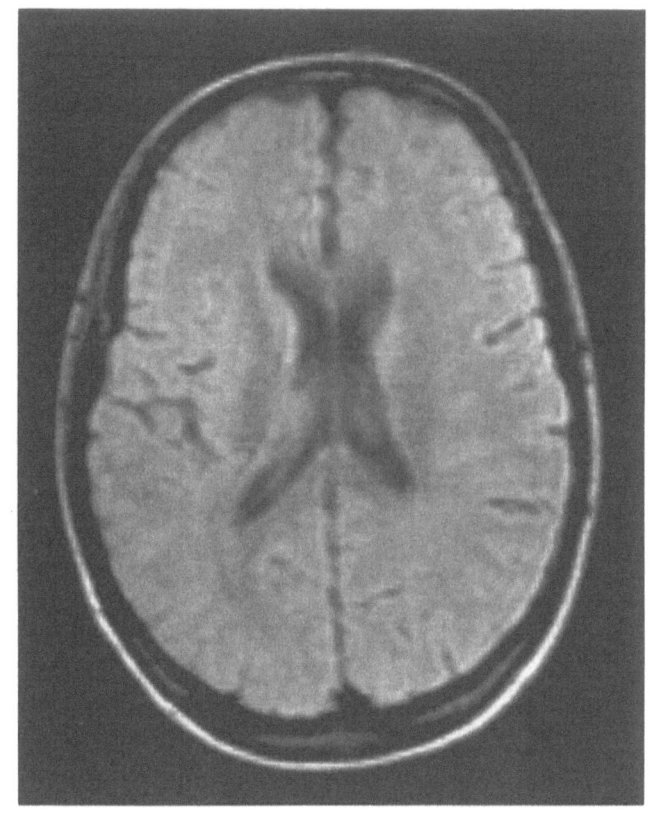

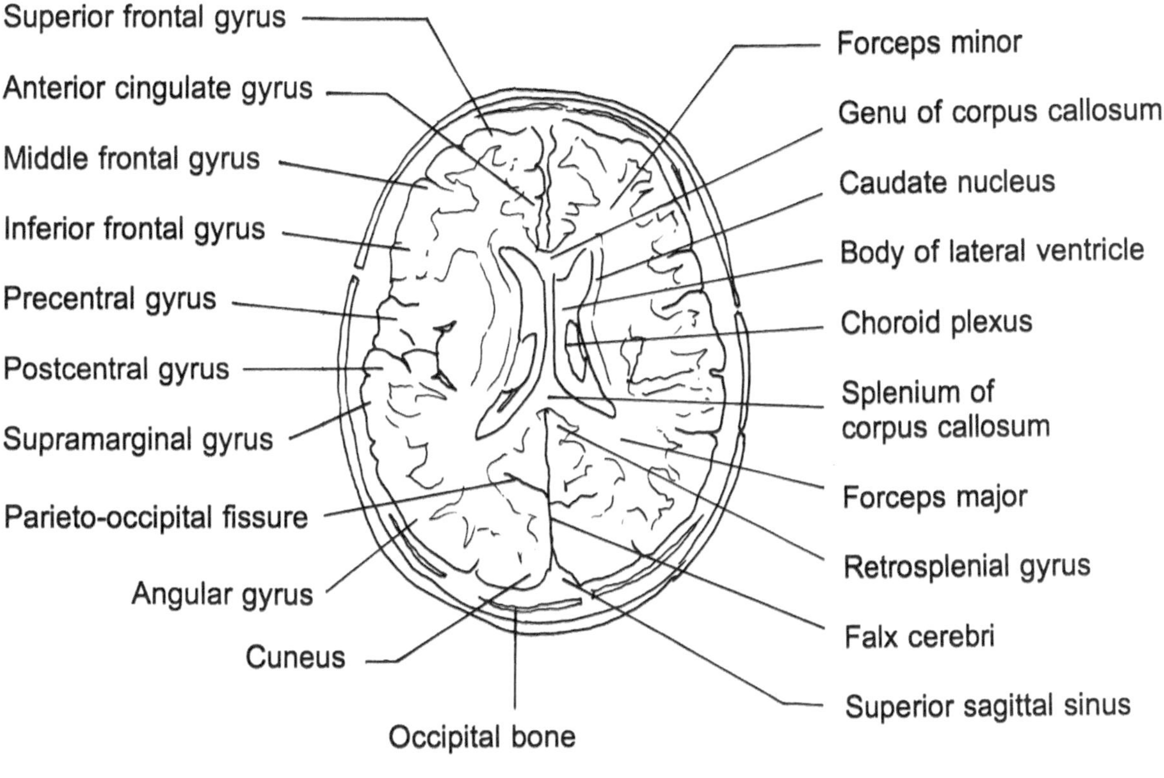

Superior frontal gyrus

Anterior cingulate gyrus

Middle frontal gyrus

Inferior frontal gyrus

Precentral gyrus

Postcentral gyrus

Supramarginal gyrus

Parieto-occipital fissure

Angular gyrus

Cuneus

Occipital bone

Forceps minor

Genu of corpus callosum

Caudate nucleus

Body of lateral ventricle

Choroid plexus

Splenium of
corpus callosum

Forceps major

Retrosplenial gyrus

Falx cerebri

Superior sagittal sinus

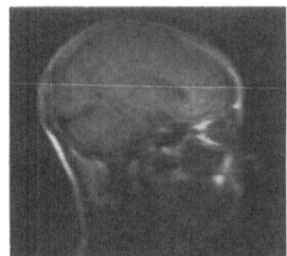

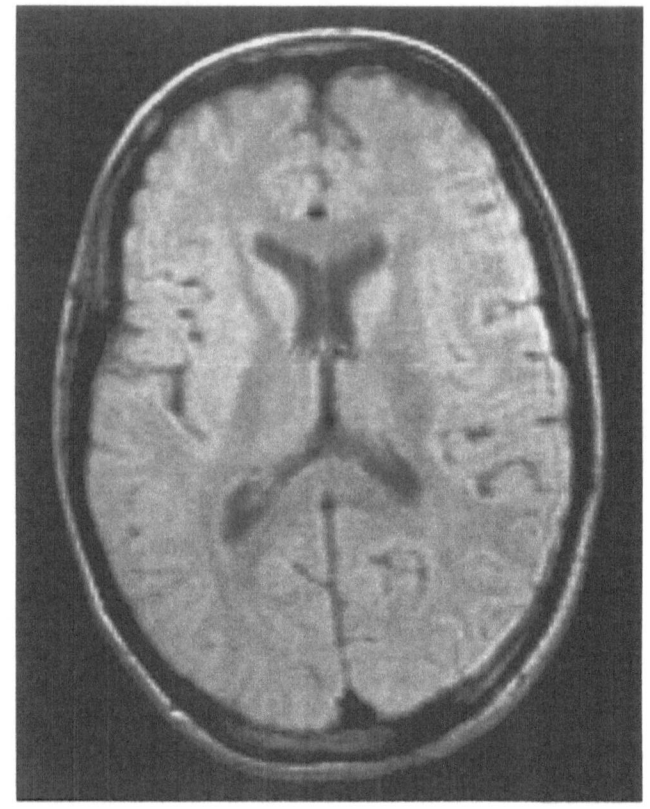

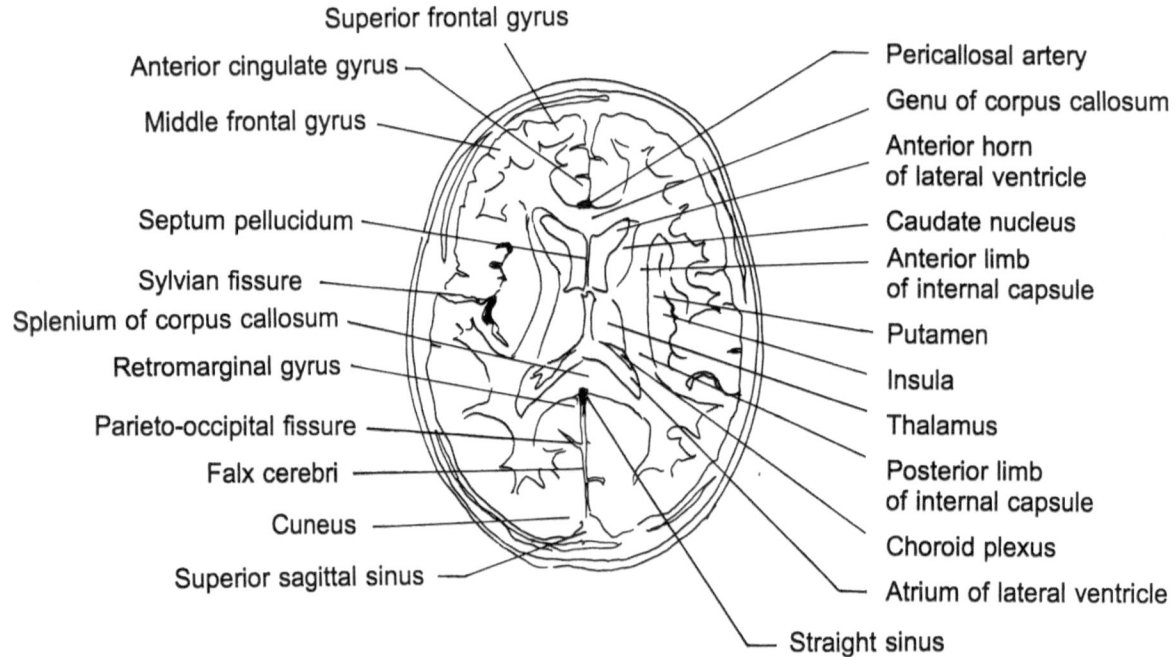

Superior frontal gyrus

Anterior cingulate gyrus

Middle frontal gyrus

Septum pellucidum

Sylvian fissure

Splenium of corpus callosum

Retromarginal gyrus

Parieto-occipital fissure

Falx cerebri

Cuneus

Superior sagittal sinus

Pericallosal artery

Genu of corpus callosum

Anterior horn
of lateral ventricle

Caudate nucleus

Anterior limb
of internal capsule

Putamen

Insula

Thalamus

Posterior limb
of internal capsule

Choroid plexus

Atrium of lateral ventricle

Straight sinus

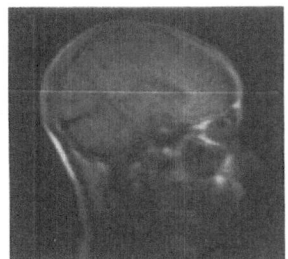

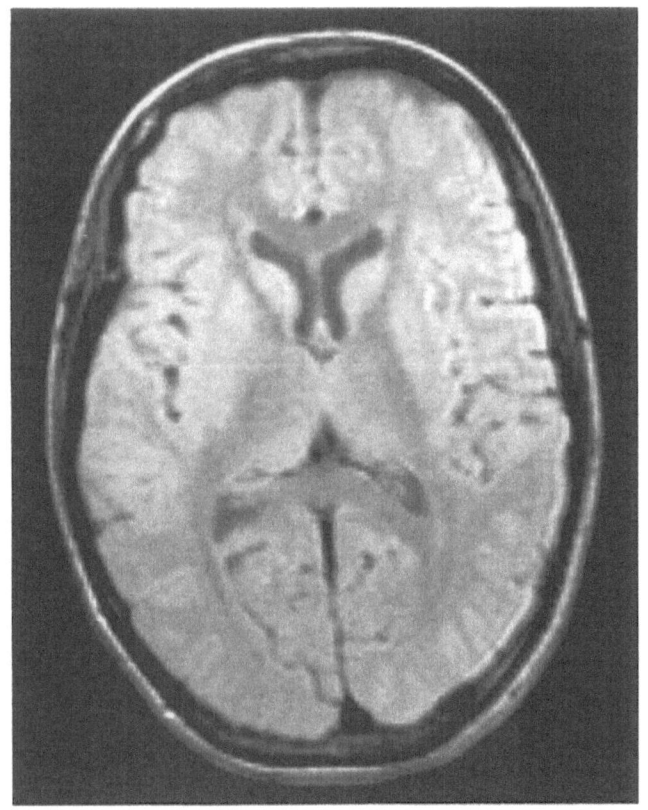

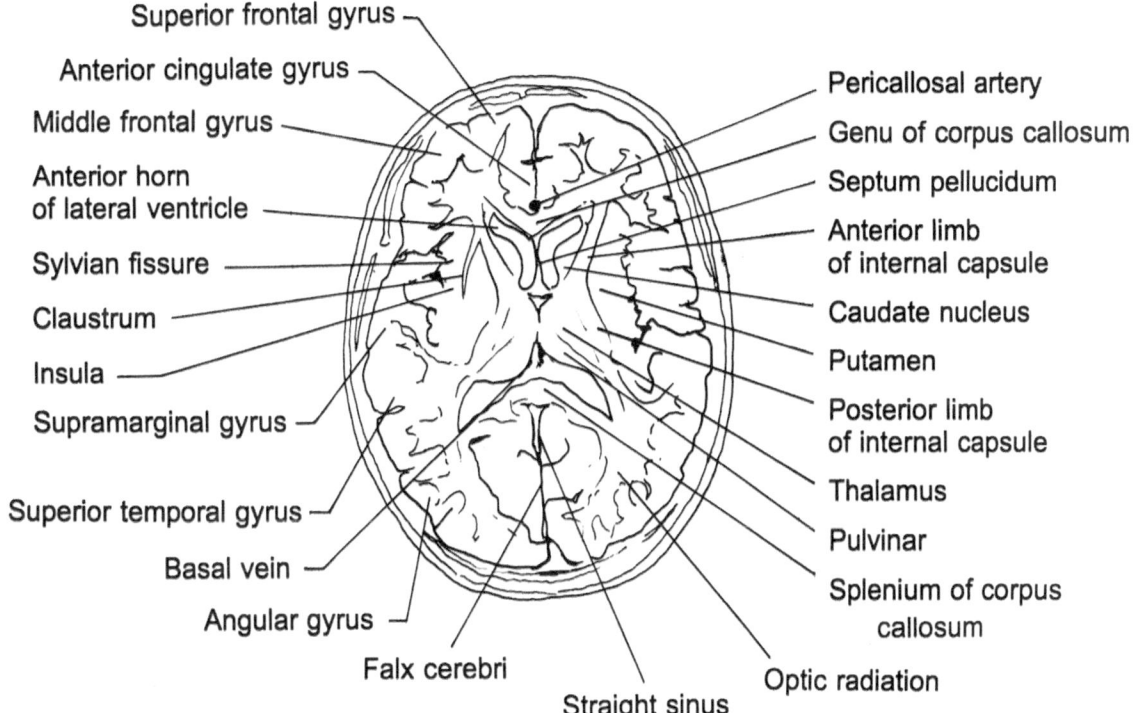

Superior frontal gyrus

Anterior cingulate gyrus

Middle frontal gyrus

Anterior horn
of lateral ventricle

Sylvian fissure

Claustrum

Insula

Supramarginal gyrus

Superior temporal gyrus

Basal vein

Angular gyrus

Falx cerebri

Straight sinus

Pericallosal artery

Genu of corpus callosum

Septum pellucidum

Anterior limb
of internal capsule

Caudate nucleus

Putamen

Posterior limb
of internal capsule

Thalamus

Pulvinar

Splenium of corpus
callosum

Optic radiation

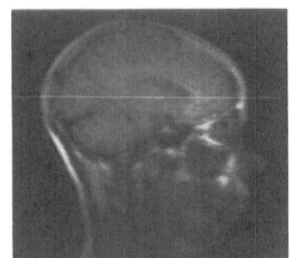

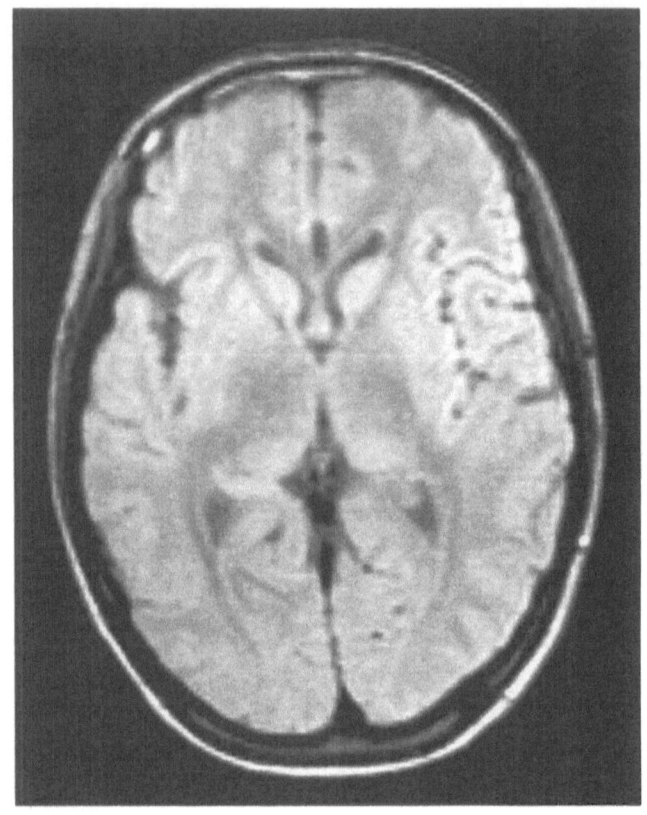

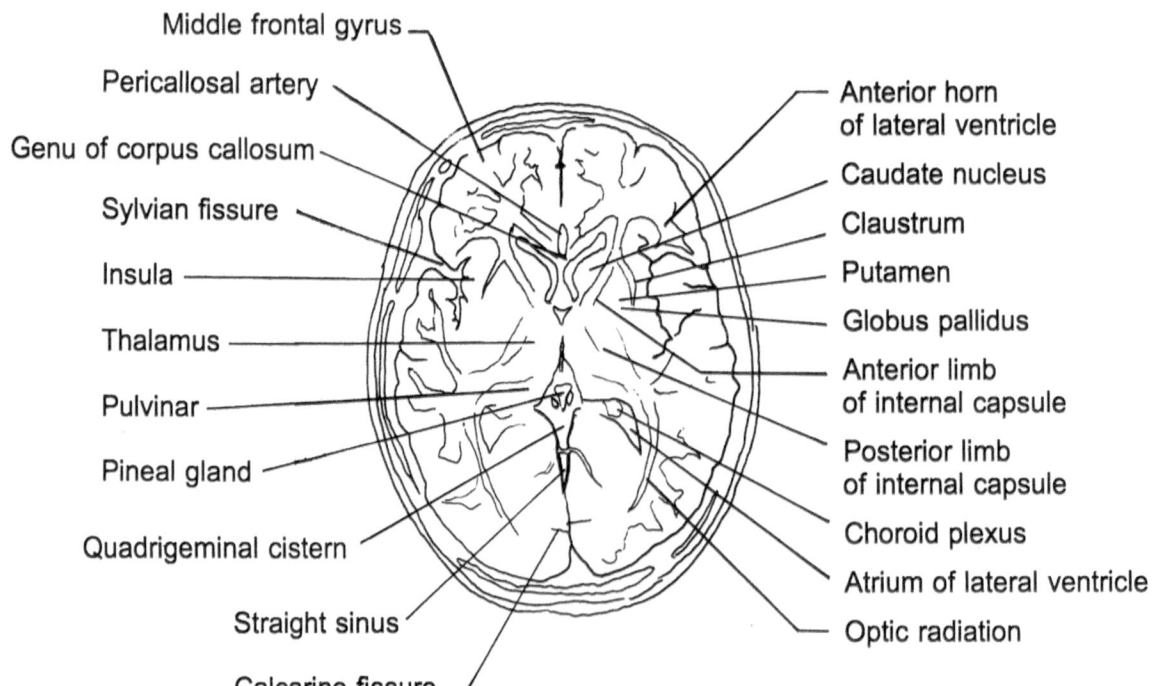

Middle frontal gyrus

Pericallosal artery

Genu of corpus callosum

Sylvian fissure

Insula

Thalamus

Pulvinar

Pineal gland

Quadrigeminal cistern

Straight sinus

Calcarine fissure

Anterior horn
of lateral ventricle

Caudate nucleus

Claustrum

Putamen

Globus pallidus

Anterior limb
of internal capsule

Posterior limb
of internal capsule

Choroid plexus

Atrium of lateral ventricle

Optic radiation

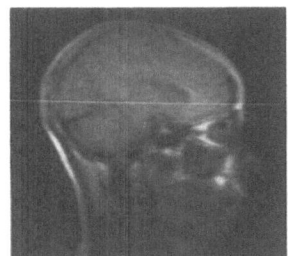

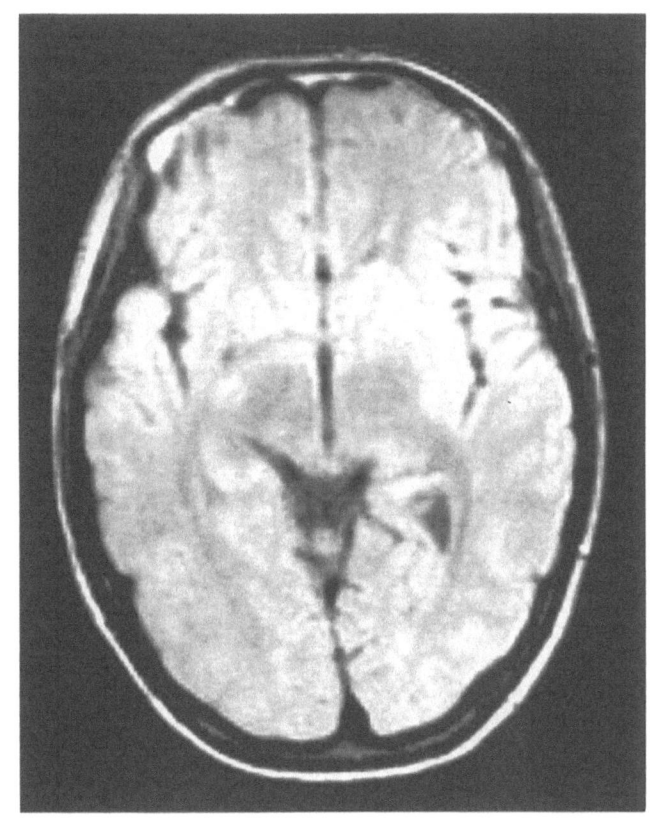

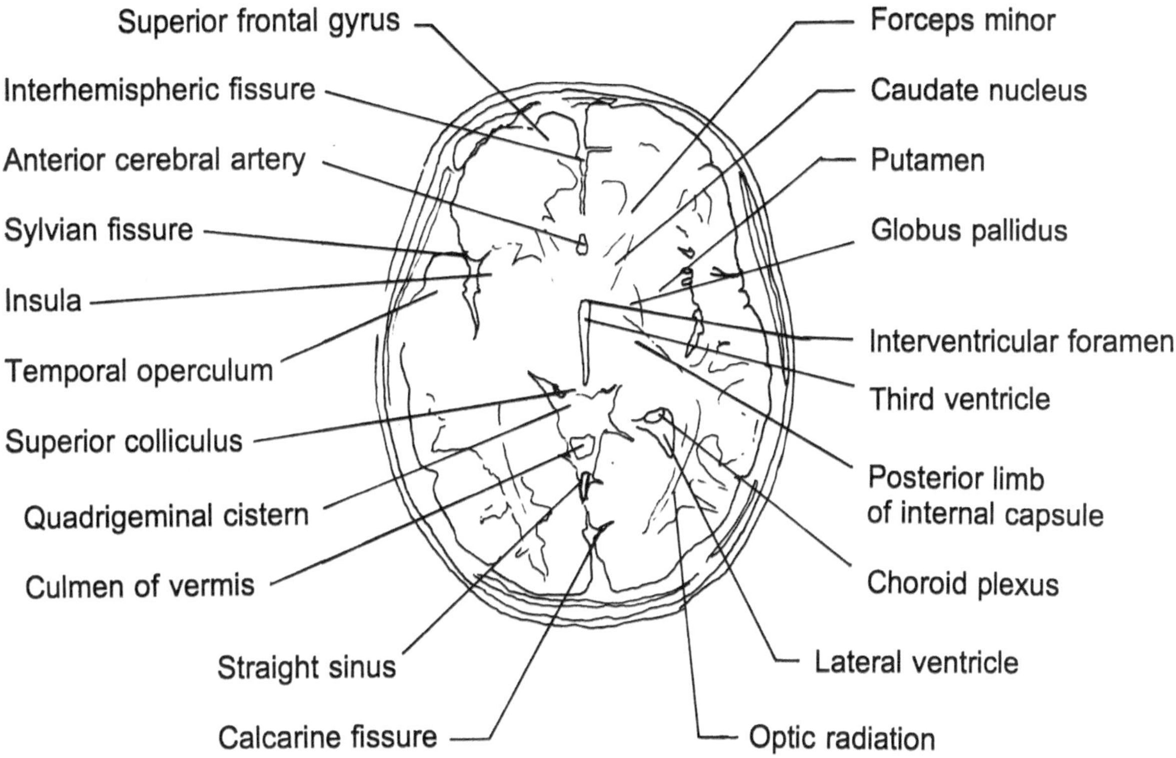

Superior frontal gyrus	Forceps minor
Interhemispheric fissure	Caudate nucleus
Anterior cerebral artery	Putamen
Sylvian fissure	Globus pallidus
Insula	Interventricular foramen
Temporal operculum	Third ventricle
Superior colliculus	Posterior limb of internal capsule
Quadrigeminal cistern	Choroid plexus
Culmen of vermis	Lateral ventricle
Straight sinus	Optic radiation
Calcarine fissure	

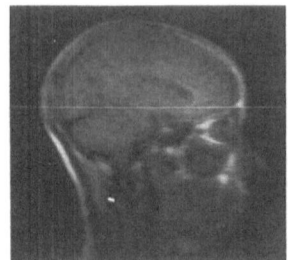

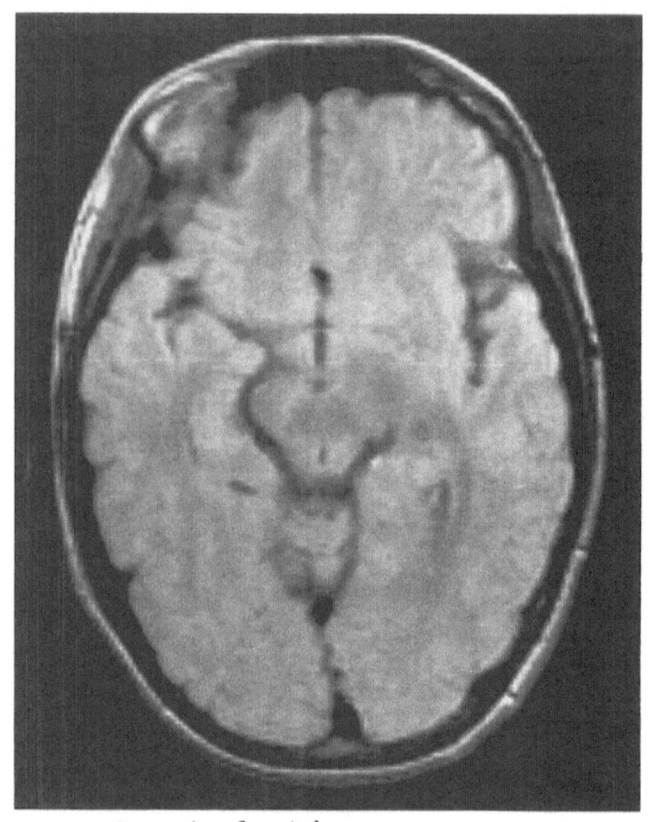

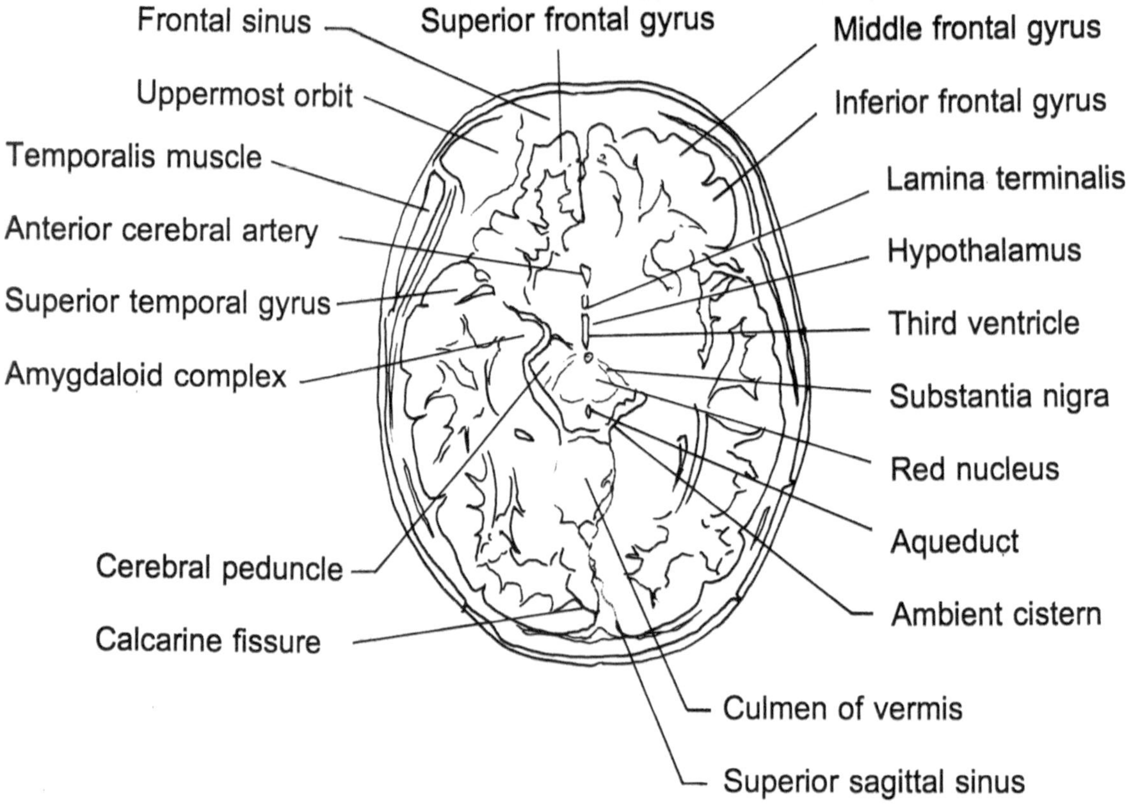

Frontal sinus	Superior frontal gyrus	Middle frontal gyrus
Uppermost orbit		Inferior frontal gyrus
Temporalis muscle		Lamina terminalis
Anterior cerebral artery		Hypothalamus
Superior temporal gyrus		Third ventricle
Amygdaloid complex		Substantia nigra
		Red nucleus
		Aqueduct
Cerebral peduncle		Ambient cistern
Calcarine fissure		
	Culmen of vermis	
	Superior sagittal sinus	

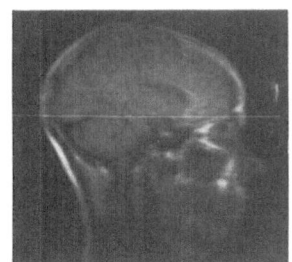

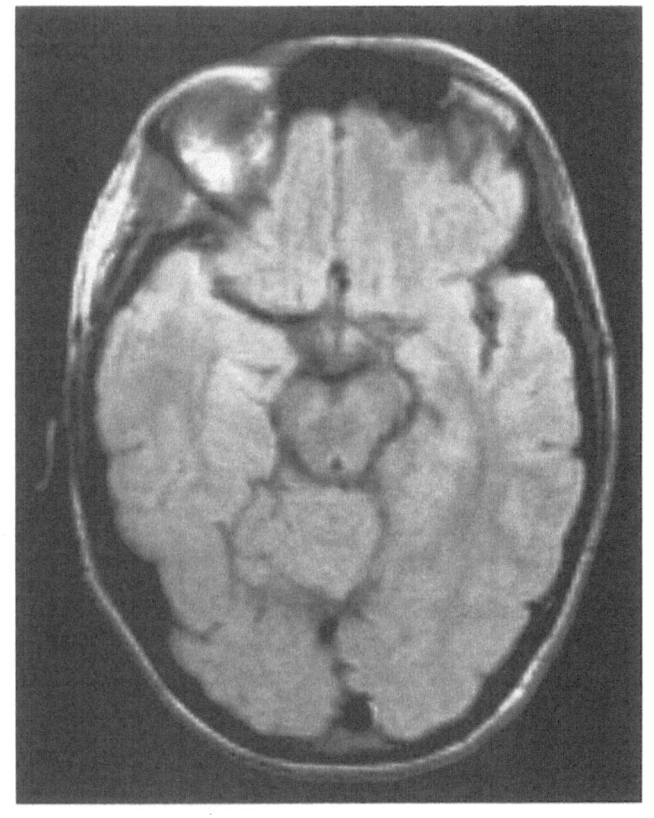

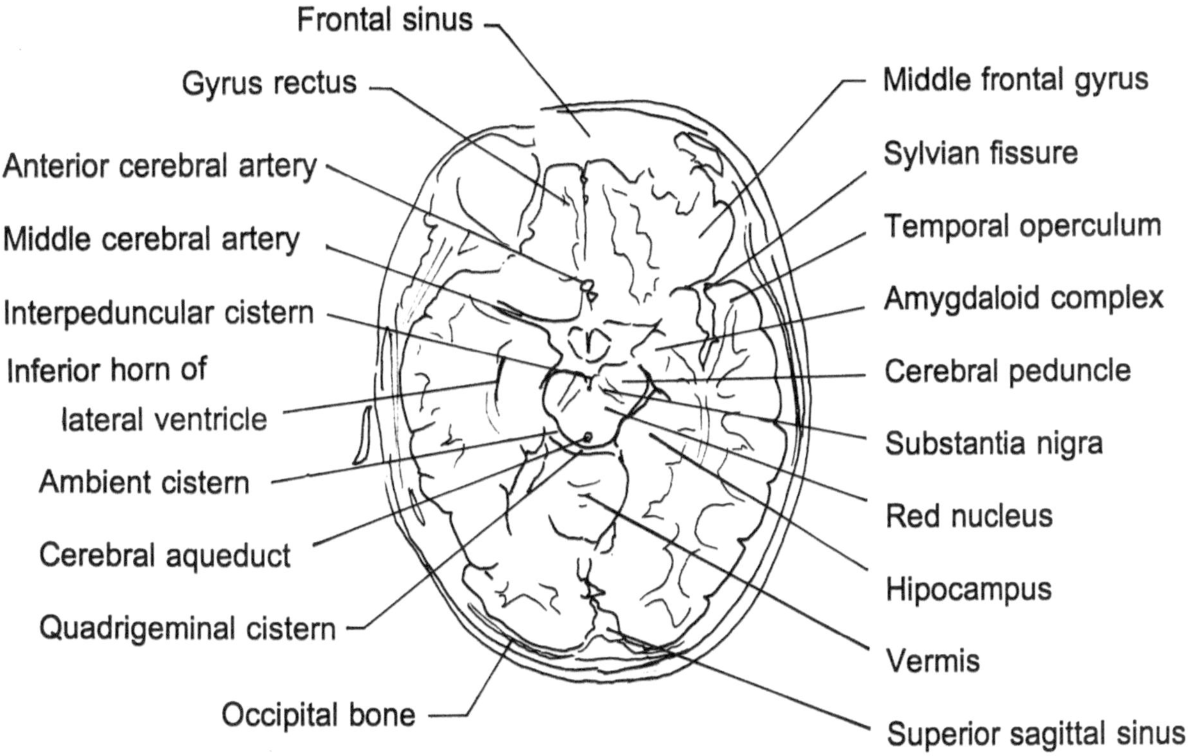

Frontal sinus

Gyrus rectus

Anterior cerebral artery

Middle cerebral artery

Interpeduncular cistern

Inferior horn of
lateral ventricle

Ambient cistern

Cerebral aqueduct

Quadrigeminal cistern

Occipital bone

Middle frontal gyrus

Sylvian fissure

Temporal operculum

Amygdaloid complex

Cerebral peduncle

Substantia nigra

Red nucleus

Hipocampus

Vermis

Superior sagittal sinus

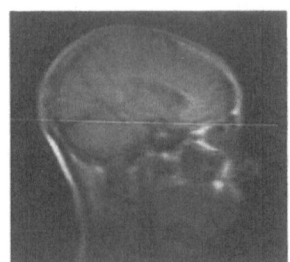

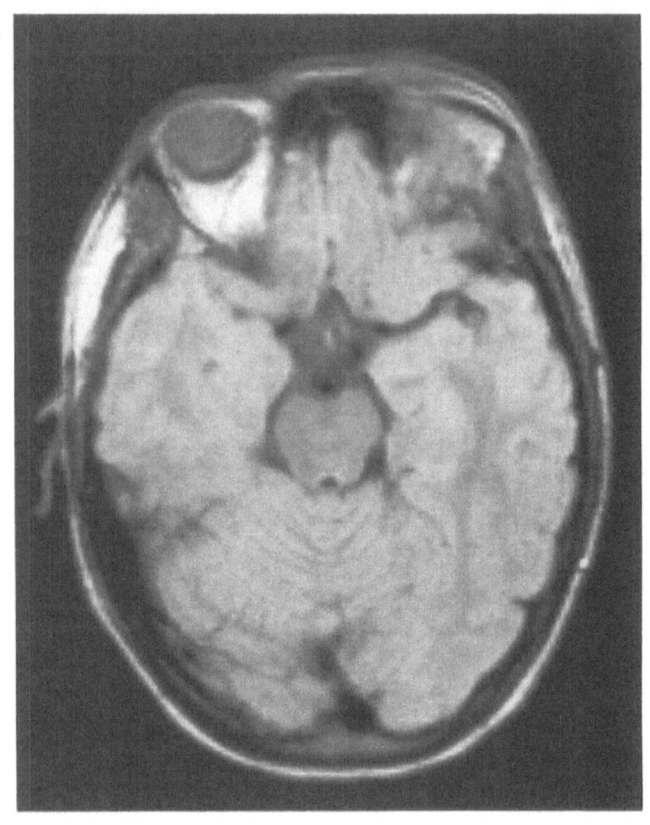

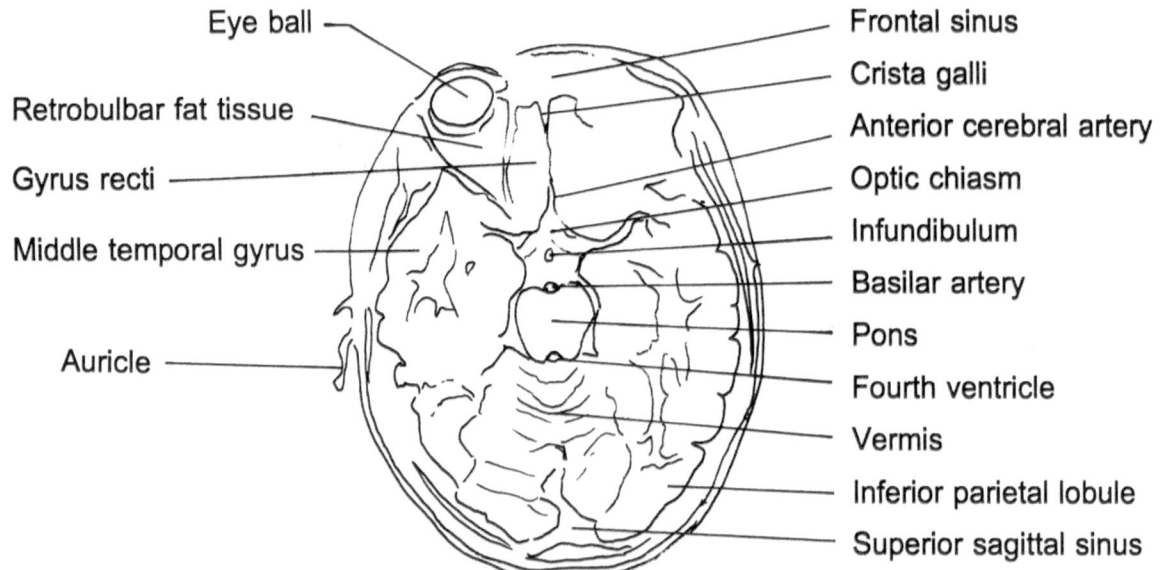

Eye ball	Frontal sinus
Retrobulbar fat tissue	Crista galli
Gyrus recti	Anterior cerebral artery
Middle temporal gyrus	Optic chiasm
Auricle	Infundibulum
	Basilar artery
	Pons
	Fourth ventricle
	Vermis
	Inferior parietal lobule
	Superior sagittal sinus

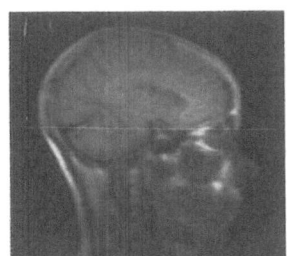

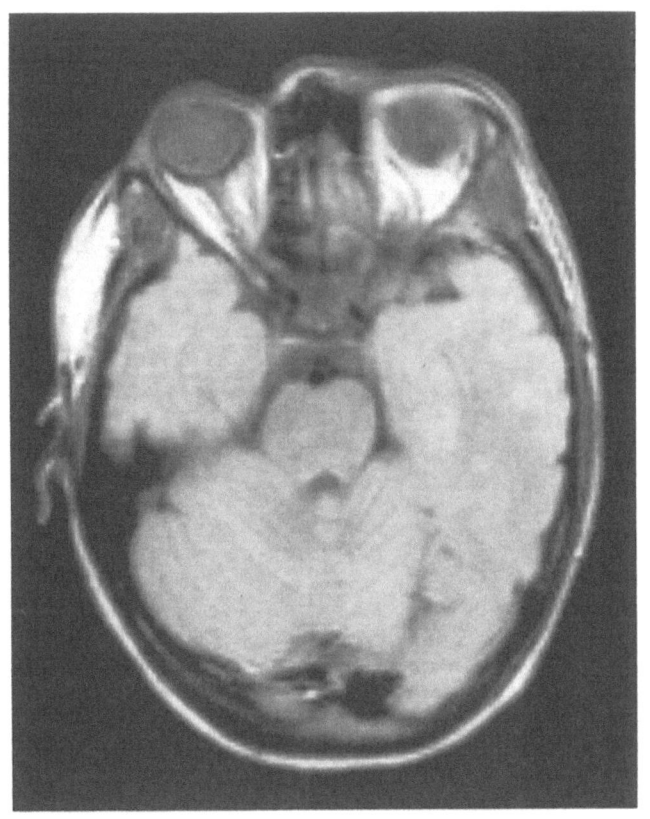

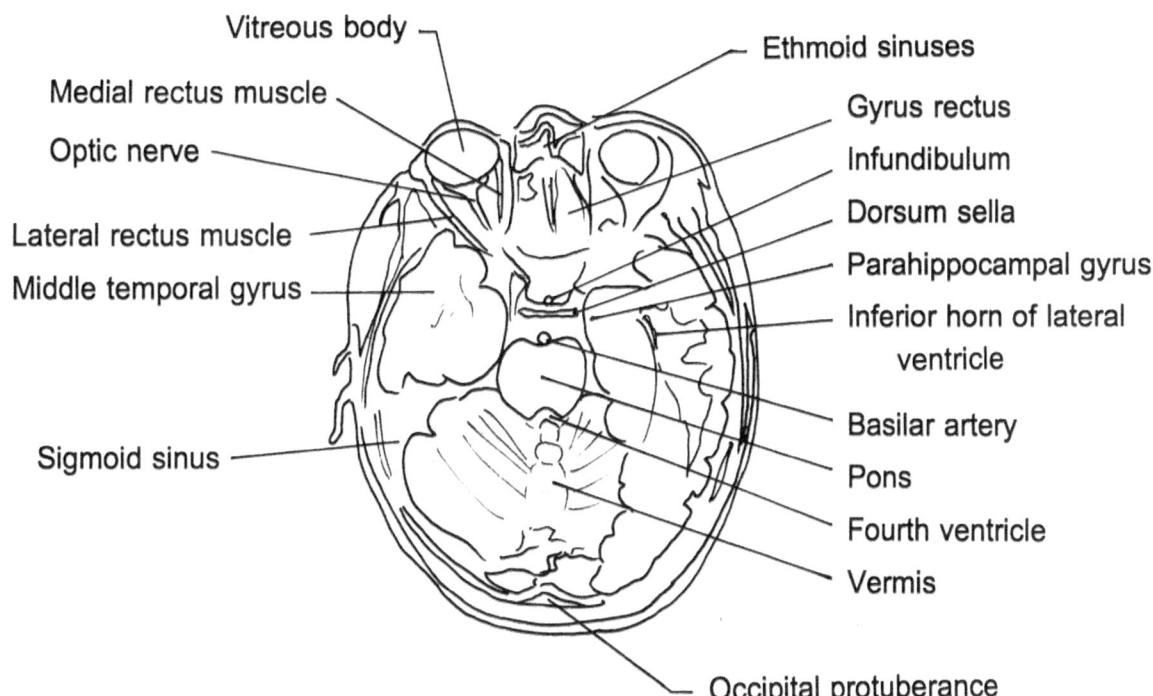

Vitreous body

Medial rectus muscle

Optic nerve

Lateral rectus muscle

Middle temporal gyrus

Sigmoid sinus

Ethmoid sinuses

Gyrus rectus

Infundibulum

Dorsum sella

Parahippocampal gyrus

Inferior horn of lateral ventricle

Basilar artery

Pons

Fourth ventricle

Vermis

Occipital protuberance

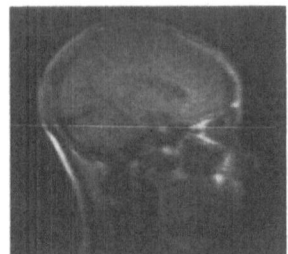

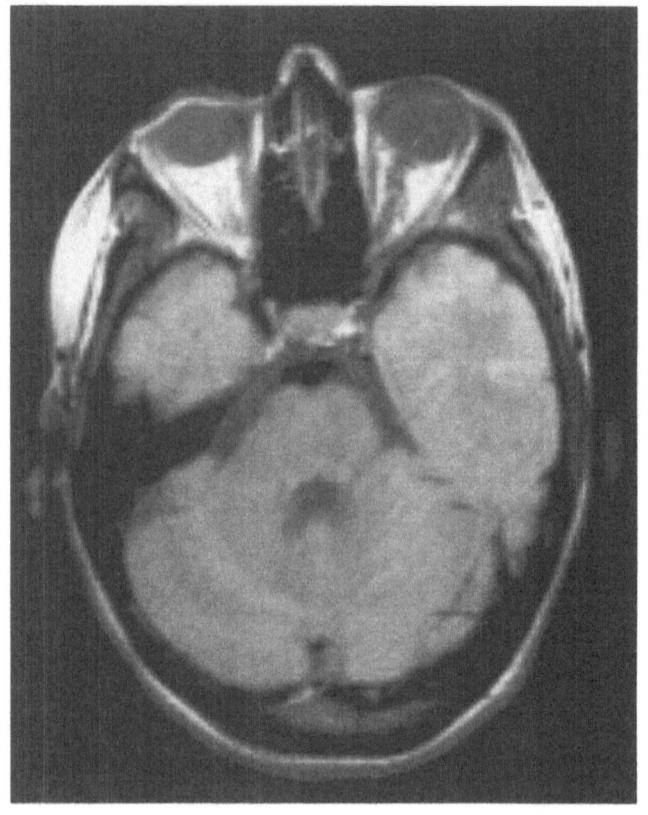

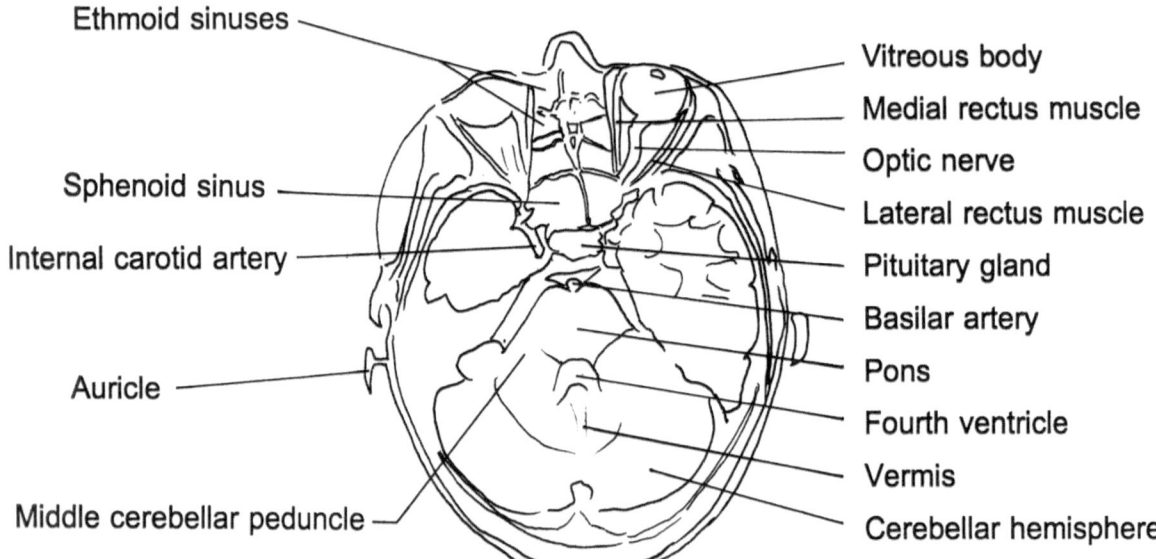

Ethmoid sinuses

Sphenoid sinus

Internal carotid artery

Auricle

Middle cerebellar peduncle

Vitreous body

Medial rectus muscle

Optic nerve

Lateral rectus muscle

Pituitary gland

Basilar artery

Pons

Fourth ventricle

Vermis

Cerebellar hemisphere

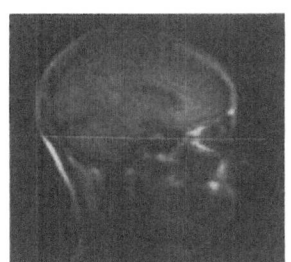

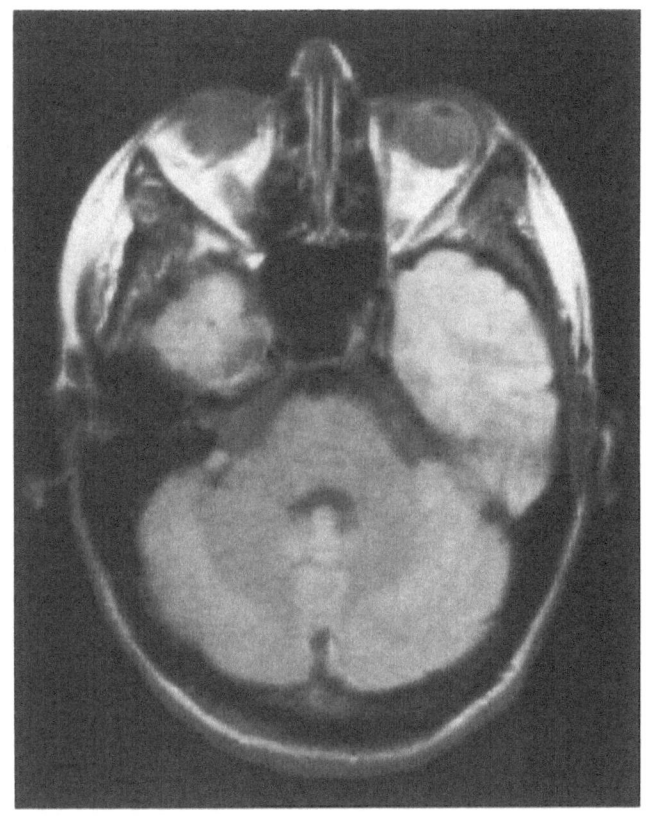

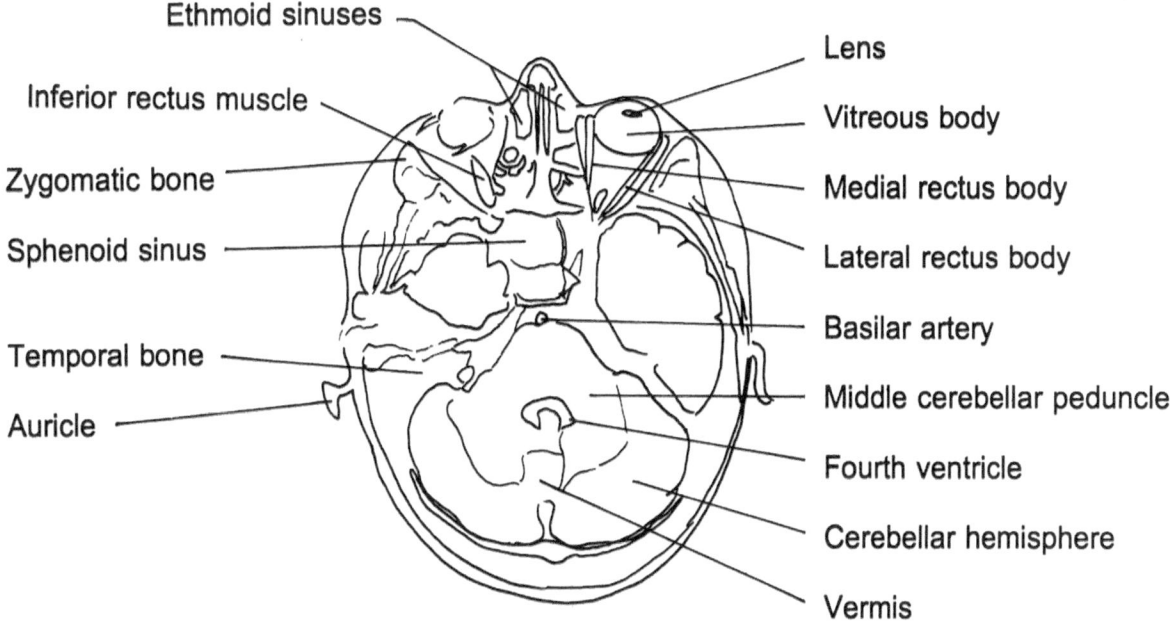

Ethmoid sinuses

Lens

Inferior rectus muscle

Vitreous body

Zygomatic bone

Medial rectus body

Sphenoid sinus

Lateral rectus body

Basilar artery

Temporal bone

Middle cerebellar peduncle

Auricle

Fourth ventricle

Cerebellar hemisphere

Vermis

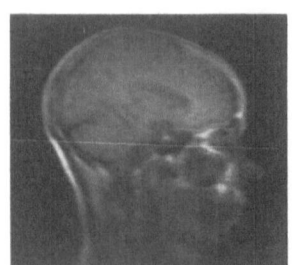

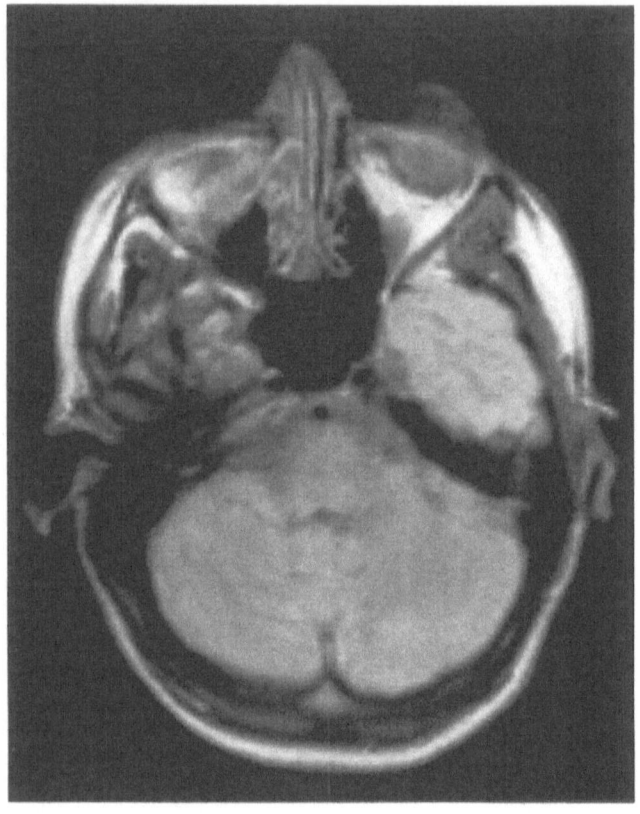

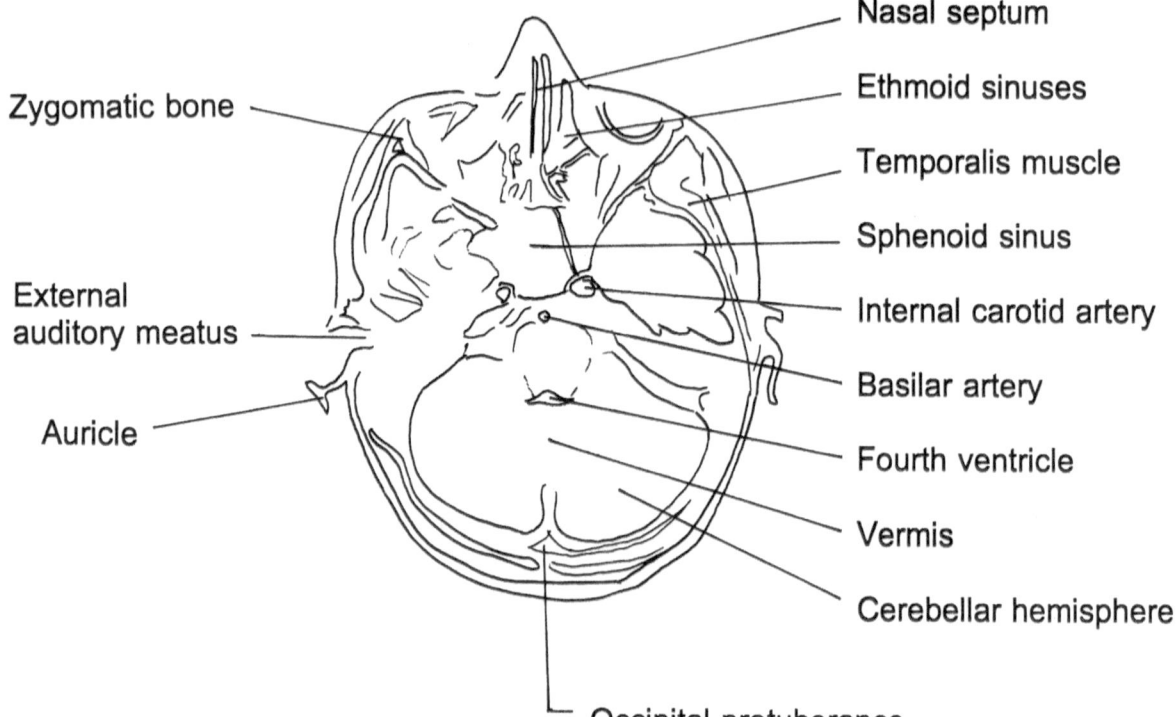

Zygomatic bone

External auditory meatus

Auricle

Nasal septum

Ethmoid sinuses

Temporalis muscle

Sphenoid sinus

Internal carotid artery

Basilar artery

Fourth ventricle

Vermis

Cerebellar hemisphere

Occipital protuberance

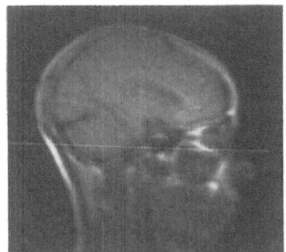

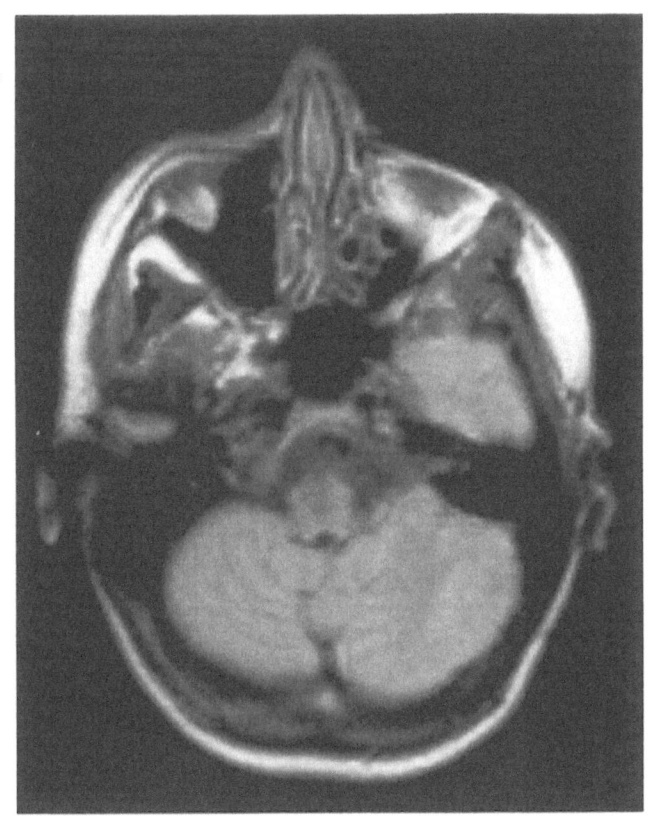

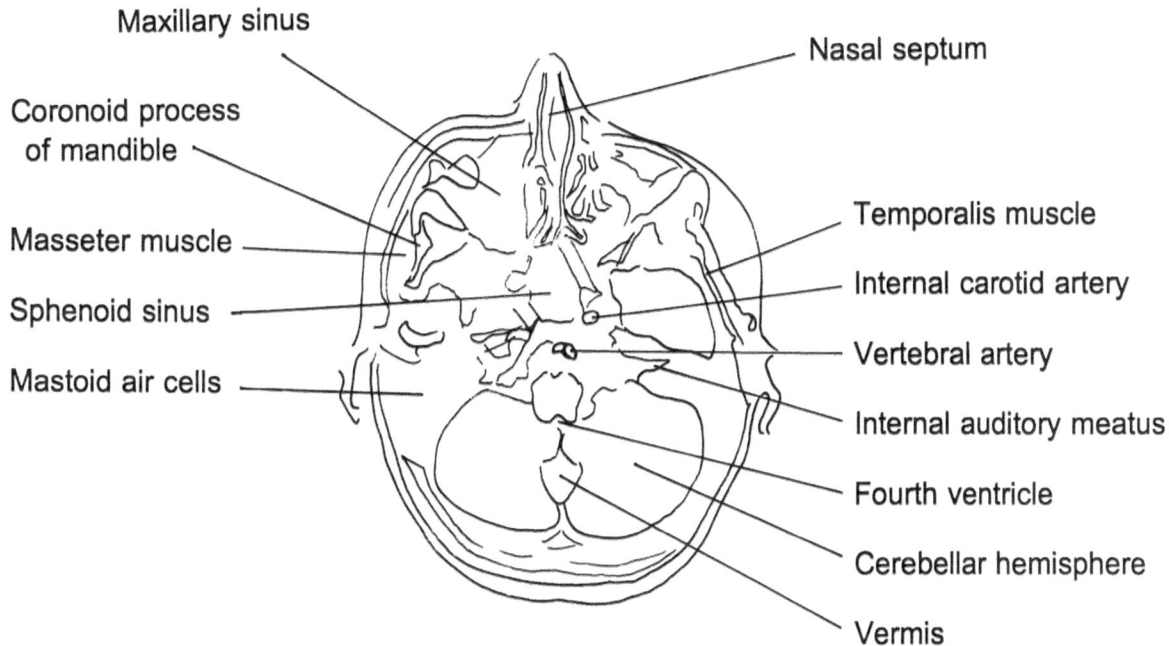

Maxillary sinus

Nasal septum

Coronoid process
of mandible

Temporalis muscle

Masseter muscle

Internal carotid artery

Sphenoid sinus

Vertebral artery

Mastoid air cells

Internal auditory meatus

Fourth ventricle

Cerebellar hemisphere

Vermis

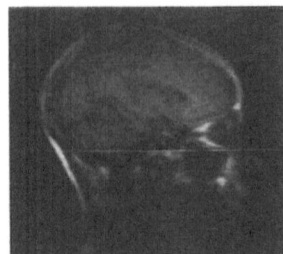

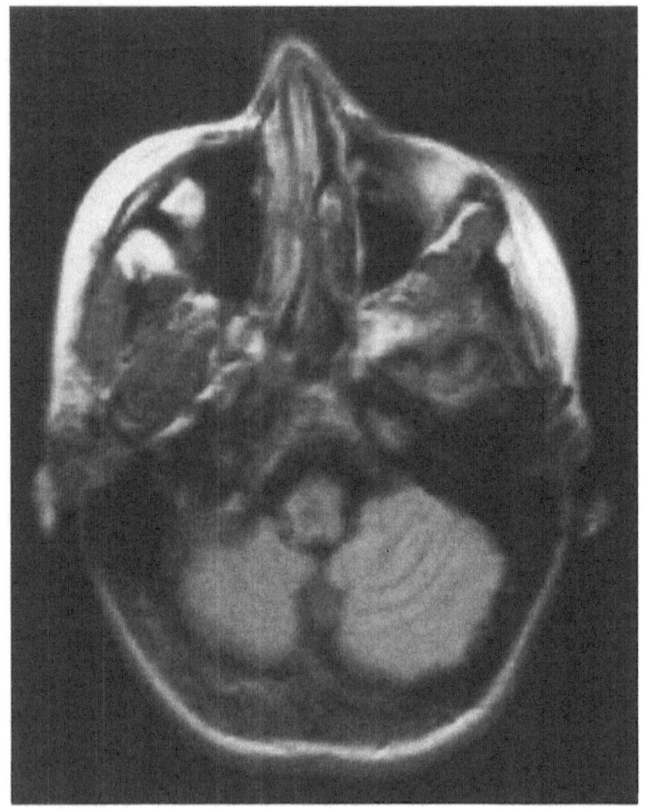

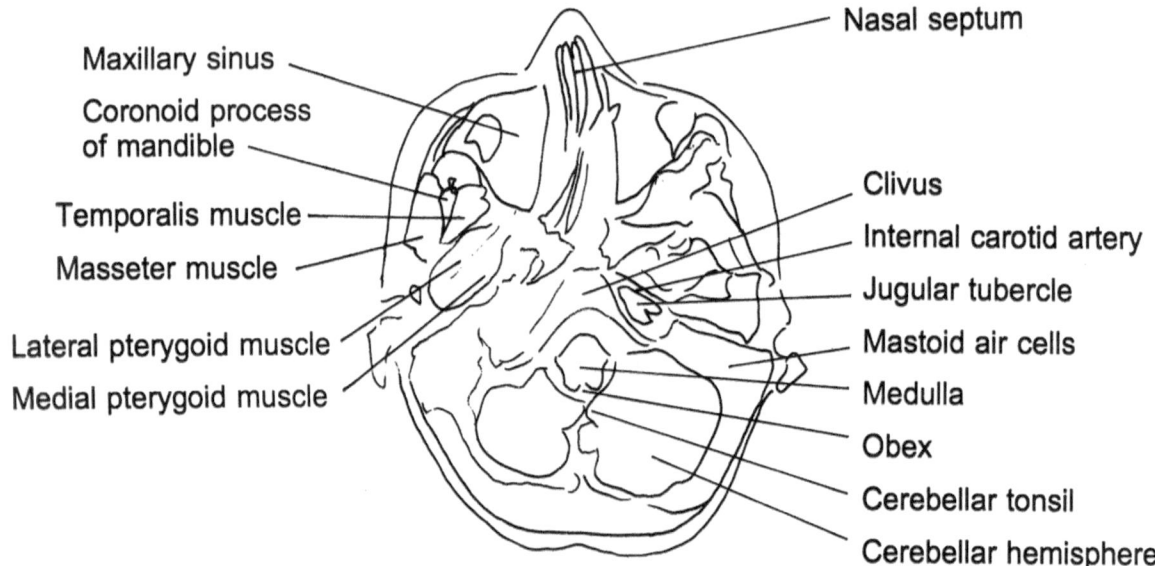

Nasal septum

Maxillary sinus

Coronoid process
of mandible

Temporalis muscle

Masseter muscle

Lateral pterygoid muscle

Medial pterygoid muscle

Clivus

Internal carotid artery

Jugular tubercle

Mastoid air cells

Medulla

Obex

Cerebellar tonsil

Cerebellar hemisphere

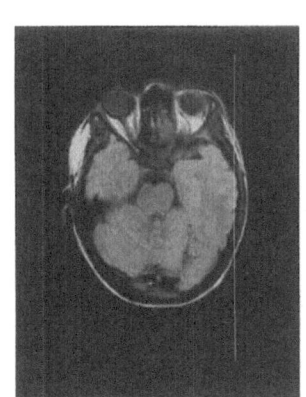

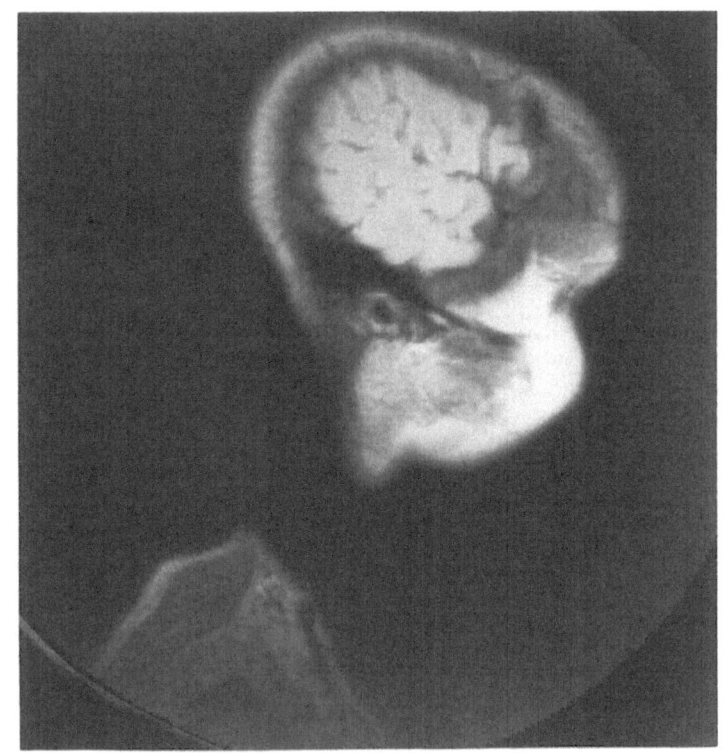

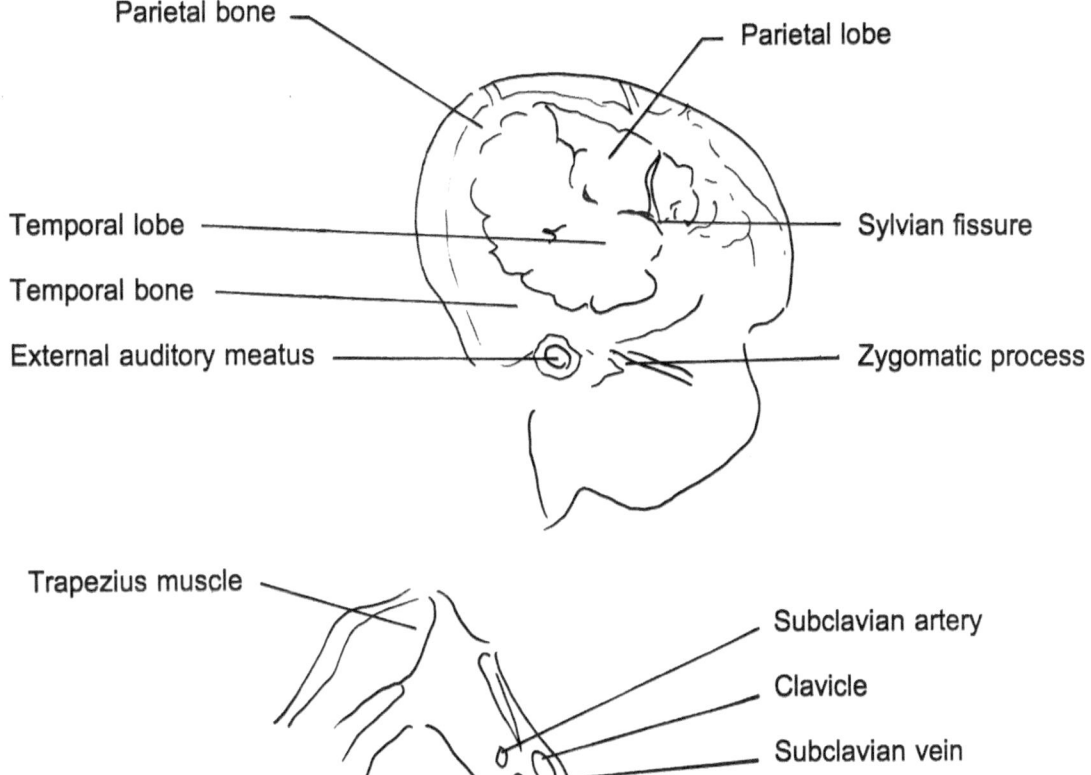

Parietal bone

Parietal lobe

Temporal lobe

Sylvian fissure

Temporal bone

External auditory meatus

Zygomatic process

Trapezius muscle

Subclavian artery

Clavicle

Subclavian vein

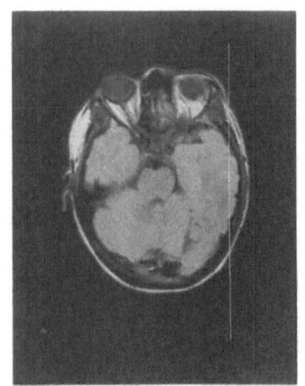

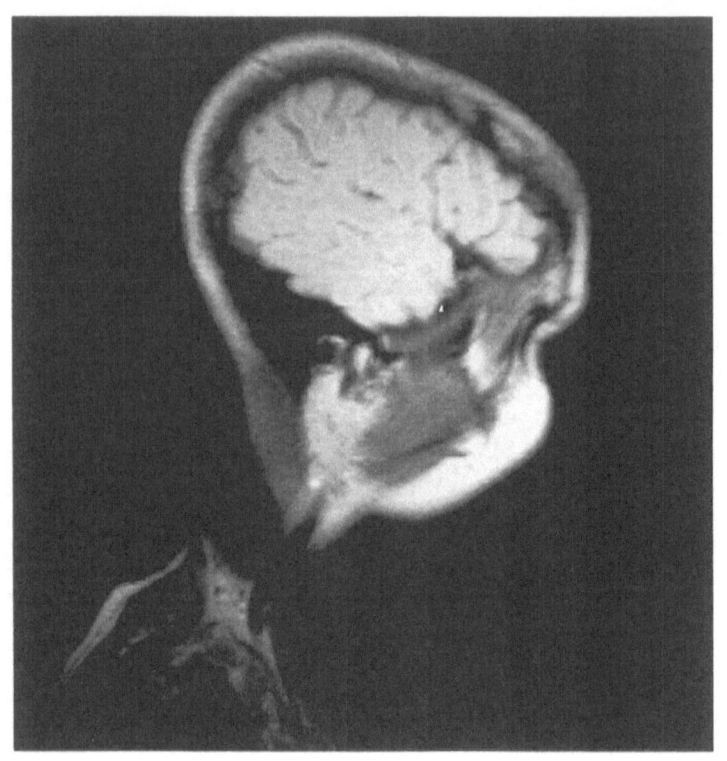

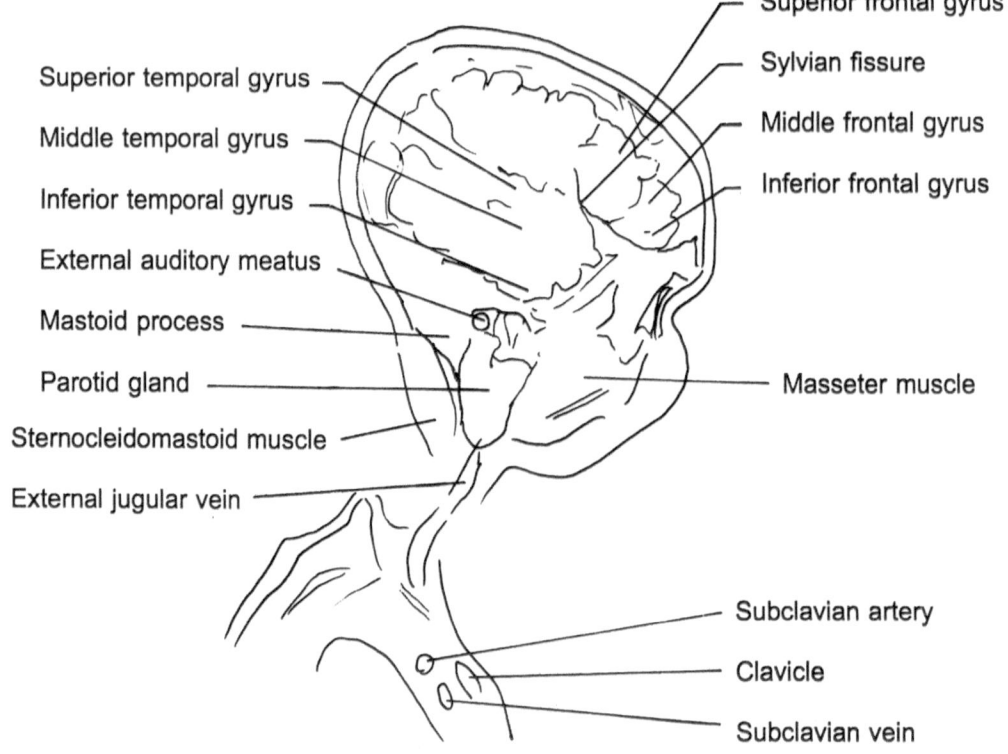

Superior temporal gyrus

Middle temporal gyrus

Inferior temporal gyrus

External auditory meatus

Mastoid process

Parotid gland

Sternocleidomastoid muscle

External jugular vein

Superior frontal gyrus

Sylvian fissure

Middle frontal gyrus

Inferior frontal gyrus

Masseter muscle

Subclavian artery

Clavicle

Subclavian vein

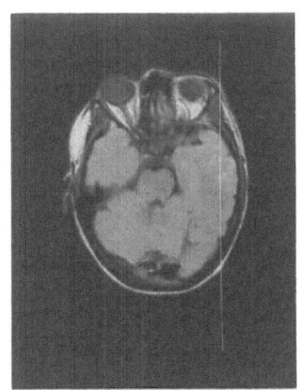

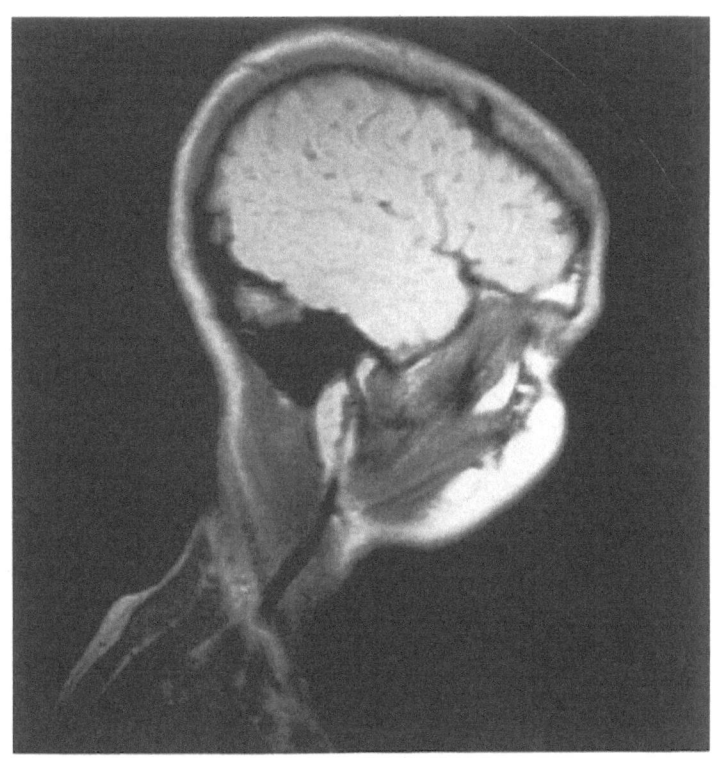

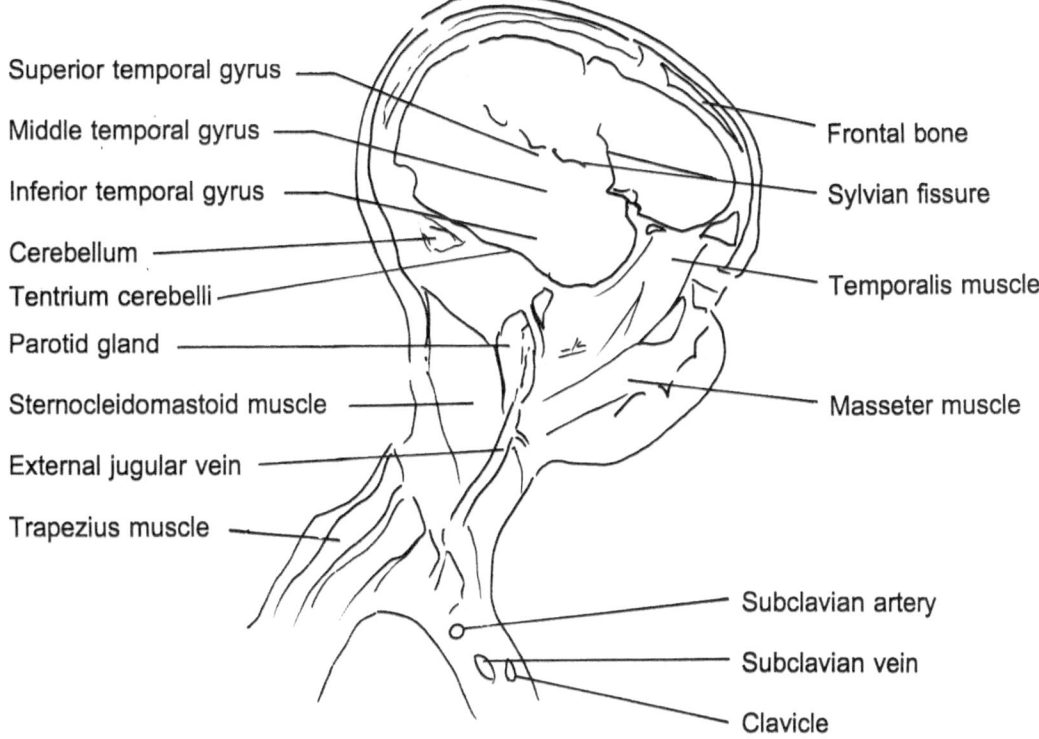

Superior temporal gyrus

Middle temporal gyrus

Inferior temporal gyrus

Cerebellum

Tentrium cerebelli

Parotid gland

Sternocleidomastoid muscle

External jugular vein

Trapezius muscle

Frontal bone

Sylvian fissure

Temporalis muscle

Masseter muscle

Subclavian artery

Subclavian vein

Clavicle

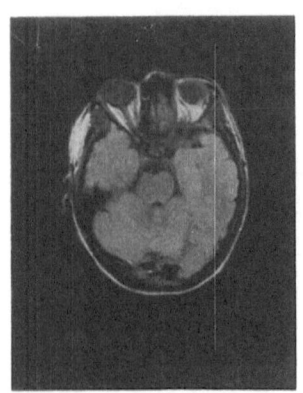

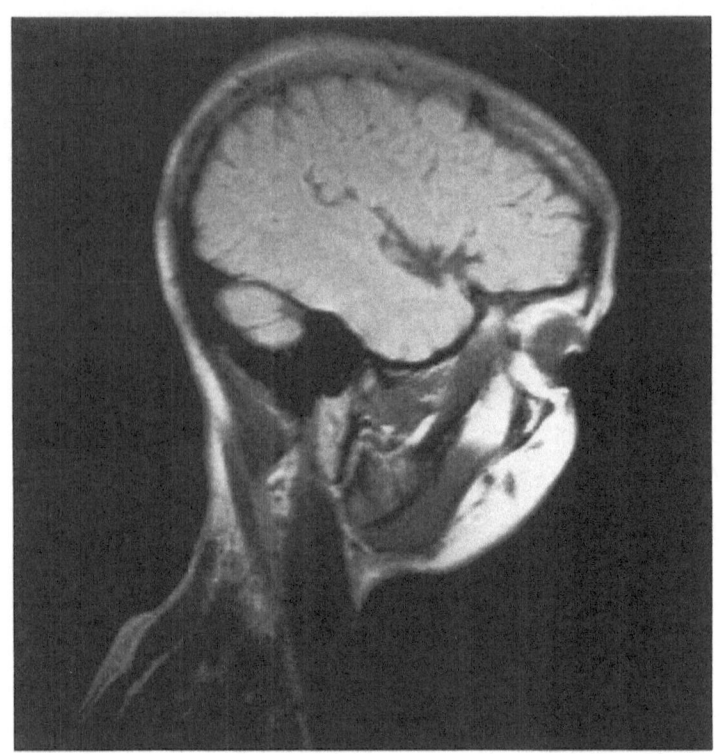

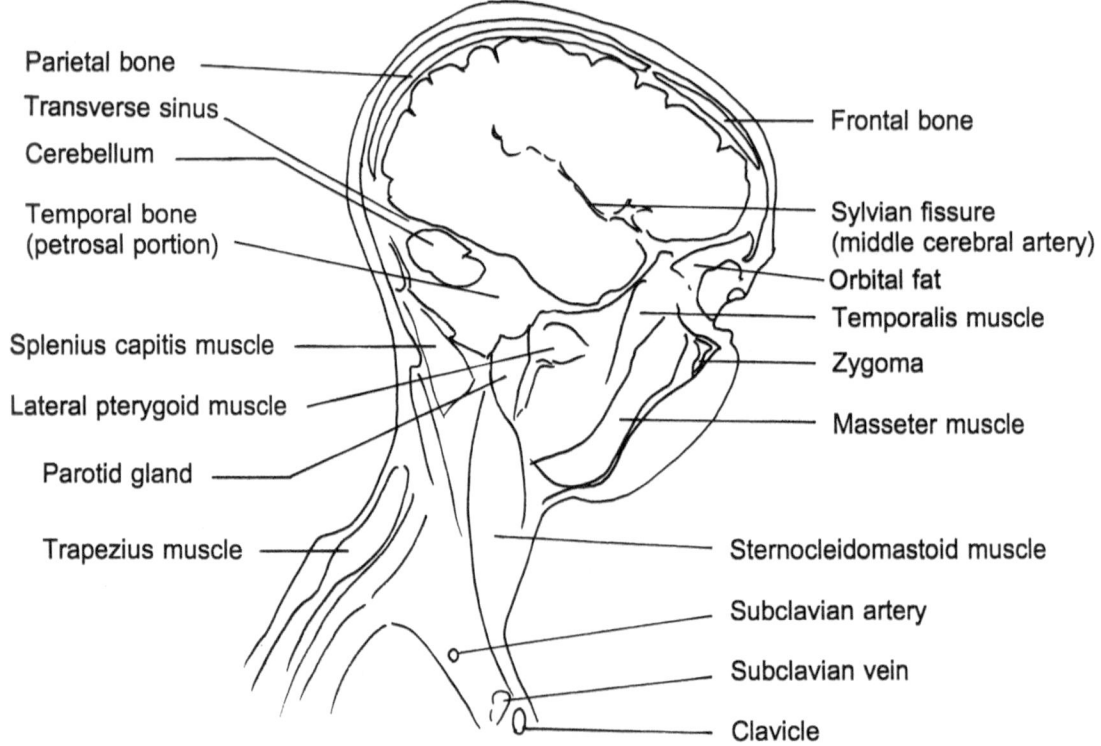

Parietal bone

Transverse sinus

Cerebellum

Temporal bone
(petrosal portion)

Splenius capitis muscle

Lateral pterygoid muscle

Parotid gland

Trapezius muscle

Frontal bone

Sylvian fissure
(middle cerebral artery)

Orbital fat

Temporalis muscle

Zygoma

Masseter muscle

Sternocleidomastoid muscle

Subclavian artery

Subclavian vein

Clavicle

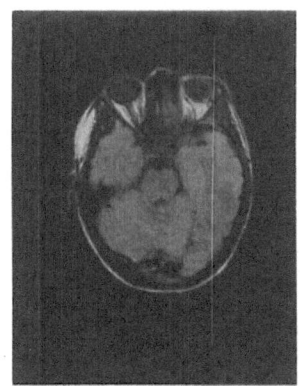

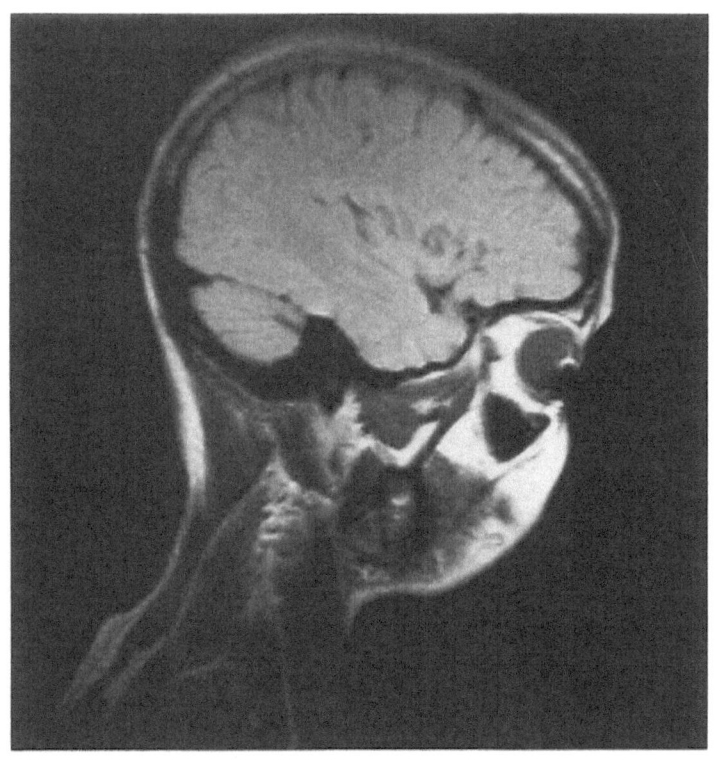

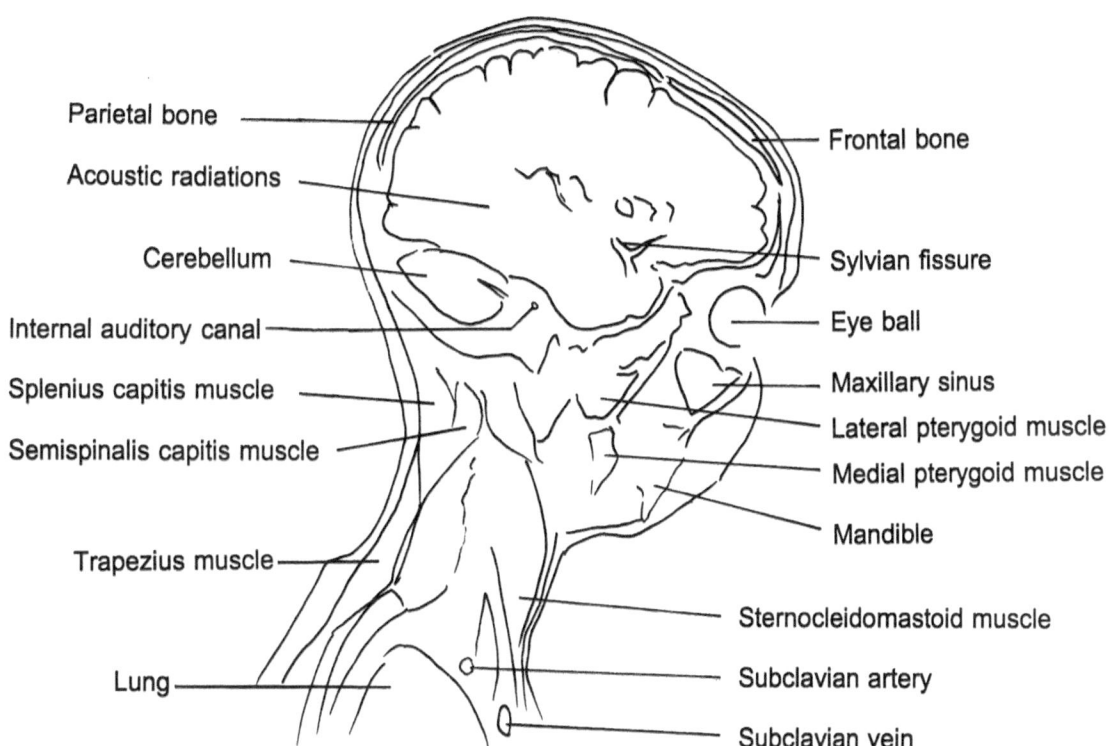

Parietal bone — Frontal bone

Acoustic radiations — Sylvian fissure

Cerebellum — Eye ball

Internal auditory canal — Maxillary sinus

Splenius capitis muscle — Lateral pterygoid muscle

Semispinalis capitis muscle — Medial pterygoid muscle

— Mandible

Trapezius muscle — Sternocleidomastoid muscle

Lung — Subclavian artery

— Subclavian vein

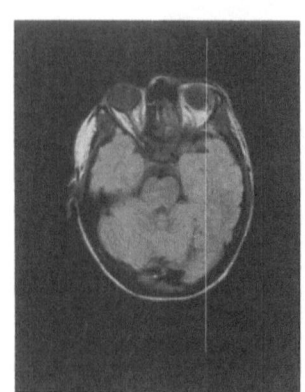

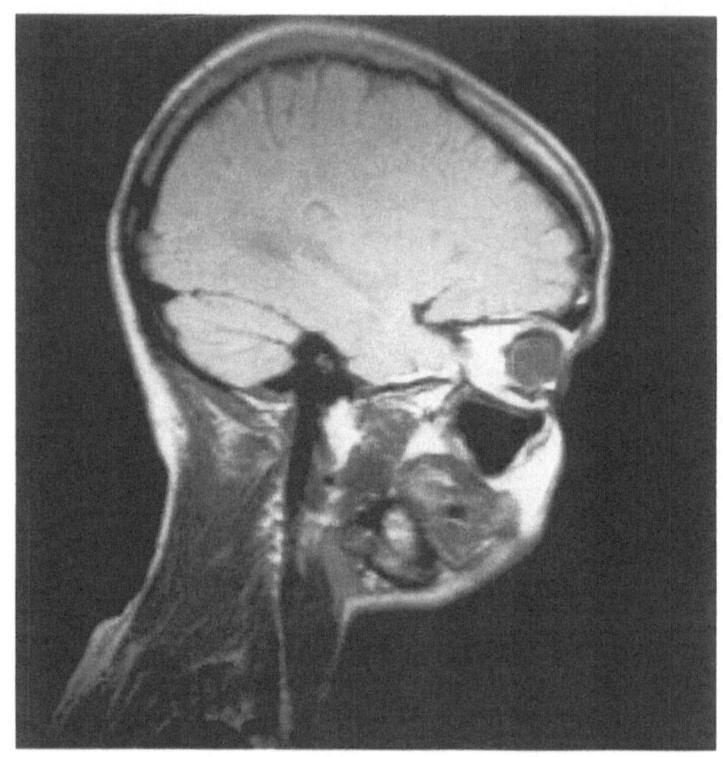

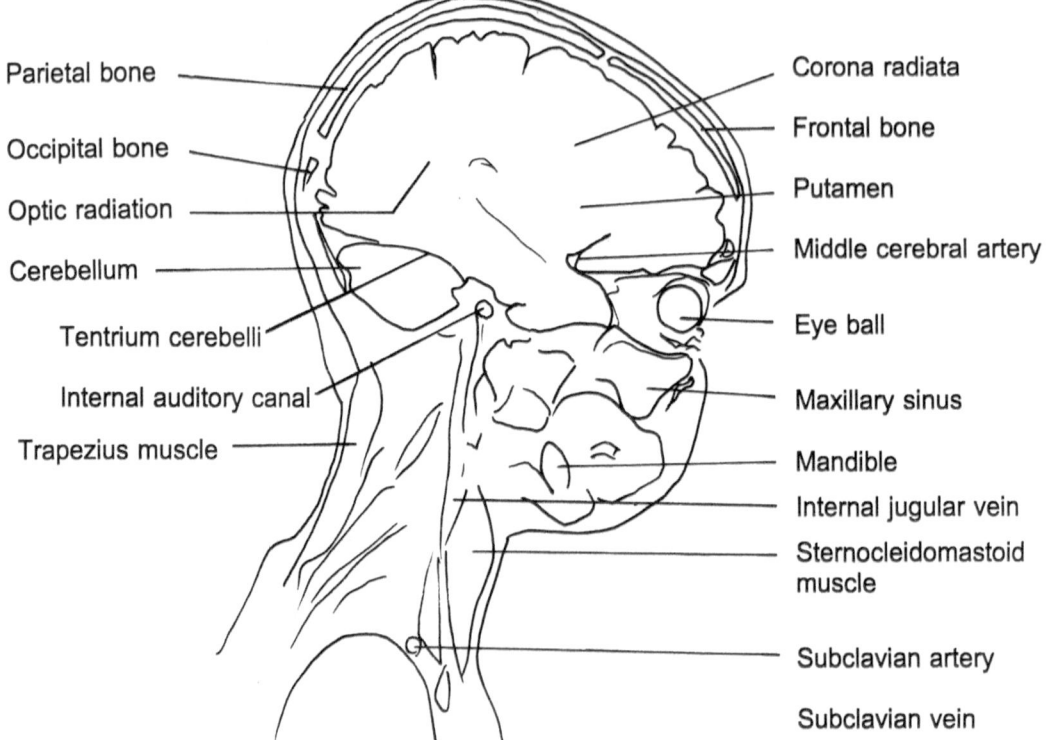

Parietal bone

Occipital bone

Optic radiation

Cerebellum

Tentrium cerebelli

Internal auditory canal

Trapezius muscle

Corona radiata

Frontal bone

Putamen

Middle cerebral artery

Eye ball

Maxillary sinus

Mandible

Internal jugular vein

Sternocleidomastoid muscle

Subclavian artery

Subclavian vein

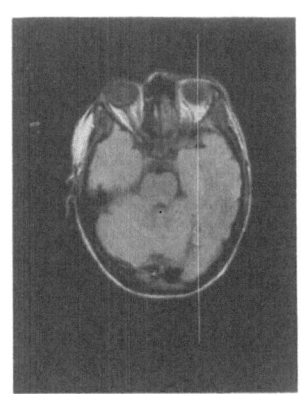

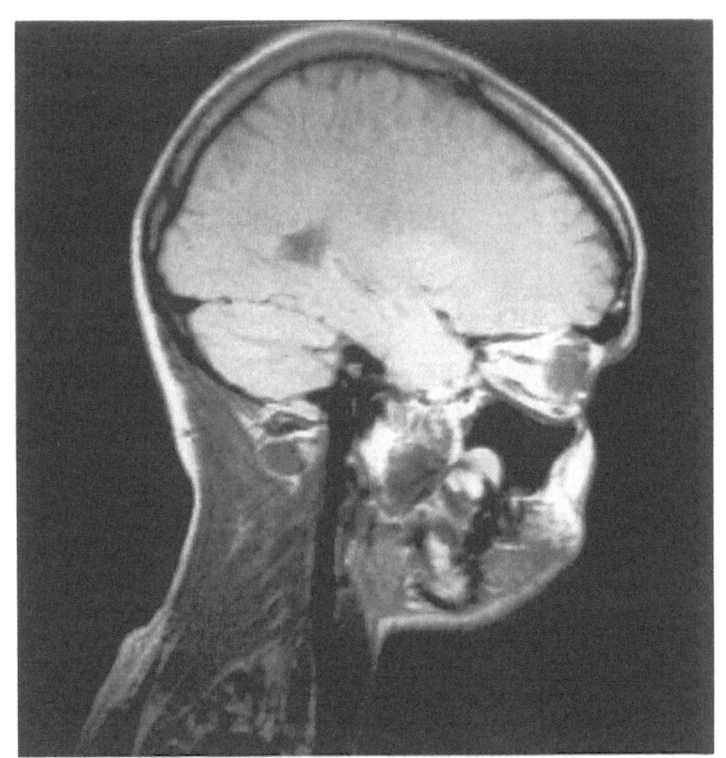

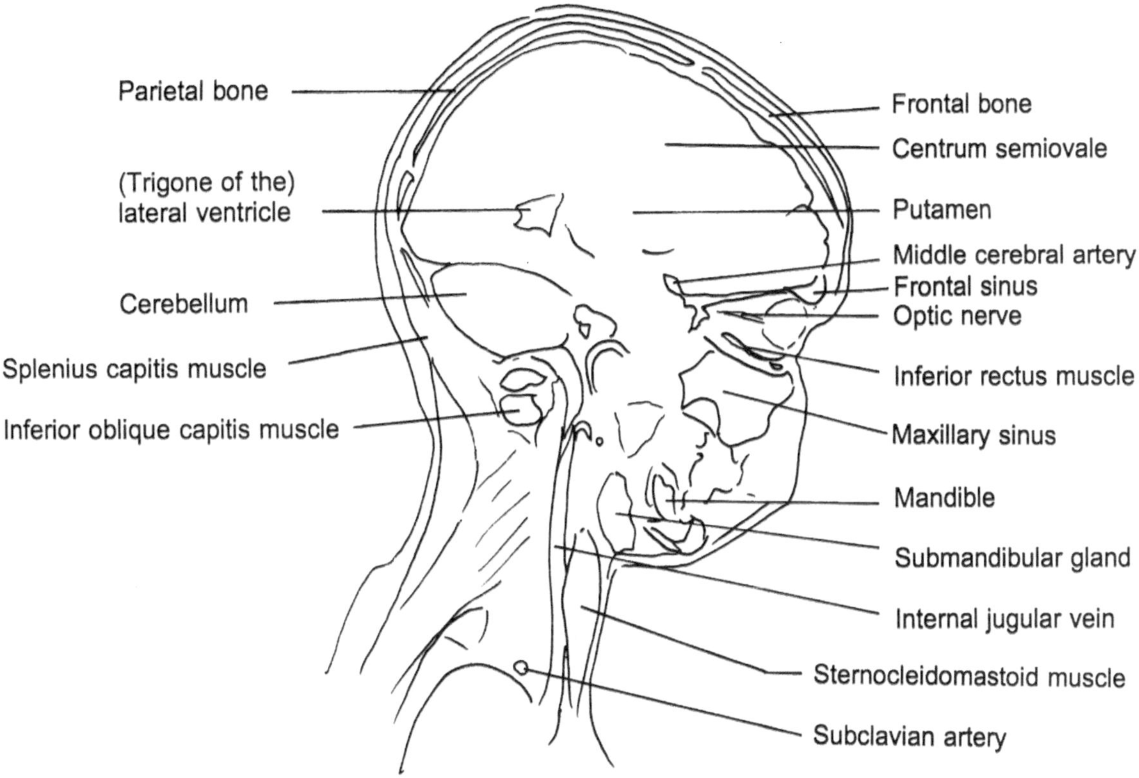

Parietal bone

(Trigone of the) lateral ventricle

Cerebellum

Splenius capitis muscle

Inferior oblique capitis muscle

Frontal bone

Centrum semiovale

Putamen

Middle cerebral artery

Frontal sinus

Optic nerve

Inferior rectus muscle

Maxillary sinus

Mandible

Submandibular gland

Internal jugular vein

Sternocleidomastoid muscle

Subclavian artery

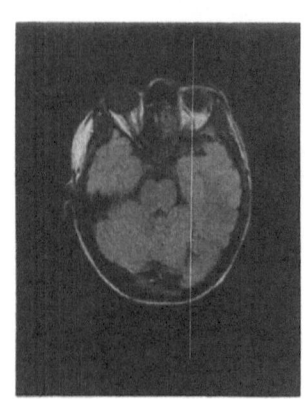

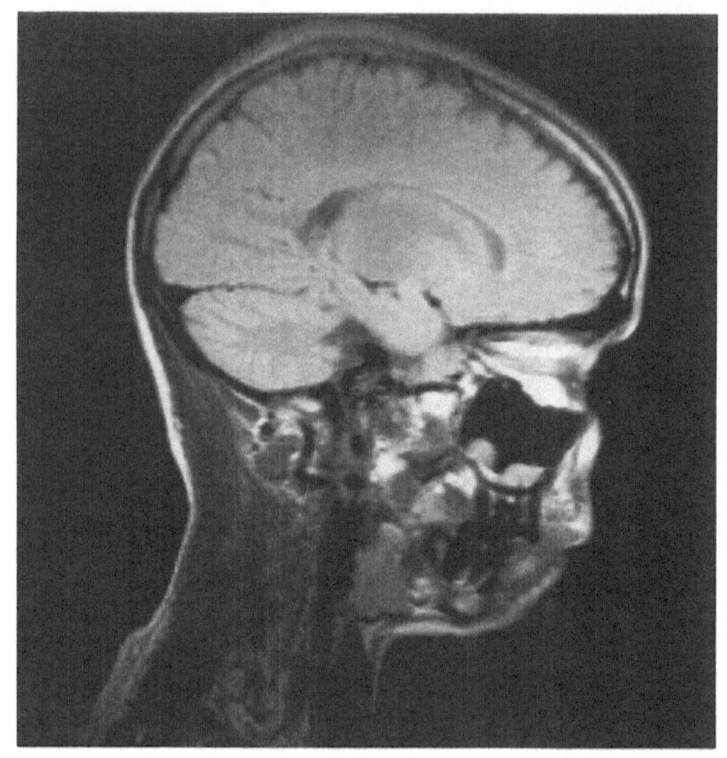

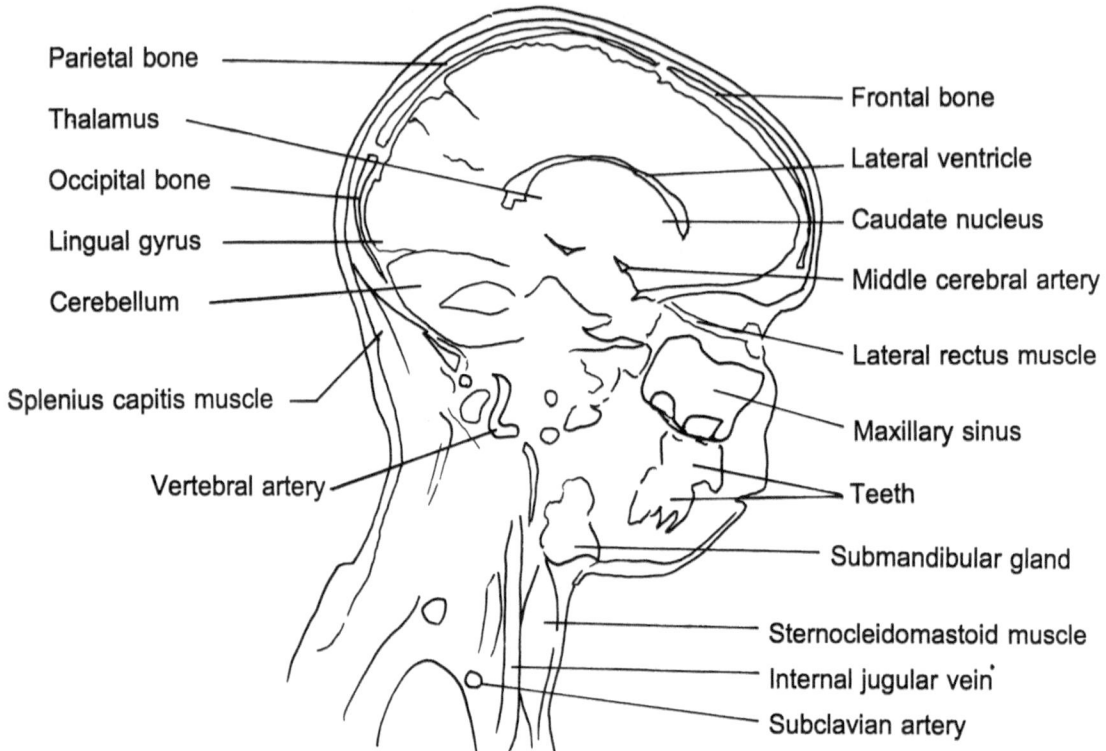

Parietal bone

Thalamus

Occipital bone

Lingual gyrus

Cerebellum

Splenius capitis muscle

Vertebral artery

Frontal bone

Lateral ventricle

Caudate nucleus

Middle cerebral artery

Lateral rectus muscle

Maxillary sinus

Teeth

Submandibular gland

Sternocleidomastoid muscle

Internal jugular vein

Subclavian artery

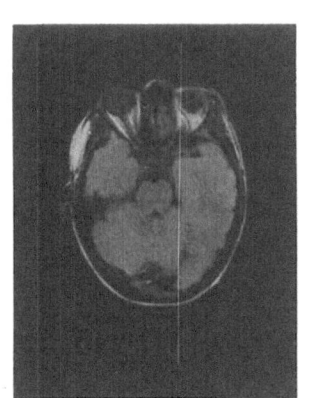

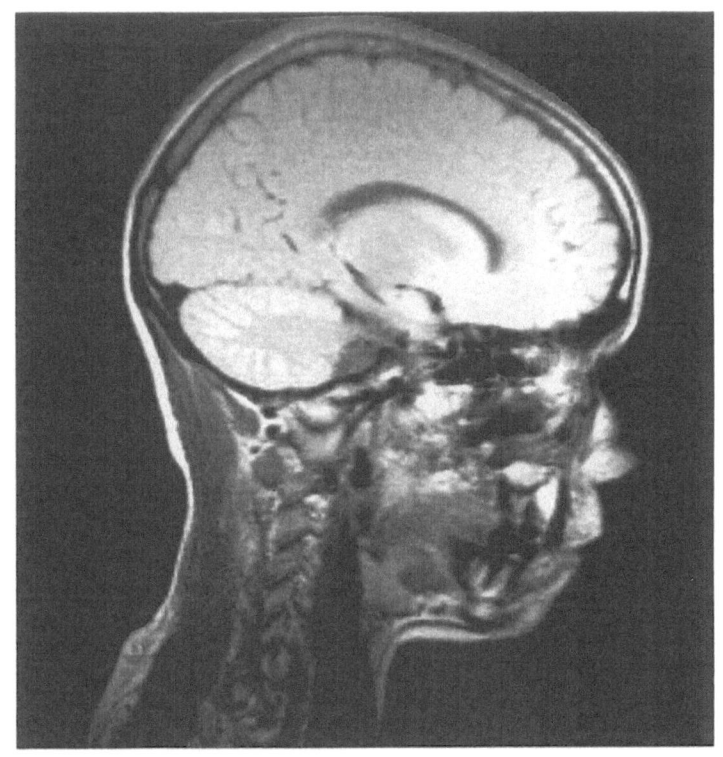

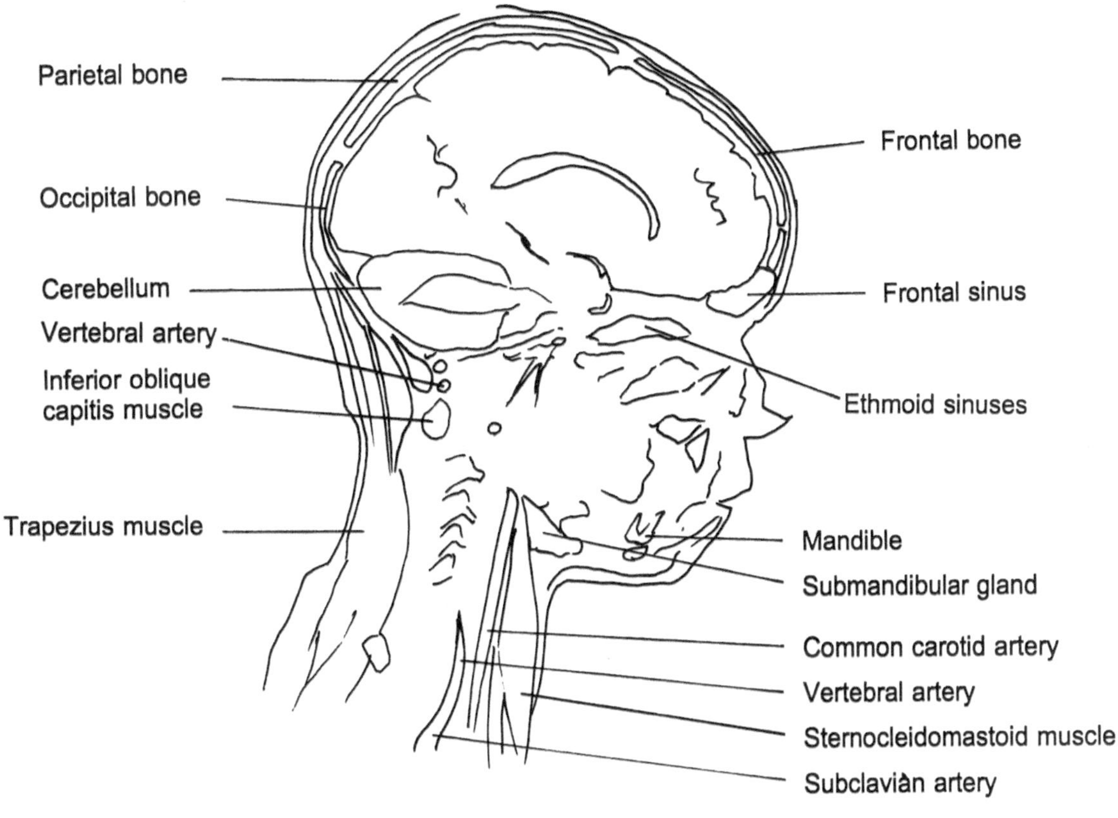

Parietal bone

Occipital bone

Cerebellum

Vertebral artery

Inferior oblique
capitis muscle

Trapezius muscle

Frontal bone

Frontal sinus

Ethmoid sinuses

Mandible

Submandibular gland

Common carotid artery

Vertebral artery

Sternocleidomastoid muscle

Subclavian artery

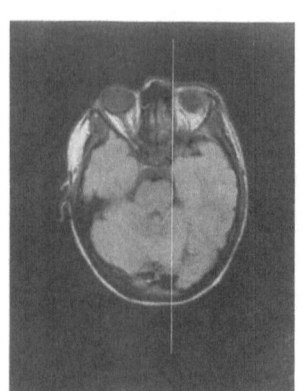

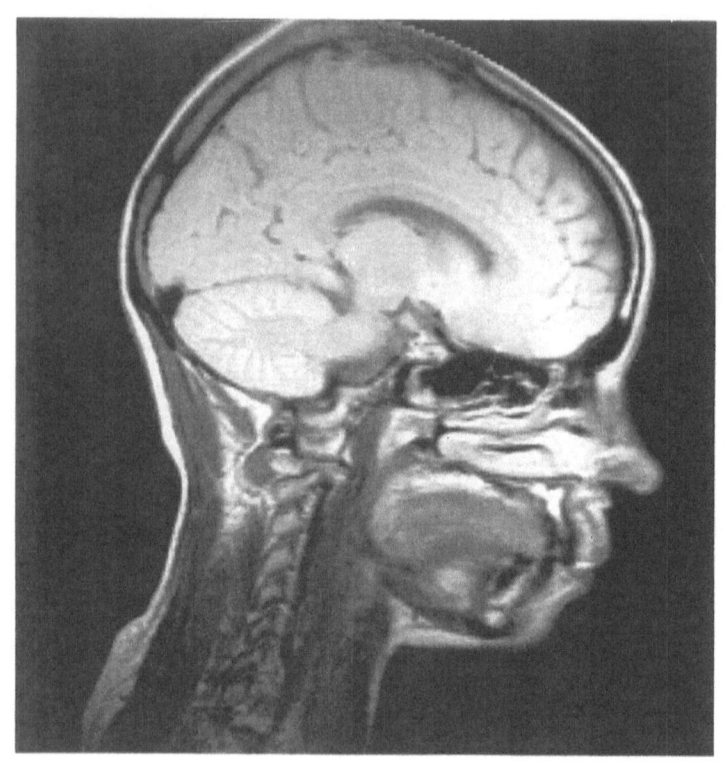

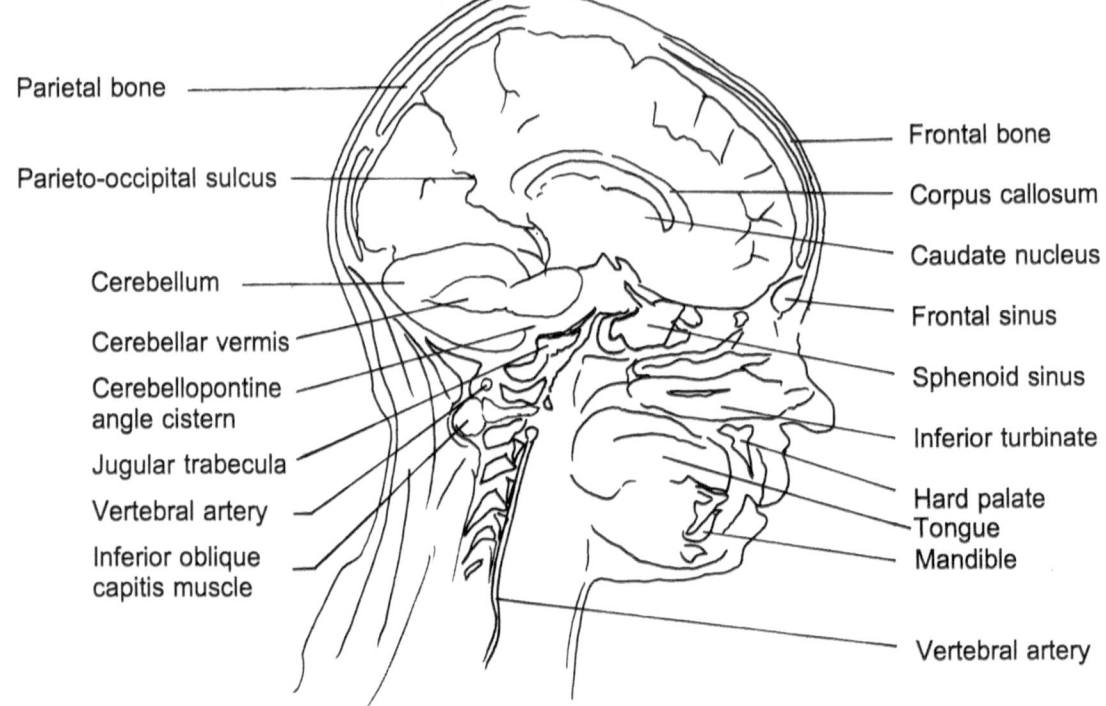

Parietal bone

Parieto-occipital sulcus

Cerebellum

Cerebellar vermis

Cerebellopontine
angle cistern

Jugular trabecula

Vertebral artery

Inferior oblique
capitis muscle

Frontal bone

Corpus callosum

Caudate nucleus

Frontal sinus

Sphenoid sinus

Inferior turbinate

Hard palate
Tongue
Mandible

Vertebral artery

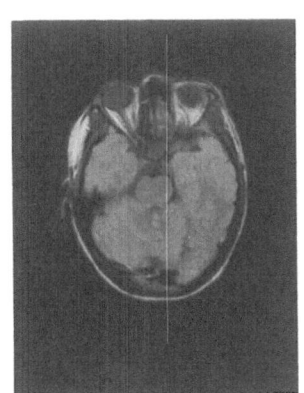

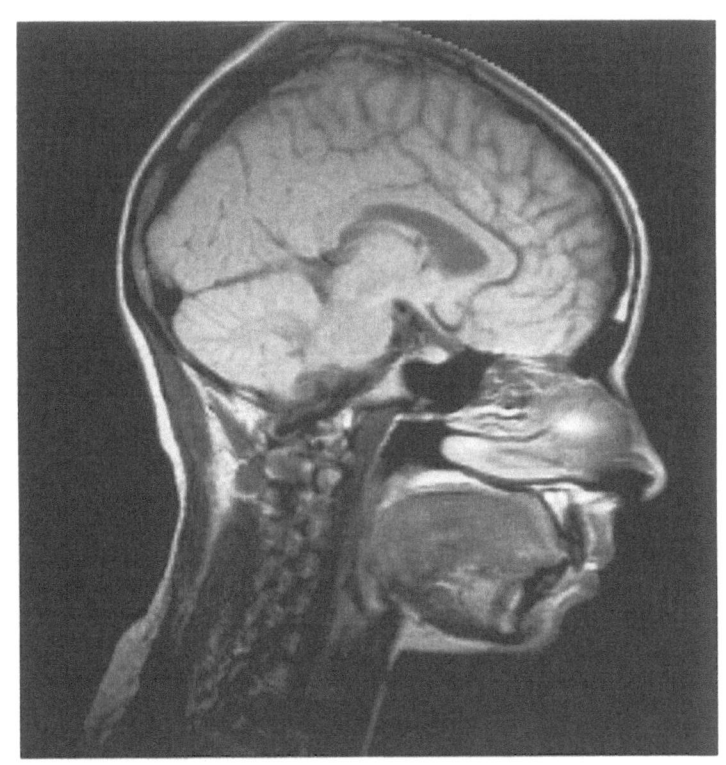

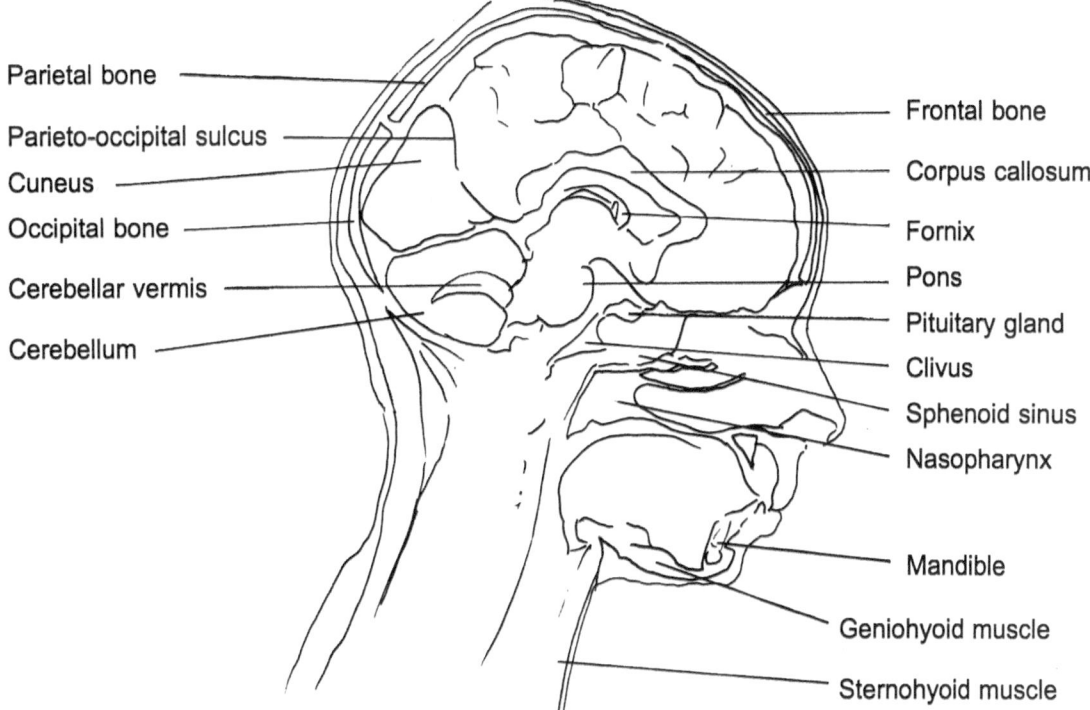

Parietal bone

Parieto-occipital sulcus

Cuneus

Occipital bone

Cerebellar vermis

Cerebellum

Frontal bone

Corpus callosum

Fornix

Pons

Pituitary gland

Clivus

Sphenoid sinus

Nasopharynx

Mandible

Geniohyoid muscle

Sternohyoid muscle

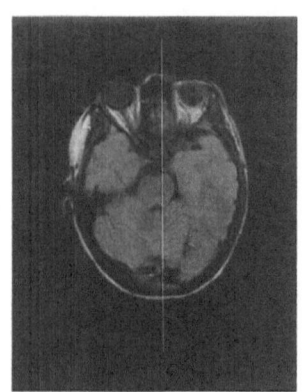

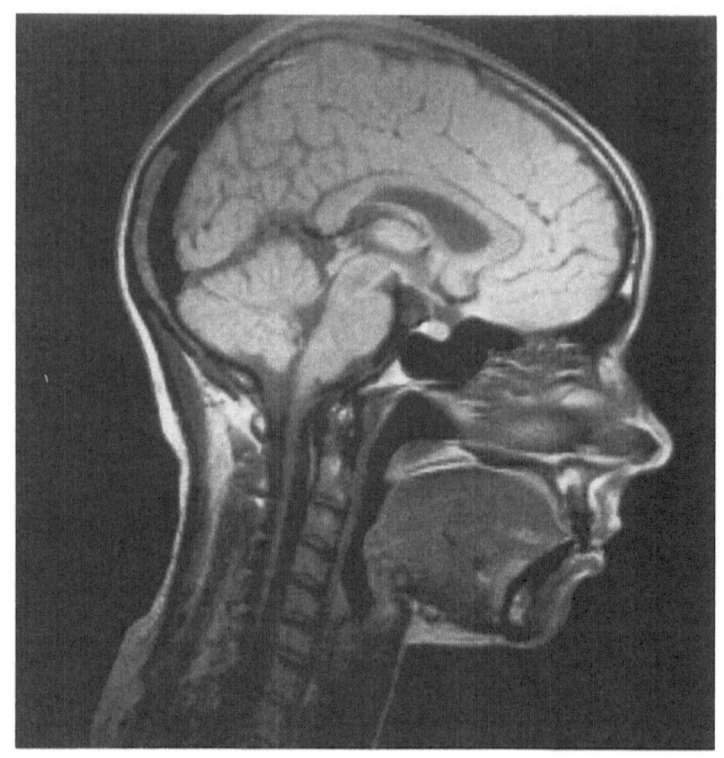

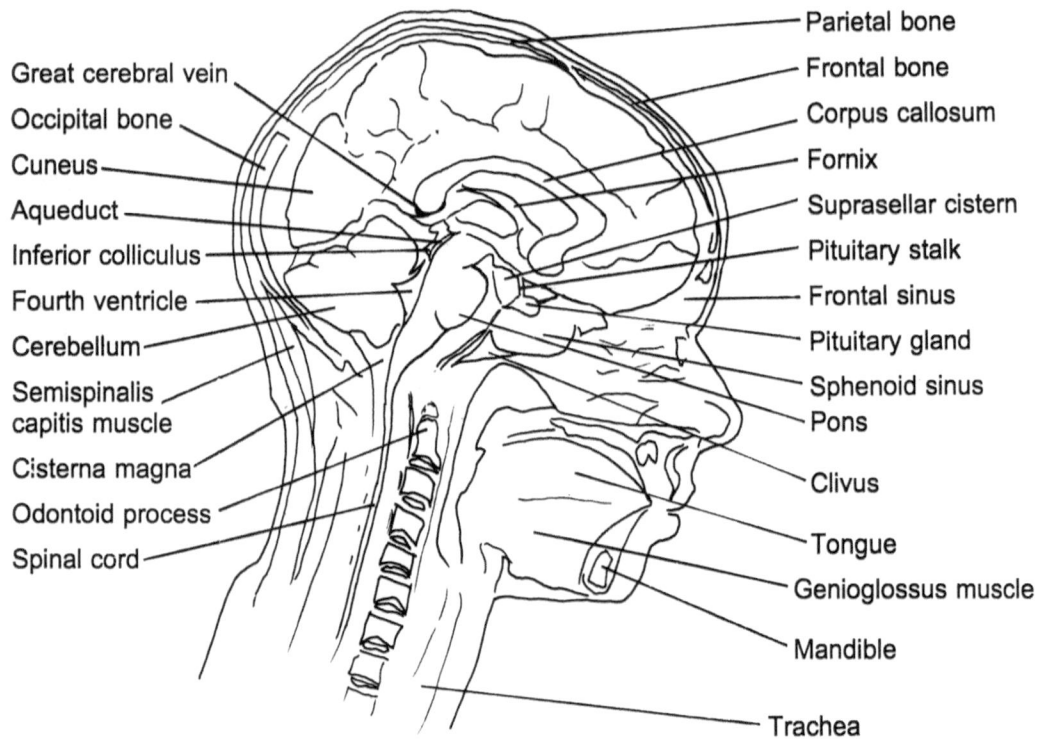

Great cerebral vein

Occipital bone

Cuneus

Aqueduct

Inferior colliculus

Fourth ventricle

Cerebellum

Semispinalis capitis muscle

Cisterna magna

Odontoid process

Spinal cord

Parietal bone

Frontal bone

Corpus callosum

Fornix

Suprasellar cistern

Pituitary stalk

Frontal sinus

Pituitary gland

Sphenoid sinus

Pons

Clivus

Tongue

Genioglossus muscle

Mandible

Trachea

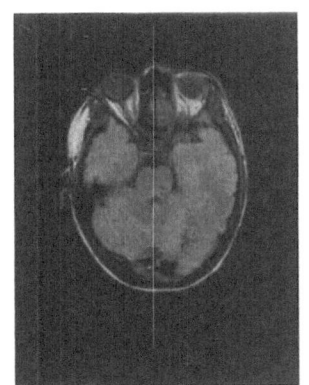
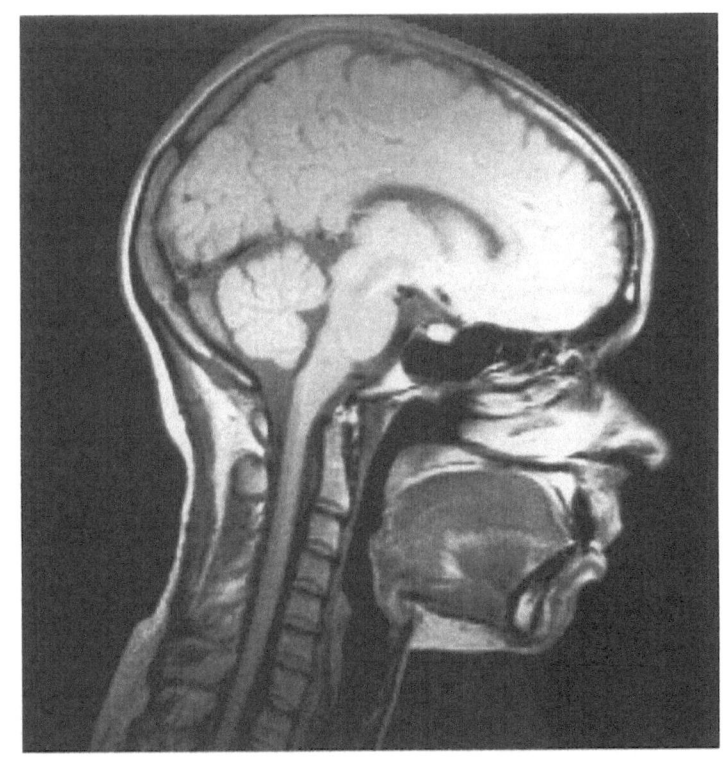

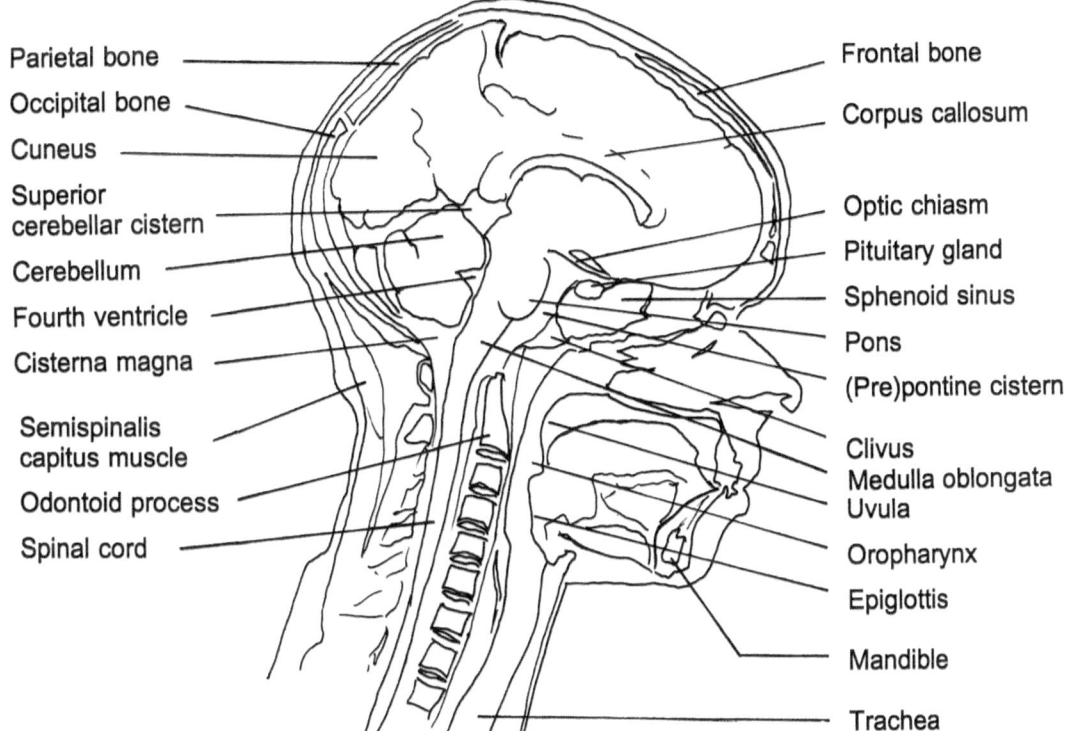

Parietal bone

Occipital bone

Cuneus

Superior
cerebellar cistern

Cerebellum

Fourth ventricle

Cisterna magna

Semispinalis
capitus muscle

Odontoid process

Spinal cord

Frontal bone

Corpus callosum

Optic chiasm

Pituitary gland

Sphenoid sinus

Pons

(Pre)pontine cistern

Clivus
Medulla oblongata
Uvula

Oropharynx

Epiglottis

Mandible

Trachea

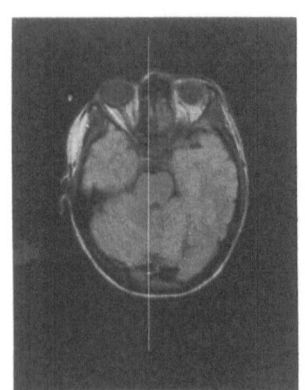

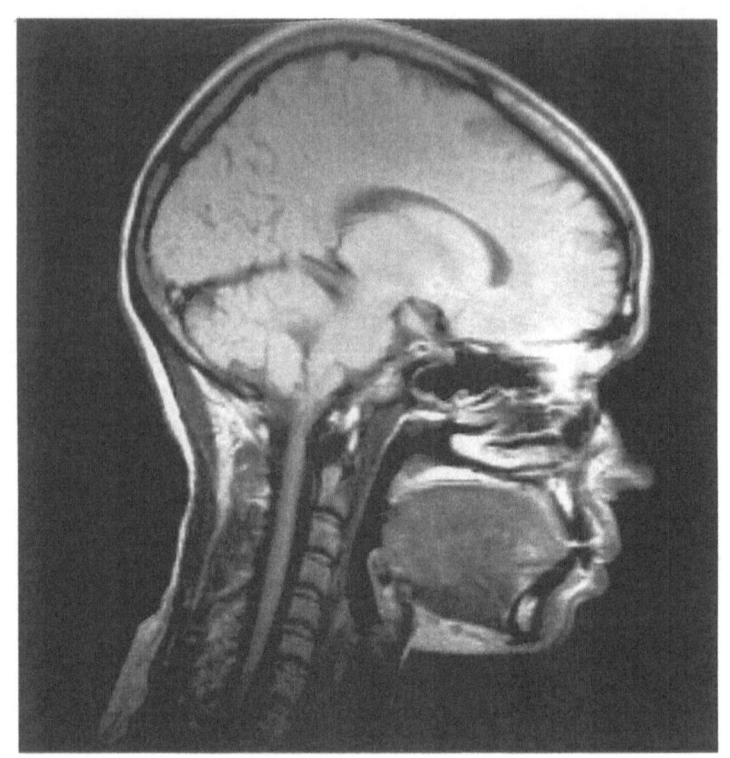

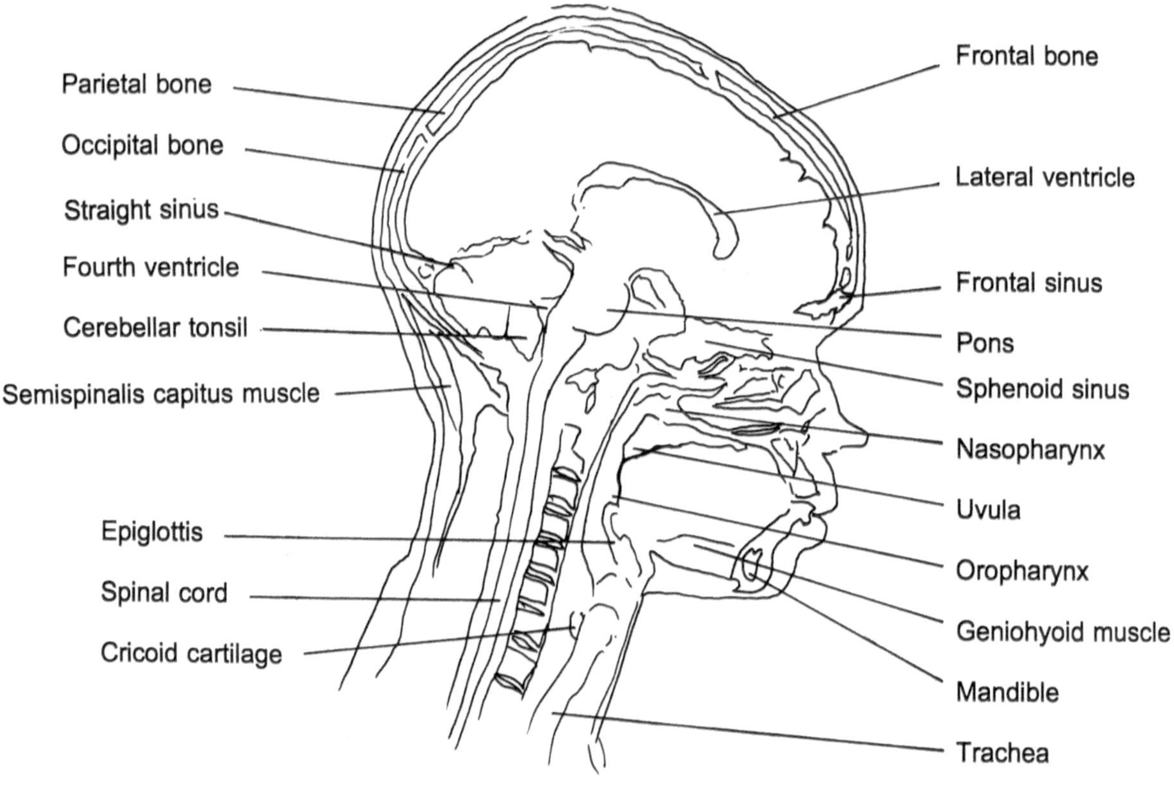

Parietal bone

Occipital bone

Straight sinus

Fourth ventricle

Cerebellar tonsil

Semispinalis capitus muscle

Epiglottis

Spinal cord

Cricoid cartilage

Frontal bone

Lateral ventricle

Frontal sinus

Pons

Sphenoid sinus

Nasopharynx

Uvula

Oropharynx

Geniohyoid muscle

Mandible

Trachea

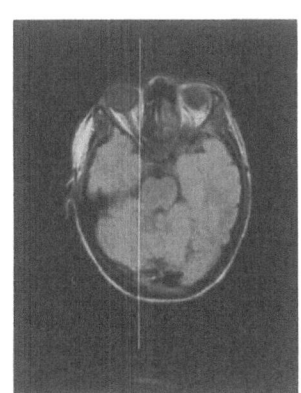

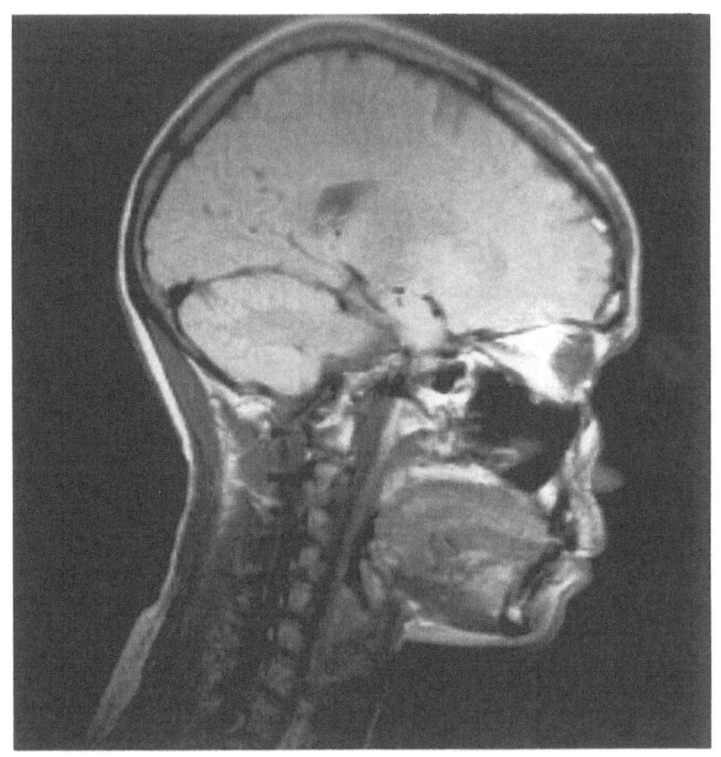

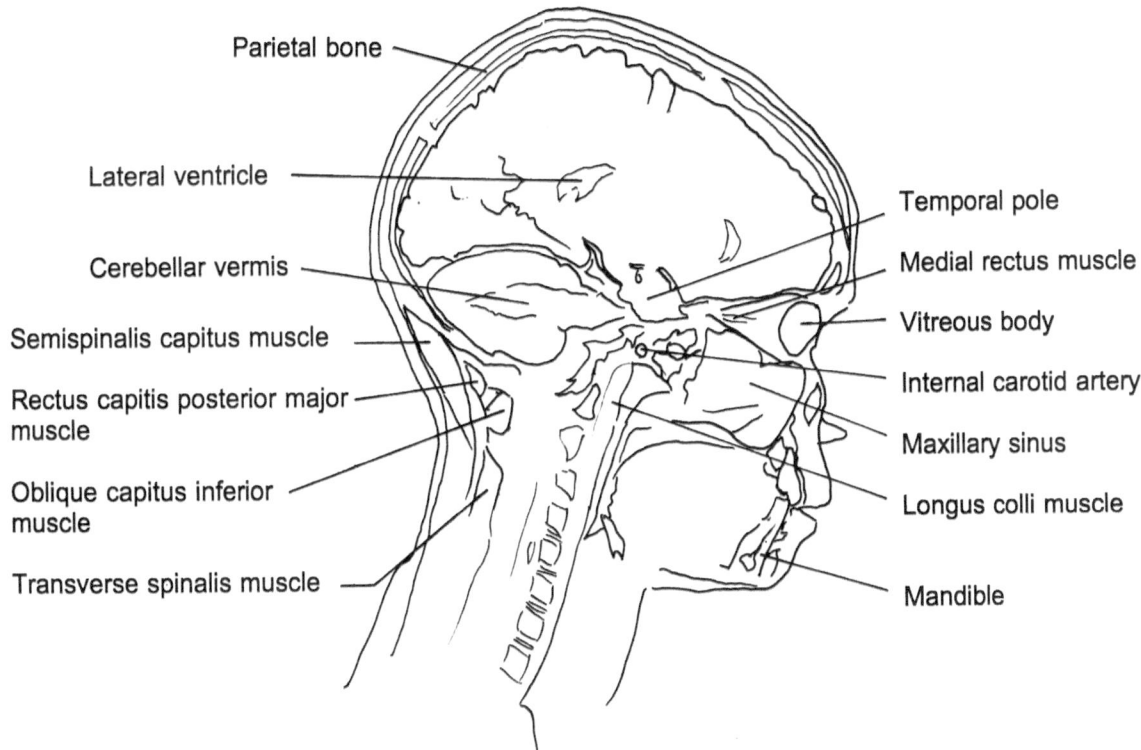

Parietal bone

Lateral ventricle

Cerebellar vermis

Semispinalis capitus muscle

Rectus capitis posterior major muscle

Oblique capitus inferior muscle

Transverse spinalis muscle

Temporal pole

Medial rectus muscle

Vitreous body

Internal carotid artery

Maxillary sinus

Longus colli muscle

Mandible

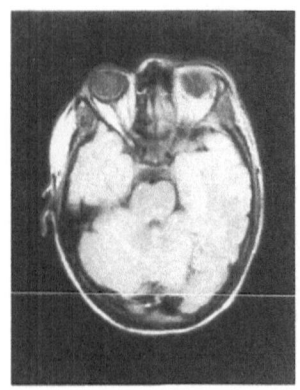

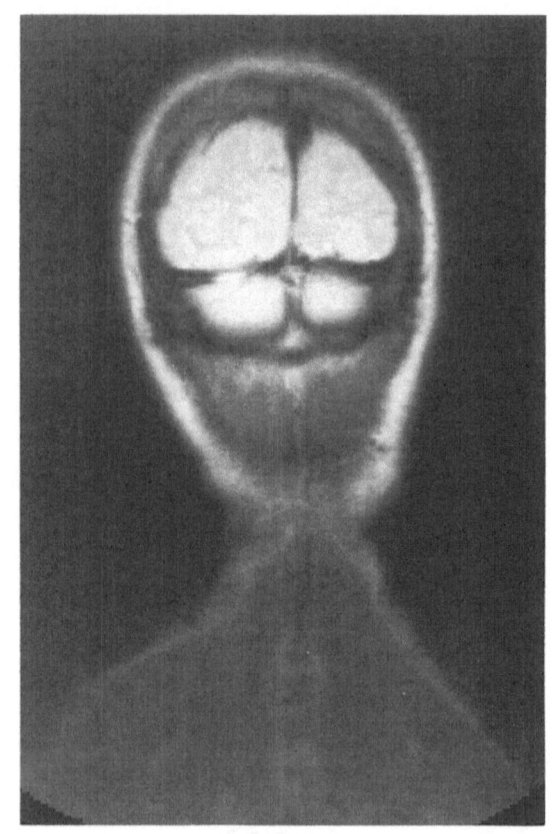

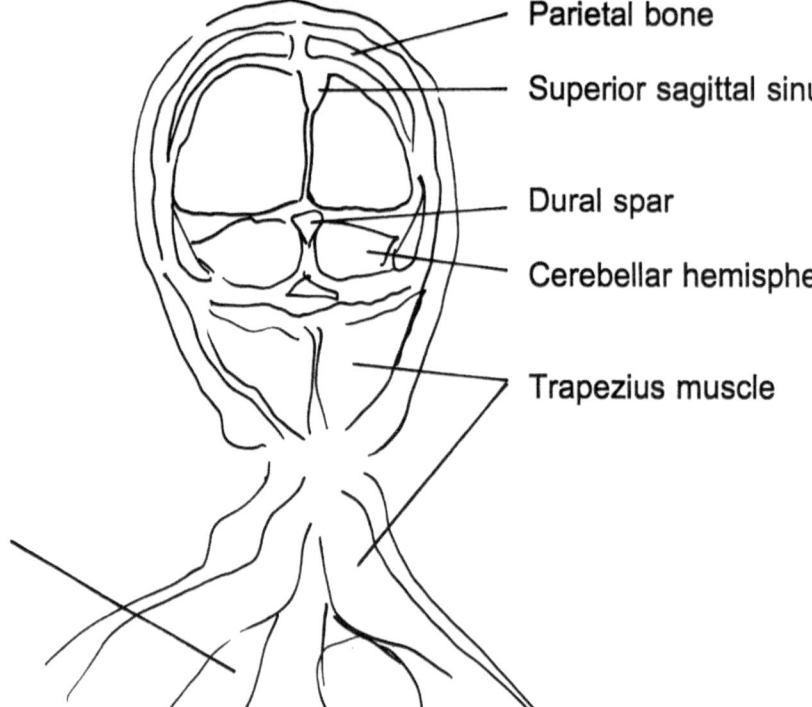

Parietal bone

Superior sagittal sinus

Dural spar

Cerebellar hemisphere

Trapezius muscle

Levator scapulae muscle

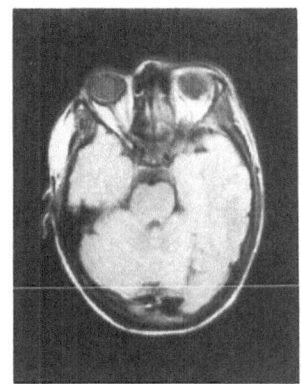

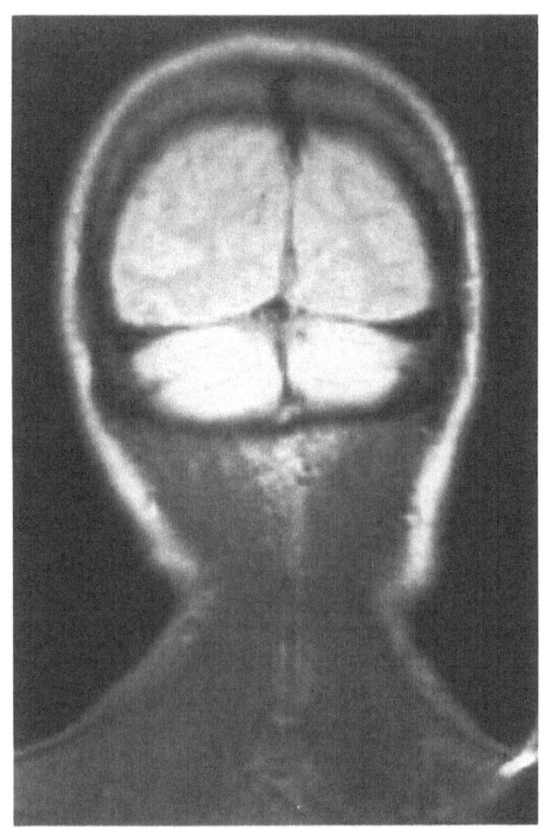

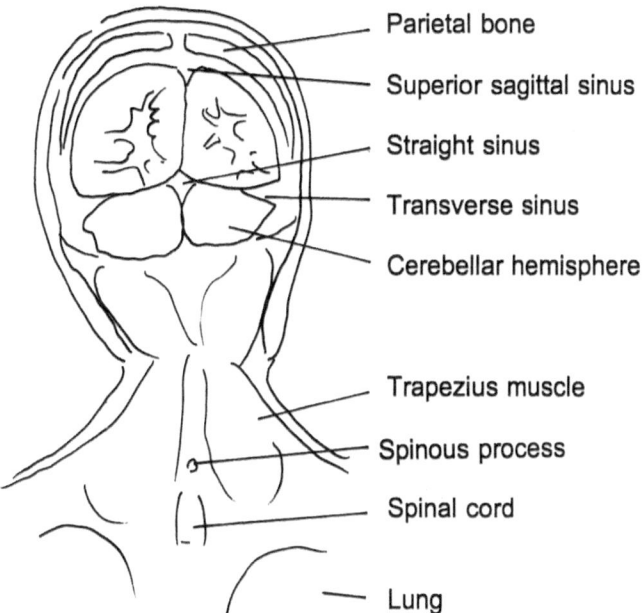

Parietal bone

Superior sagittal sinus

Straight sinus

Transverse sinus

Cerebellar hemisphere

Trapezius muscle

Spinous process

Spinal cord

Lung

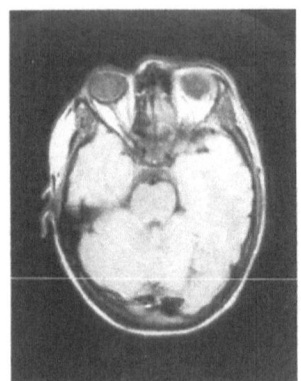

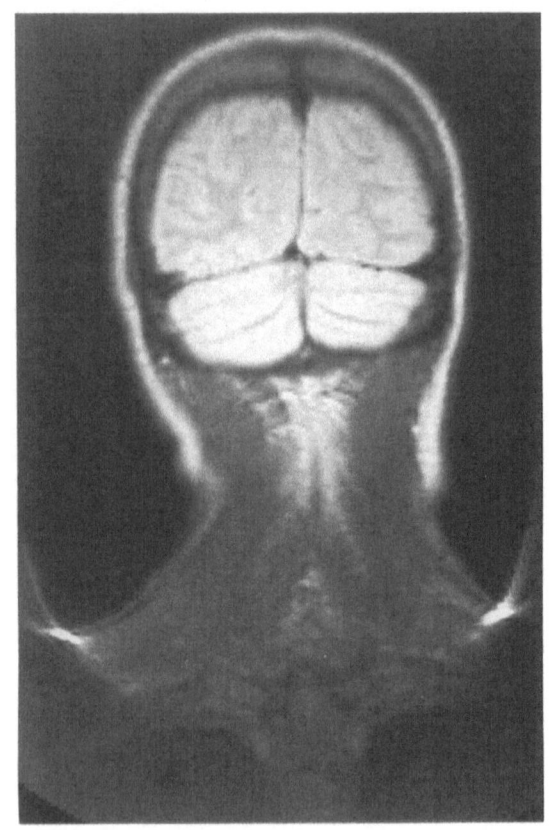

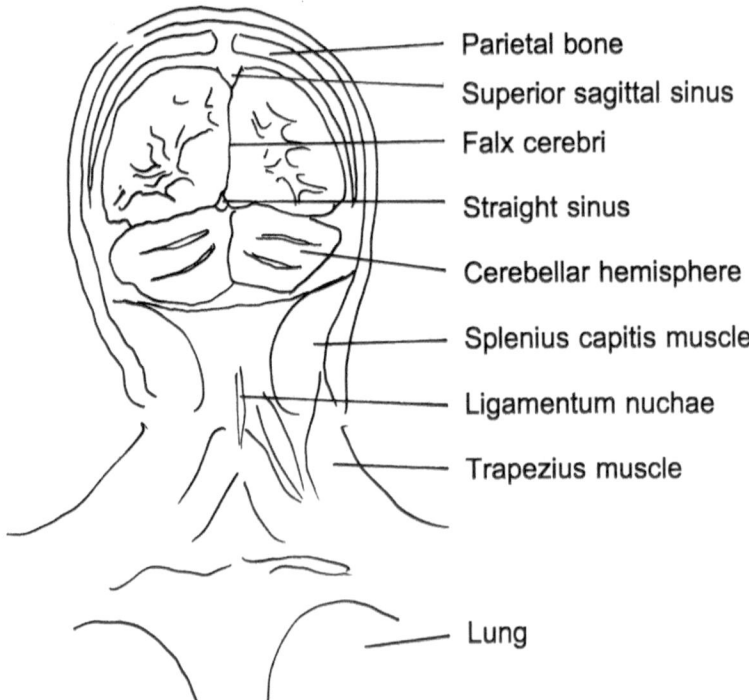

Parietal bone

Superior sagittal sinus

Falx cerebri

Straight sinus

Cerebellar hemisphere

Splenius capitis muscle

Ligamentum nuchae

Trapezius muscle

Lung

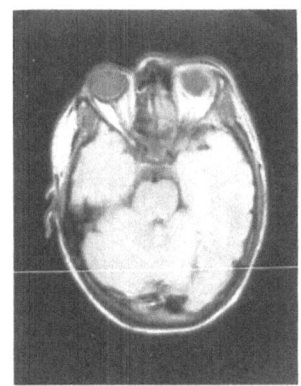

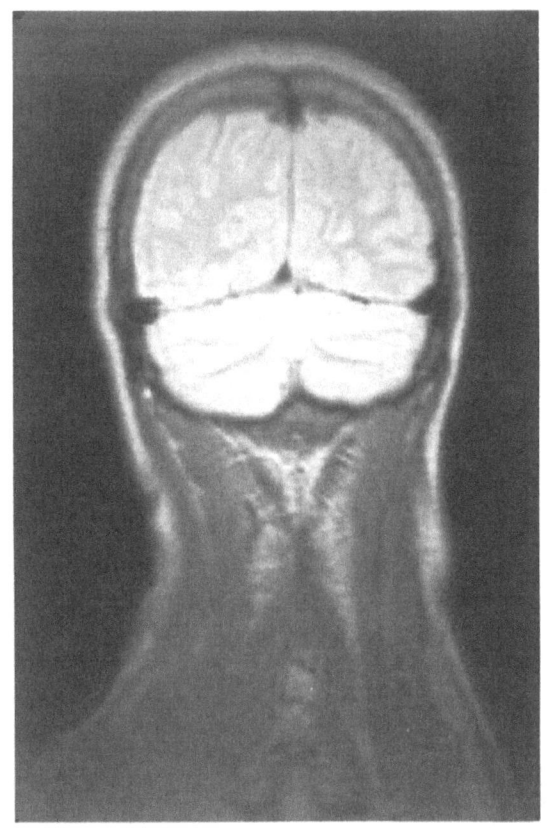

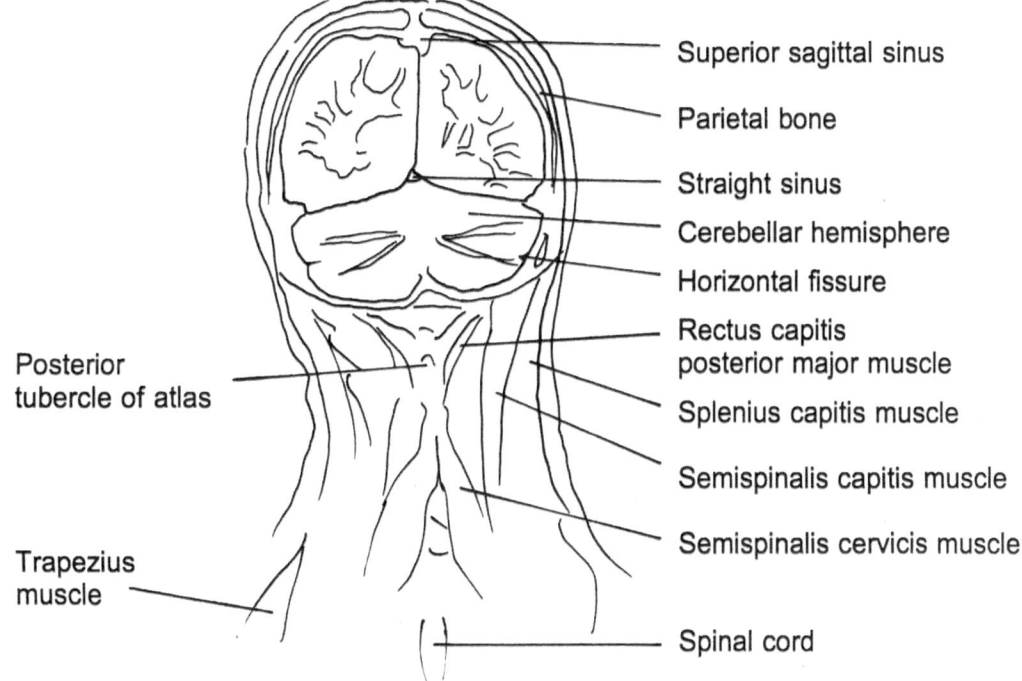

Superior sagittal sinus

Parietal bone

Straight sinus

Cerebellar hemisphere

Horizontal fissure

Rectus capitis
posterior major muscle

Splenius capitis muscle

Semispinalis capitis muscle

Semispinalis cervicis muscle

Posterior
tubercle of atlas

Trapezius
muscle

Spinal cord

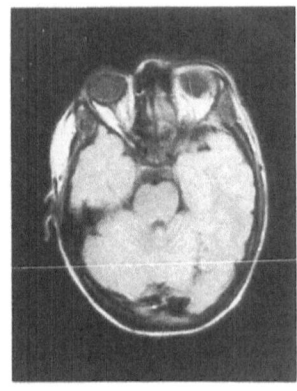

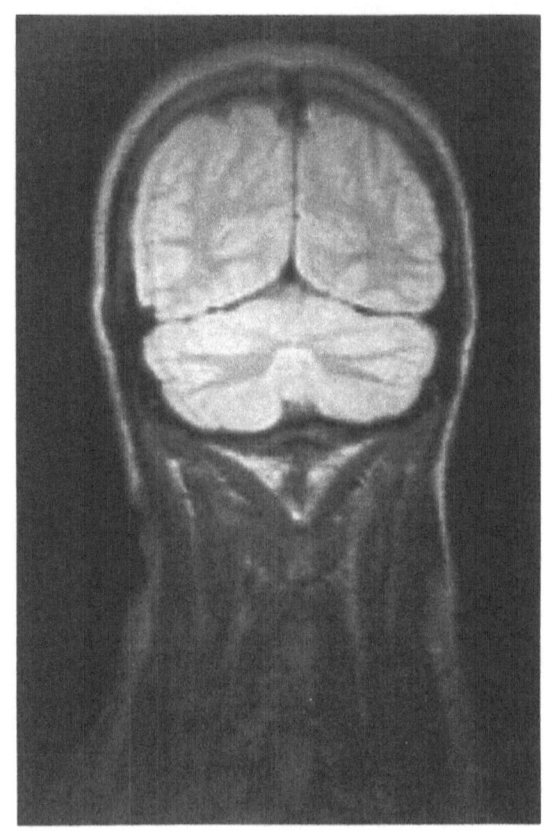

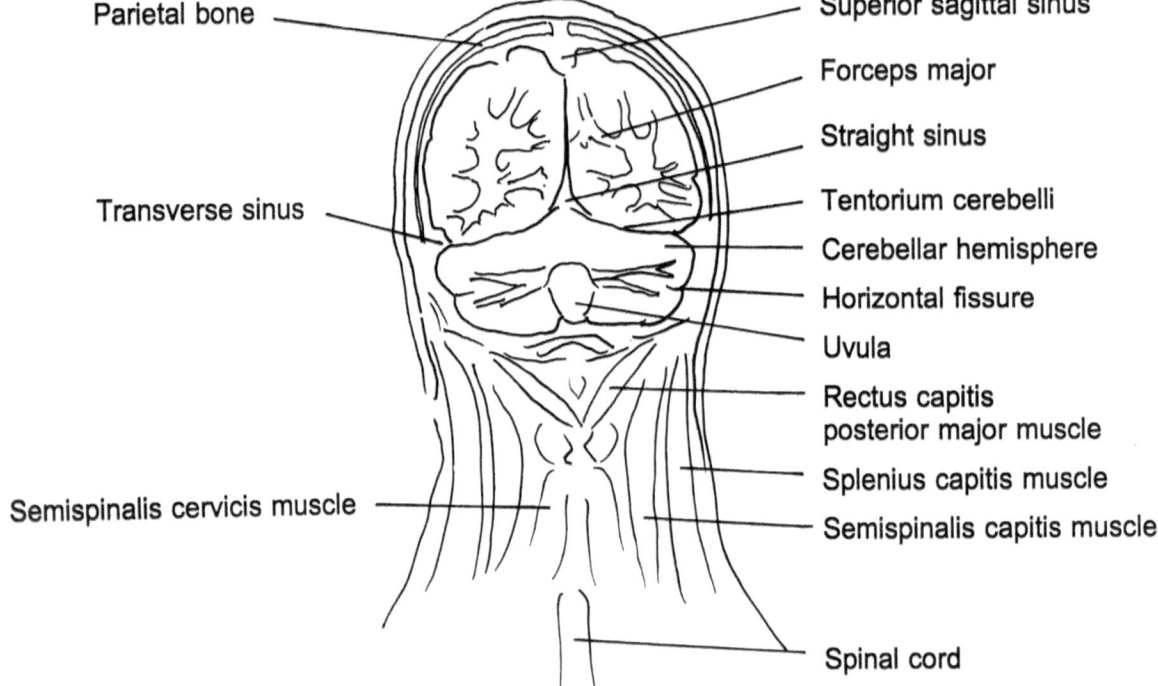

Parietal bone

Superior sagittal sinus

Forceps major

Straight sinus

Transverse sinus

Tentorium cerebelli

Cerebellar hemisphere

Horizontal fissure

Uvula

Rectus capitis
posterior major muscle

Splenius capitis muscle

Semispinalis cervicis muscle

Semispinalis capitis muscle

Spinal cord

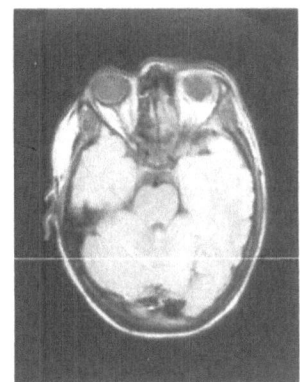

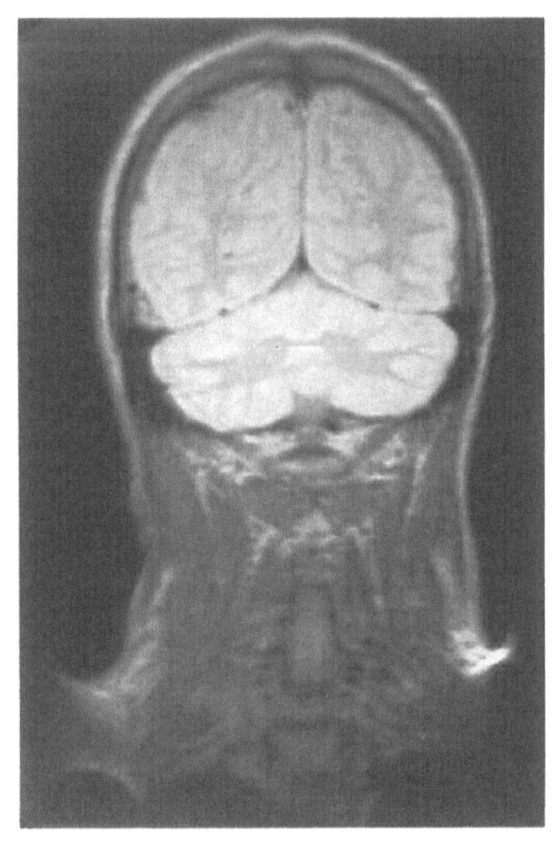

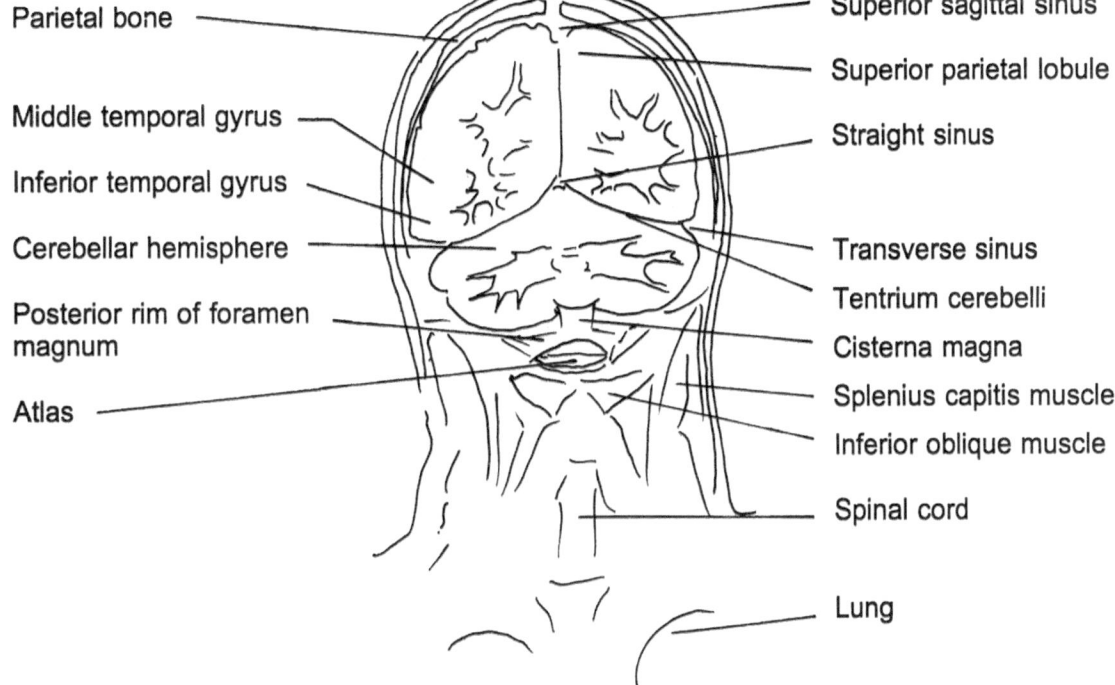

Parietal bone

Middle temporal gyrus

Inferior temporal gyrus

Cerebellar hemisphere

Posterior rim of foramen magnum

Atlas

Superior sagittal sinus

Superior parietal lobule

Straight sinus

Transverse sinus

Tentrium cerebelli

Cisterna magna

Splenius capitis muscle

Inferior oblique muscle

Spinal cord

Lung

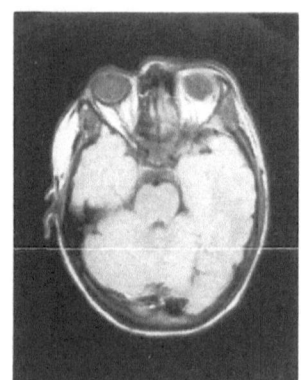

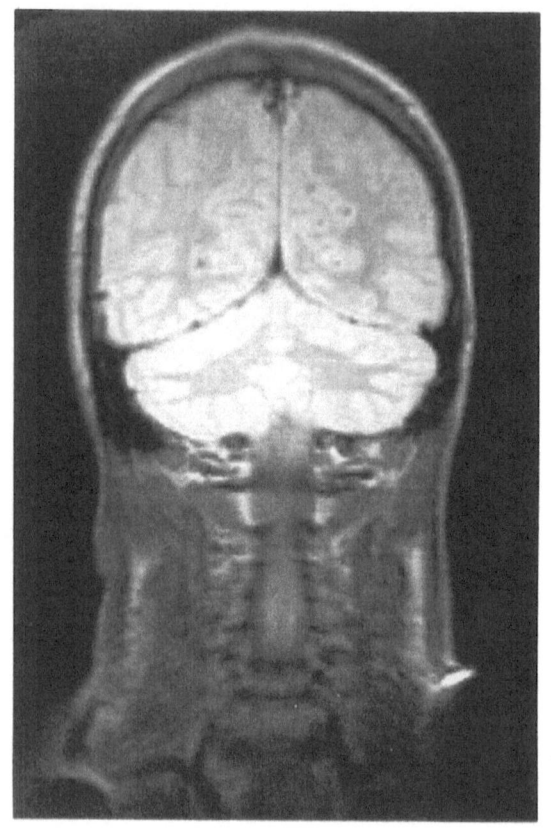

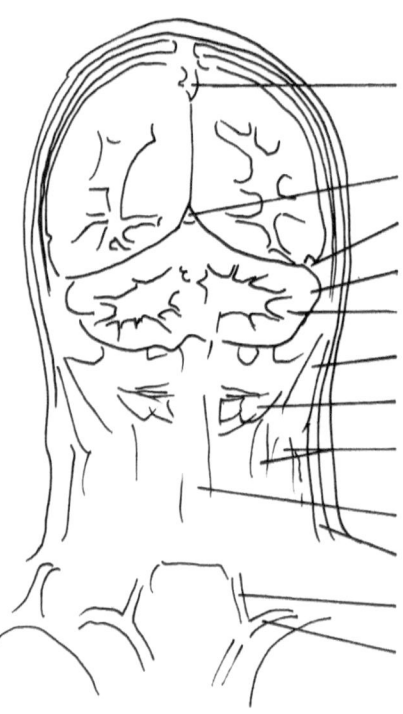

Superior sagittal sinus

Straight sinus

Transverse sinus

Cerebellar hemisphere

Horizontal sulcus

Splenius capitis muscle

Inferior oblique muscle

Levator scapulae and
semispinalis capitis muscles

Spinal cord

Sternocleidomastoid muscle

Vertebral artery

Subclavian artery

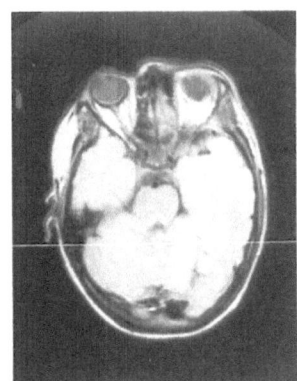

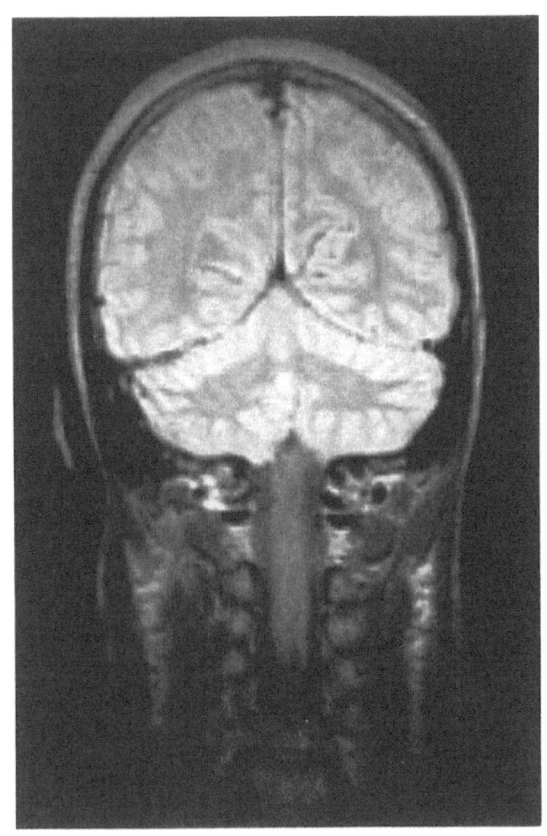

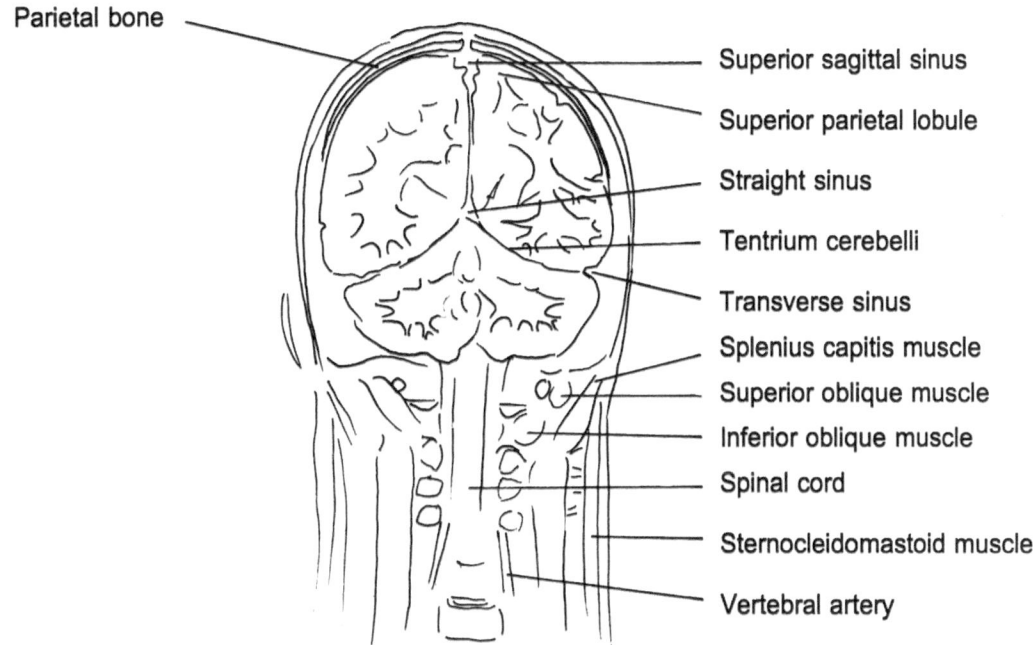

Parietal bone

Superior sagittal sinus

Superior parietal lobule

Straight sinus

Tentrium cerebelli

Transverse sinus

Splenius capitis muscle

Superior oblique muscle

Inferior oblique muscle

Spinal cord

Sternocleidomastoid muscle

Vertebral artery

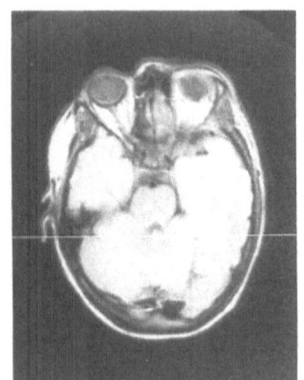

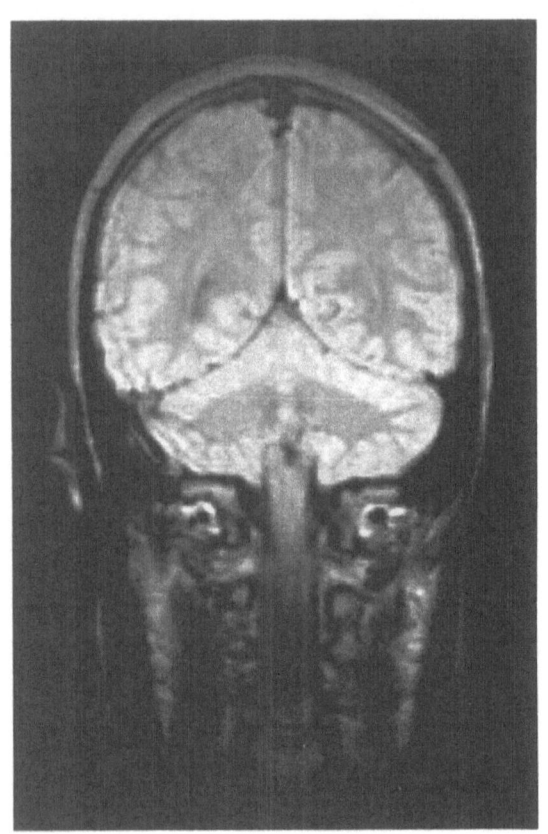

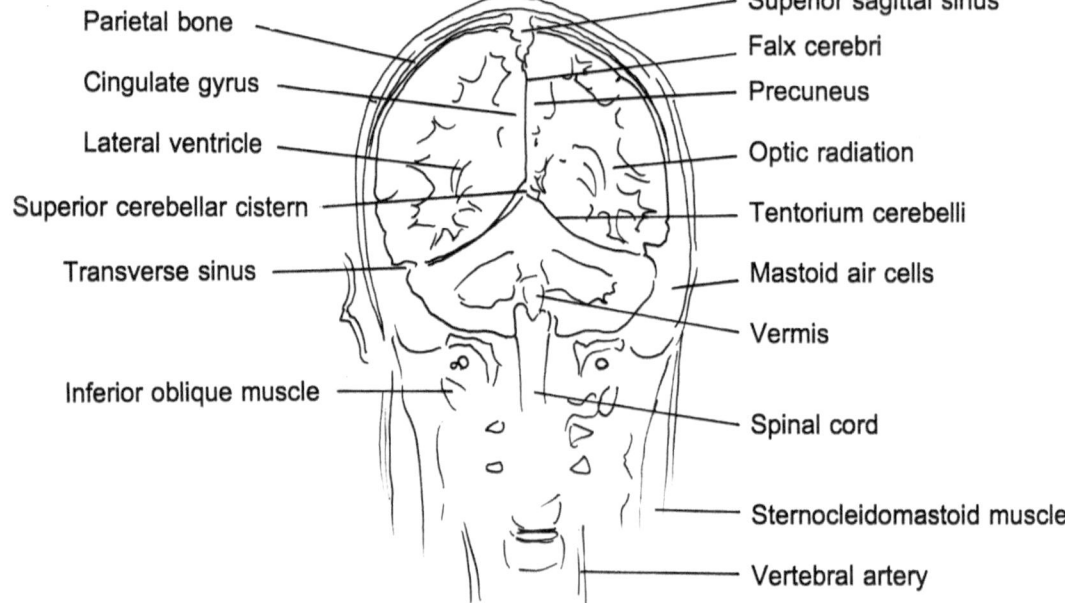

Parietal bone

Cingulate gyrus

Lateral ventricle

Superior cerebellar cistern

Transverse sinus

Inferior oblique muscle

Superior sagittal sinus

Falx cerebri

Precuneus

Optic radiation

Tentorium cerebelli

Mastoid air cells

Vermis

Spinal cord

Sternocleidomastoid muscle

Vertebral artery

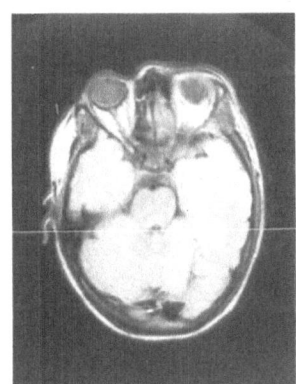

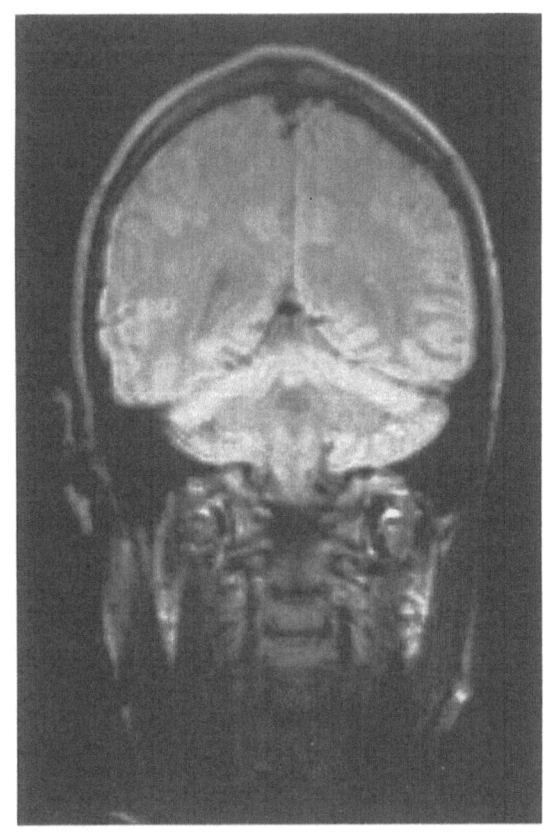

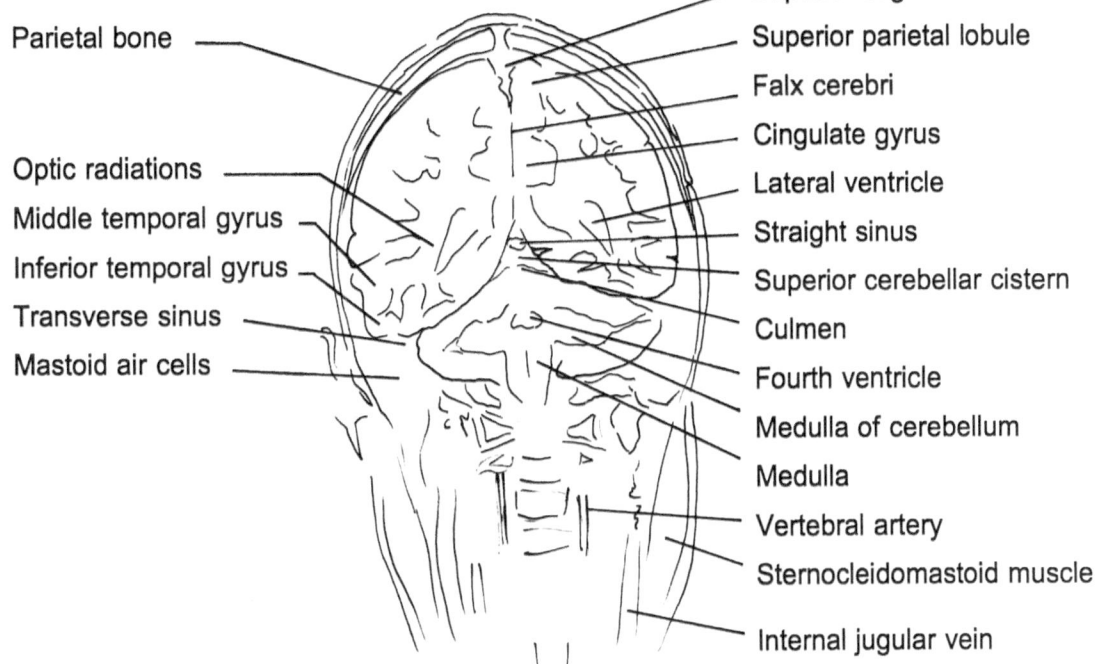

Parietal bone

Optic radiations

Middle temporal gyrus

Inferior temporal gyrus

Transverse sinus

Mastoid air cells

Superior sagittal sinus

Superior parietal lobule

Falx cerebri

Cingulate gyrus

Lateral ventricle

Straight sinus

Superior cerebellar cistern

Culmen

Fourth ventricle

Medulla of cerebellum

Medulla

Vertebral artery

Sternocleidomastoid muscle

Internal jugular vein

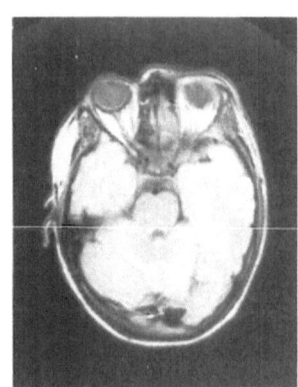

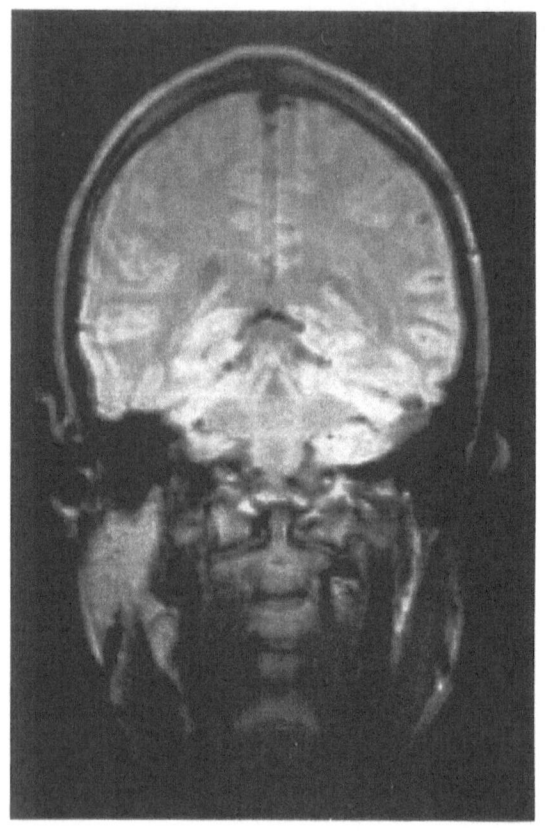

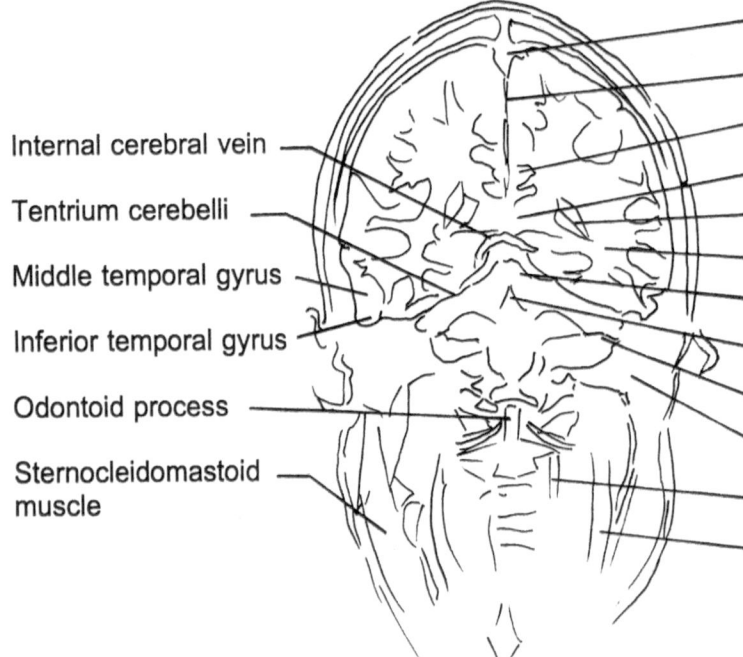

Superior sagittal sinus

Falx cerebri

Cingulate gyrus

Splenium of corpus callosum

Internal cerebral vein

Lateral ventricle

Tentrium cerebelli

Optic radiation

Middle temporal gyrus

Middle cerebellar peduncle

Inferior temporal gyrus

Fourth ventricle

Odontoid process

Horizontal fissure

Sternocleidomastoid
muscle

Mastoid air cells

Vertebral artery

Internal jugular vein

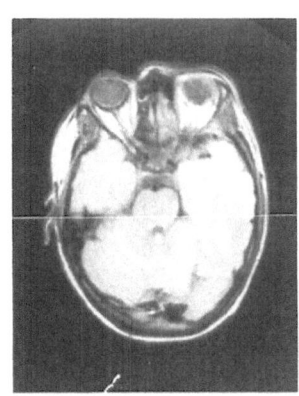

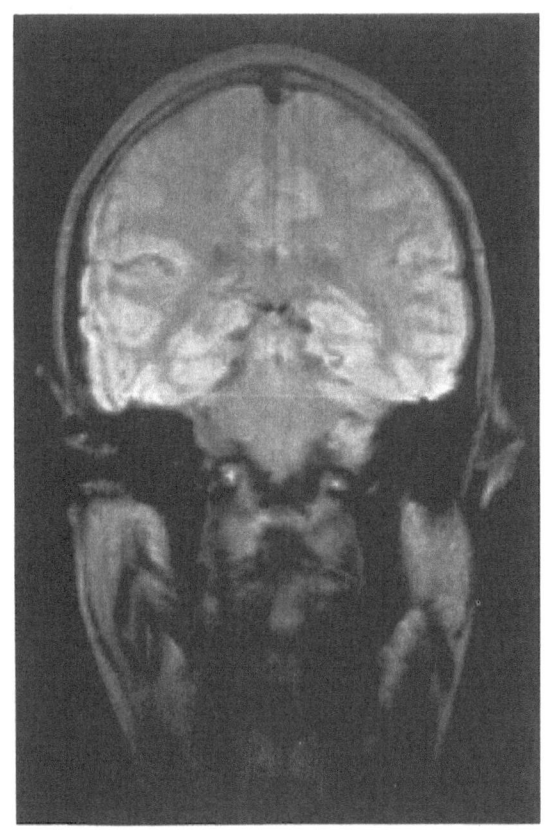

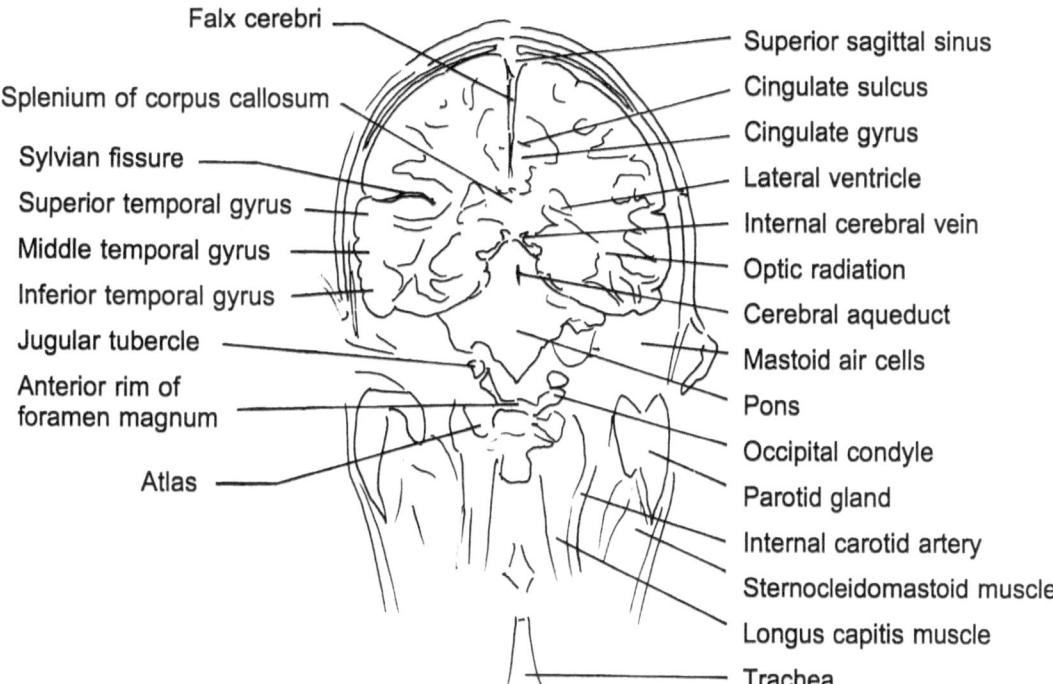

Falx cerebri

Splenium of corpus callosum

Sylvian fissure

Superior temporal gyrus

Middle temporal gyrus

Inferior temporal gyrus

Jugular tubercle

Anterior rim of foramen magnum

Atlas

Superior sagittal sinus

Cingulate sulcus

Cingulate gyrus

Lateral ventricle

Internal cerebral vein

Optic radiation

Cerebral aqueduct

Mastoid air cells

Pons

Occipital condyle

Parotid gland

Internal carotid artery

Sternocleidomastoid muscle

Longus capitis muscle

Trachea

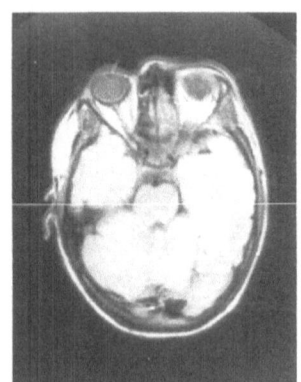

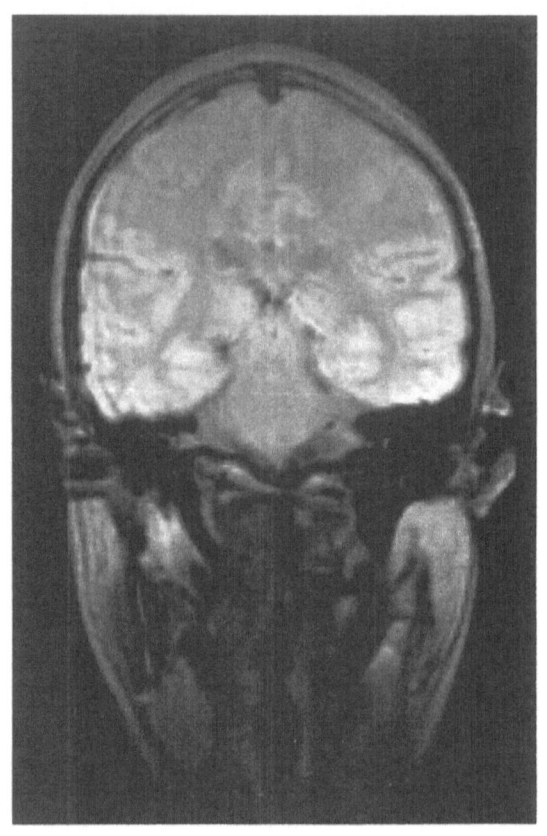

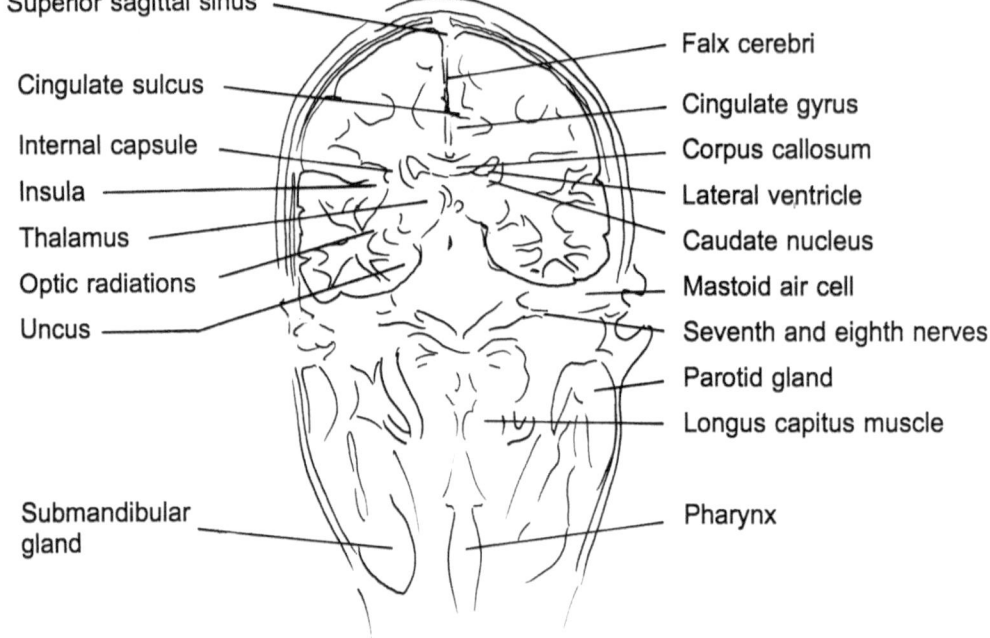

Superior sagittal sinus

Cingulate sulcus

Internal capsule

Insula

Thalamus

Optic radiations

Uncus

Submandibular gland

Falx cerebri

Cingulate gyrus

Corpus callosum

Lateral ventricle

Caudate nucleus

Mastoid air cell

Seventh and eighth nerves

Parotid gland

Longus capitus muscle

Pharynx

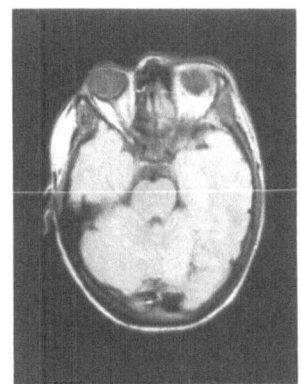

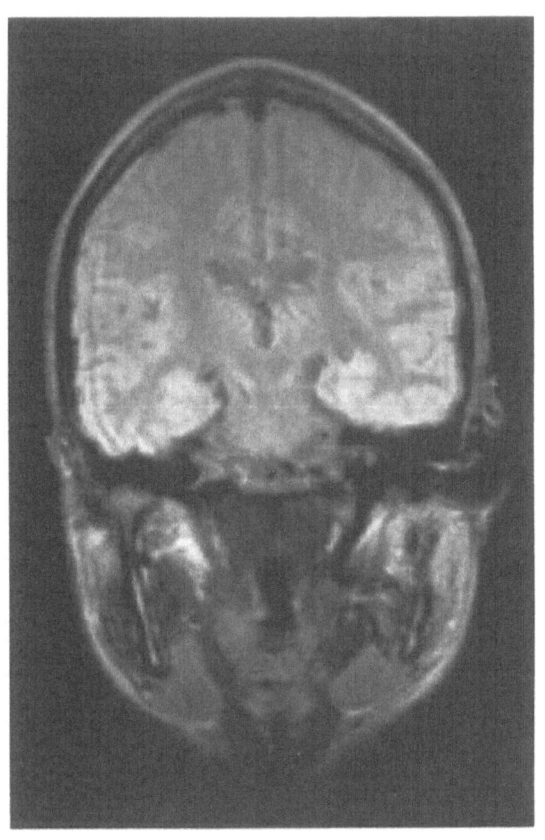

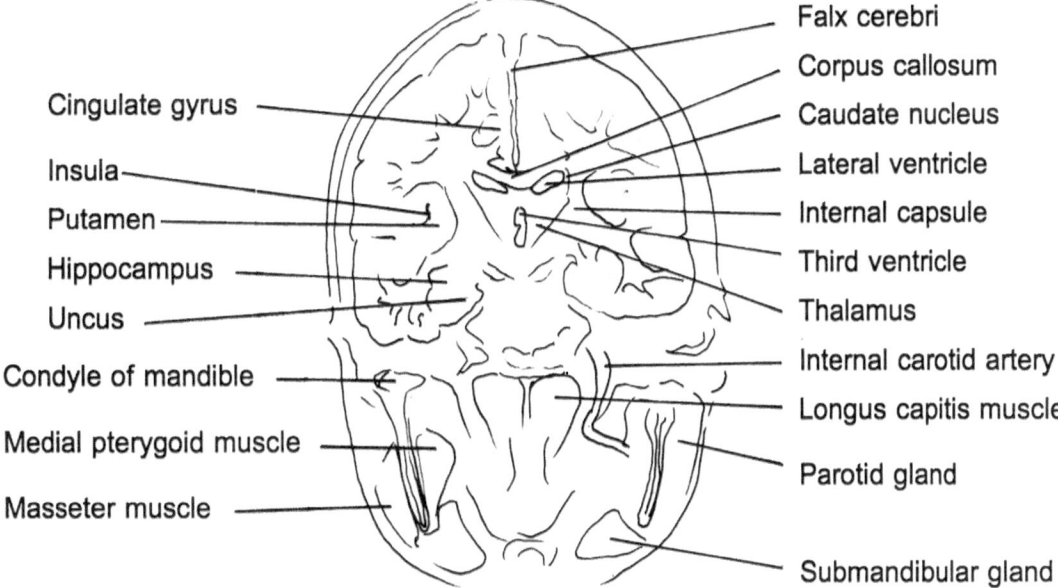

Cingulate gyrus

Insula

Putamen

Hippocampus

Uncus

Condyle of mandible

Medial pterygoid muscle

Masseter muscle

Falx cerebri

Corpus callosum

Caudate nucleus

Lateral ventricle

Internal capsule

Third ventricle

Thalamus

Internal carotid artery

Longus capitis muscle

Parotid gland

Submandibular gland

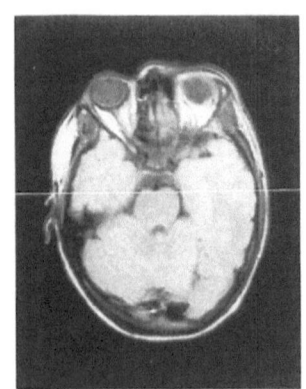

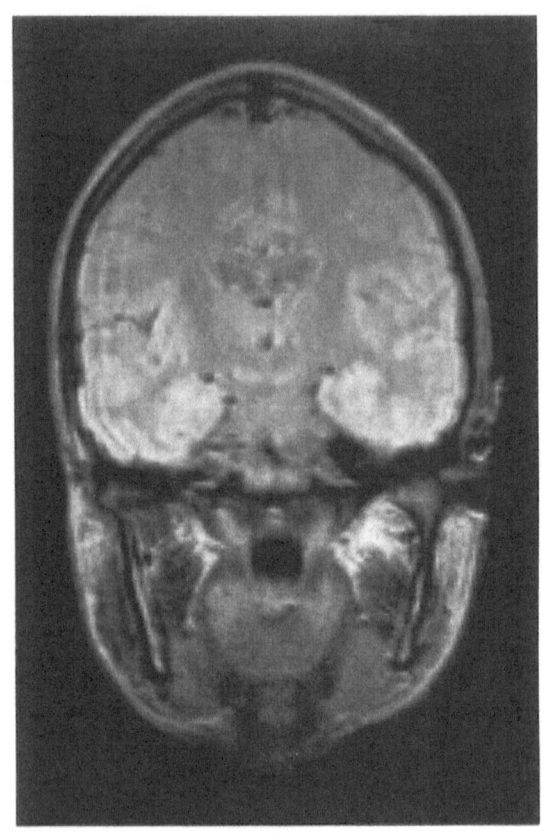

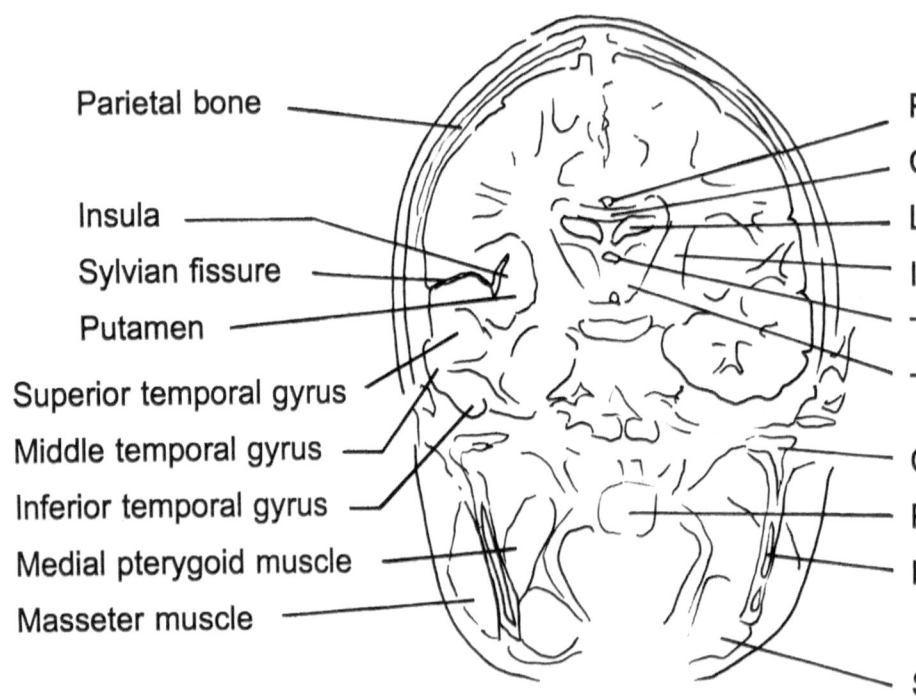

Parietal bone

Insula

Sylvian fissure

Putamen

Superior temporal gyrus

Middle temporal gyrus

Inferior temporal gyrus

Medial pterygoid muscle

Masseter muscle

Pericallosal artery

Corpus callosum

Lateral ventricle

Internal capsule

Third ventricle

Thalamus

Condyle of mandible

Pharynx

Ramus of mandible

Submandibular gland

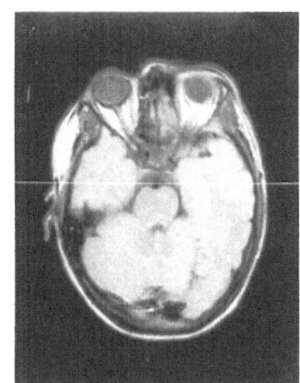
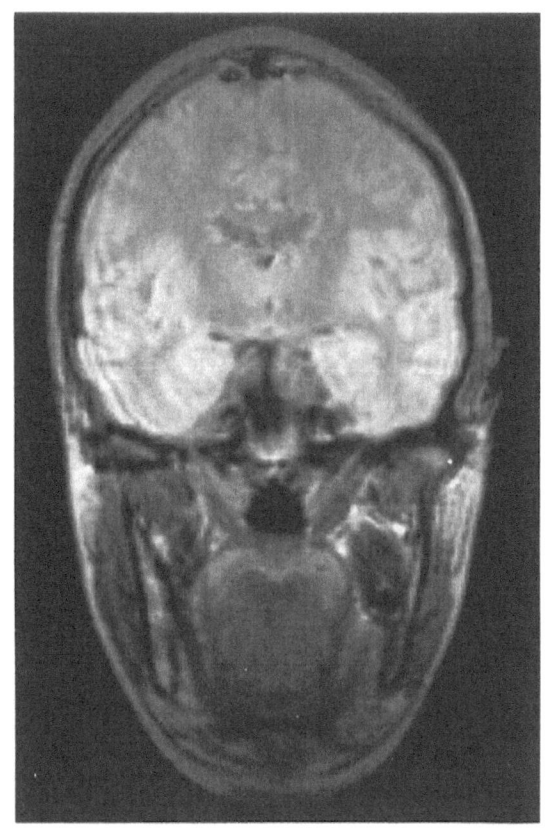

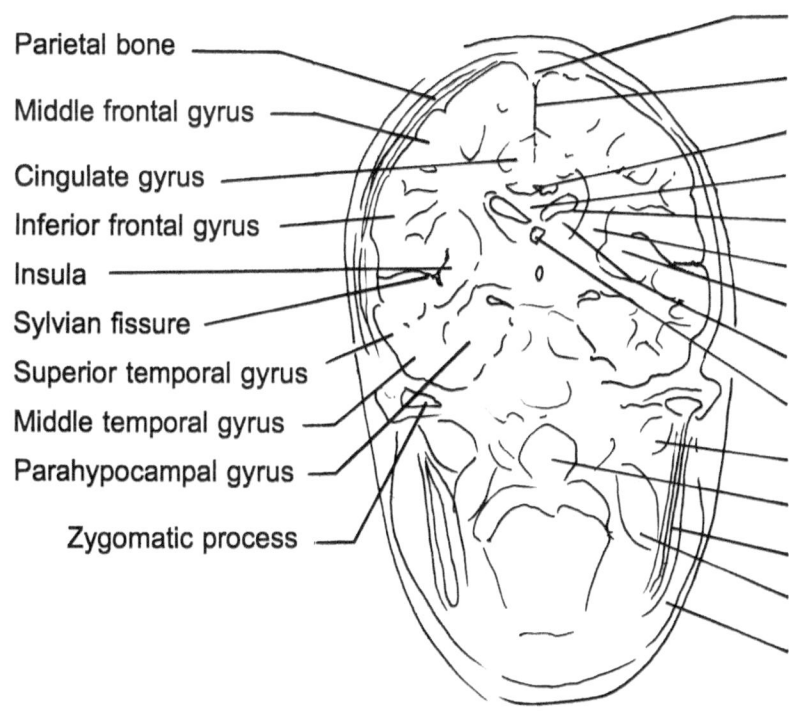

Parietal bone

Middle frontal gyrus

Cingulate gyrus

Inferior frontal gyrus

Insula

Sylvian fissure

Superior temporal gyrus

Middle temporal gyrus

Parahypocampal gyrus

Zygomatic process

Superior sagittal sinus

Falx cerebri

Pericallosal artery

Corpus callosum

Lateral ventricle

Internal capsule

Putamen

Caudate nucleus

Third ventricle

Lateral pterygoid muscle

Pharynx

Ramus of mandible

Medial pterygoid muscle

Masseter muscle

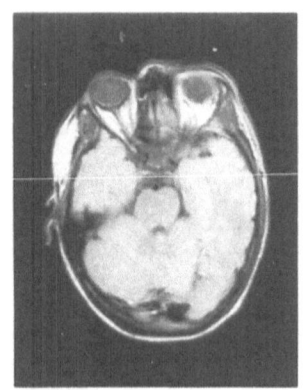

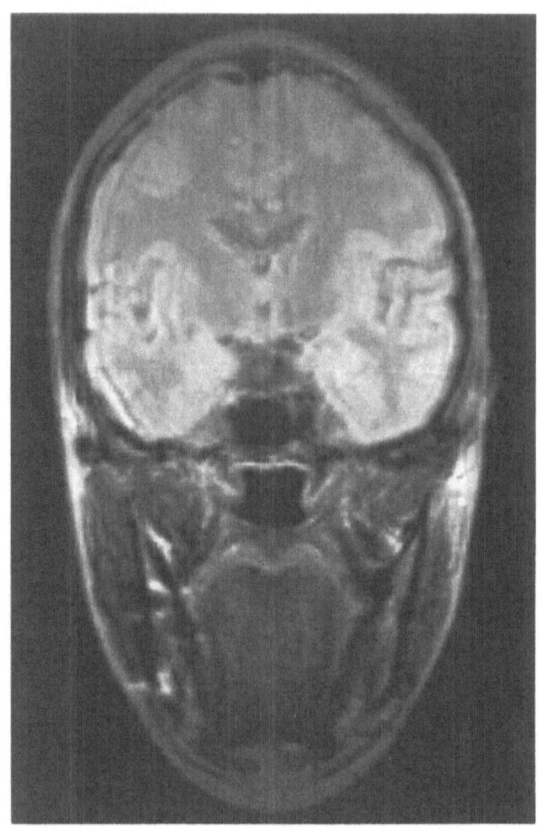

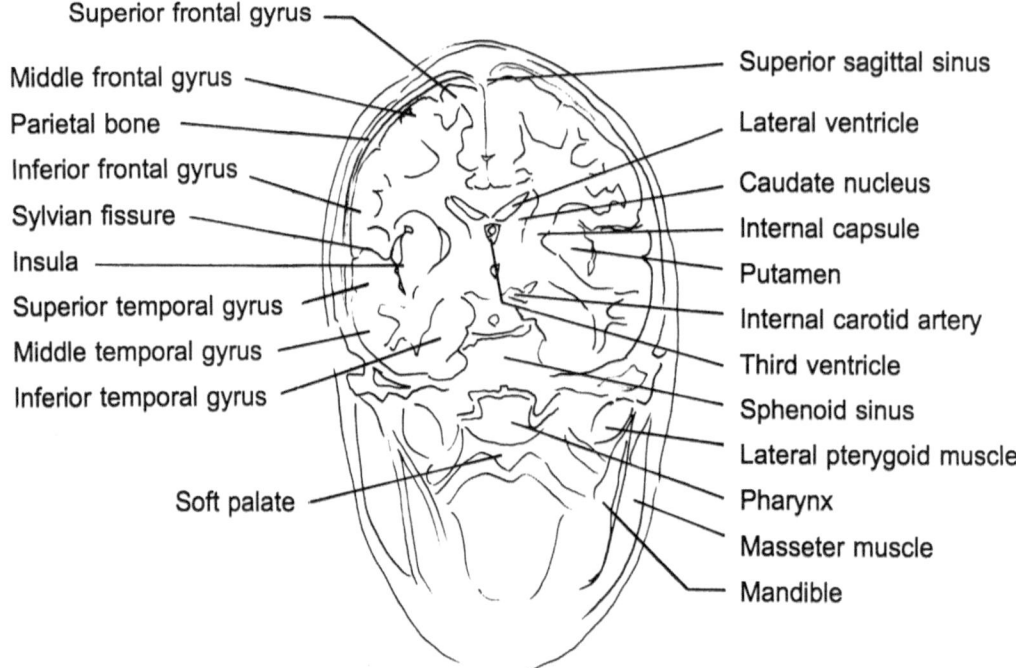

Superior frontal gyrus	Superior sagittal sinus
Middle frontal gyrus	Lateral ventricle
Parietal bone	Caudate nucleus
Inferior frontal gyrus	Internal capsule
Sylvian fissure	Putamen
Insula	Internal carotid artery
Superior temporal gyrus	Third ventricle
Middle temporal gyrus	Sphenoid sinus
Inferior temporal gyrus	Lateral pterygoid muscle
Soft palate	Pharynx
	Masseter muscle
	Mandible

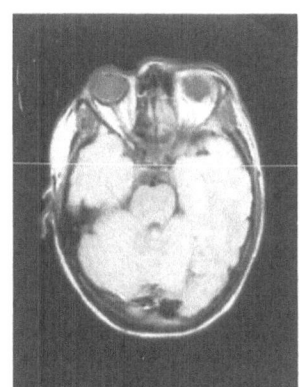

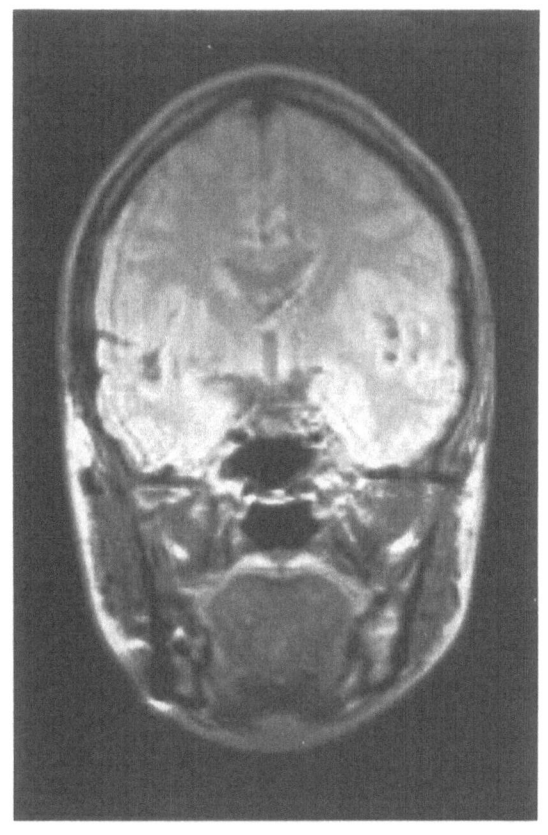

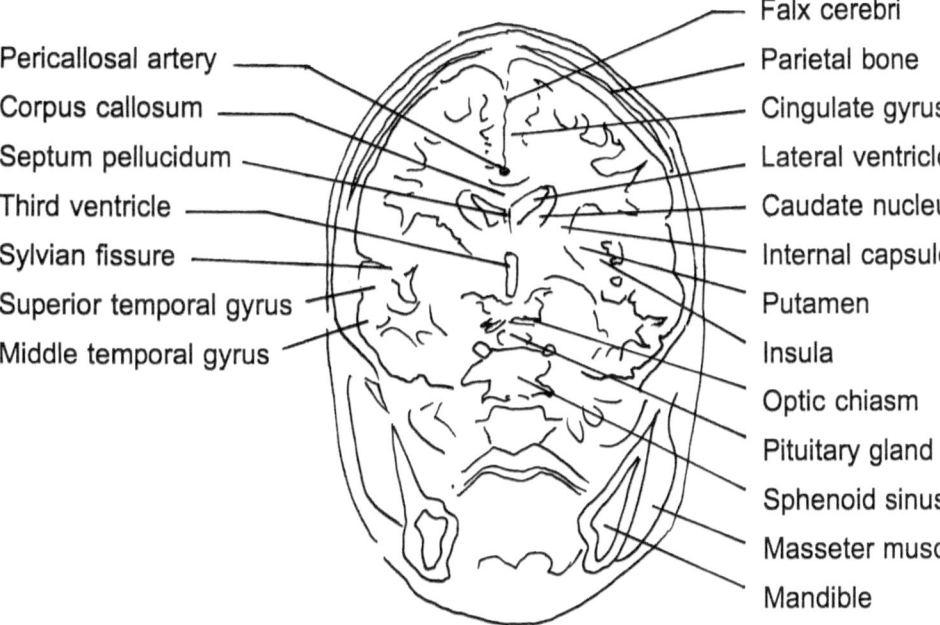

Pericallosal artery

Corpus callosum

Septum pellucidum

Third ventricle

Sylvian fissure

Superior temporal gyrus

Middle temporal gyrus

Falx cerebri

Parietal bone

Cingulate gyrus

Lateral ventricle

Caudate nucleus

Internal capsule

Putamen

Insula

Optic chiasm

Pituitary gland

Sphenoid sinus

Masseter muscle

Mandible

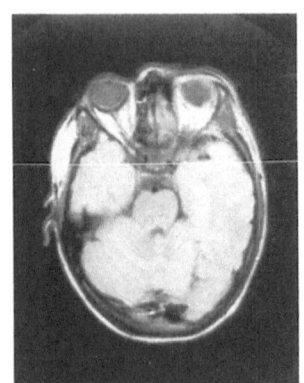

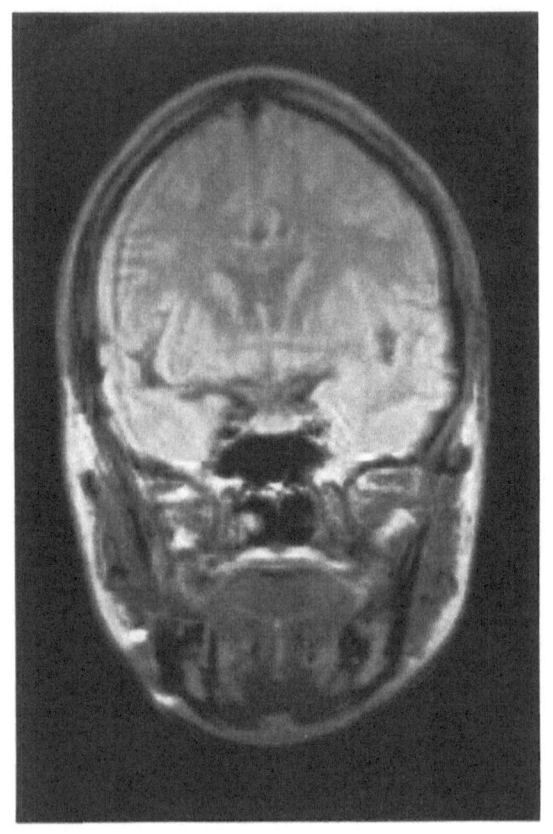

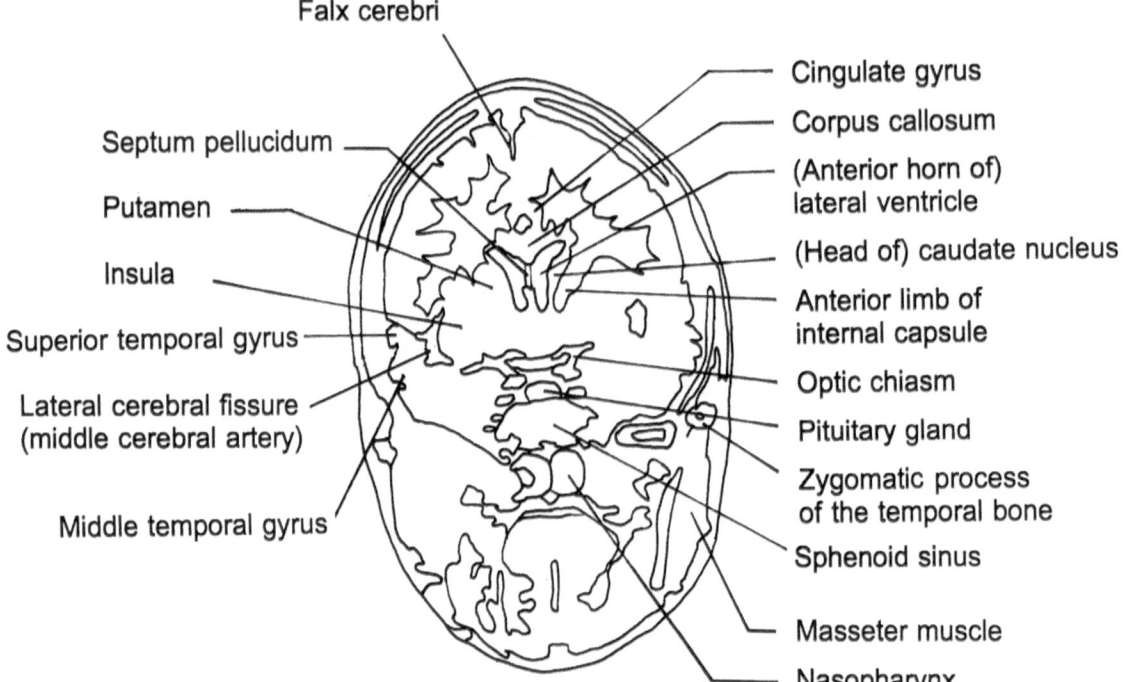

Falx cerebri

Cingulate gyrus

Corpus callosum

Septum pellucidum

(Anterior horn of) lateral ventricle

Putamen

(Head of) caudate nucleus

Insula

Anterior limb of internal capsule

Superior temporal gyrus

Optic chiasm

Lateral cerebral fissure (middle cerebral artery)

Pituitary gland

Zygomatic process of the temporal bone

Middle temporal gyrus

Sphenoid sinus

Masseter muscle

Nasopharynx

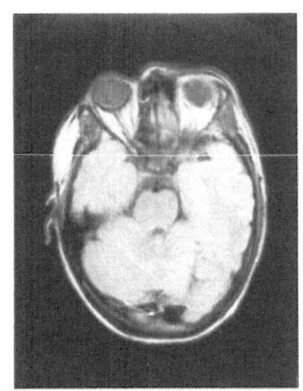

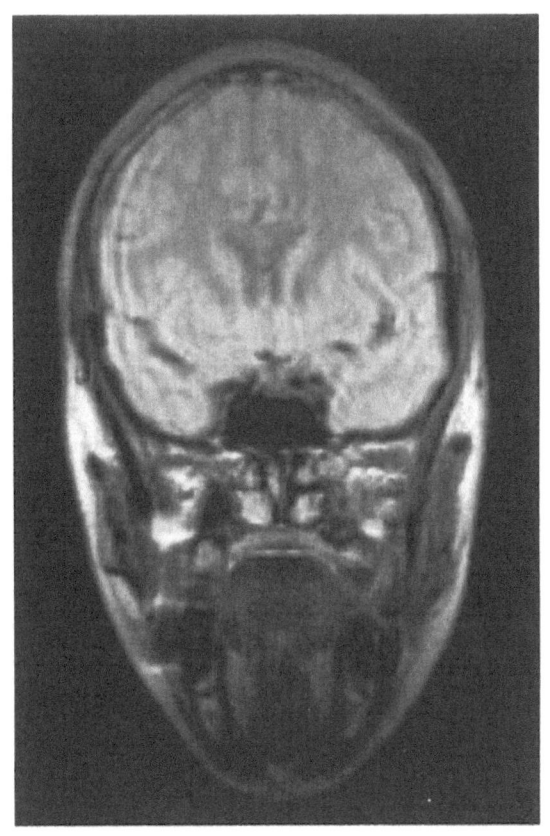

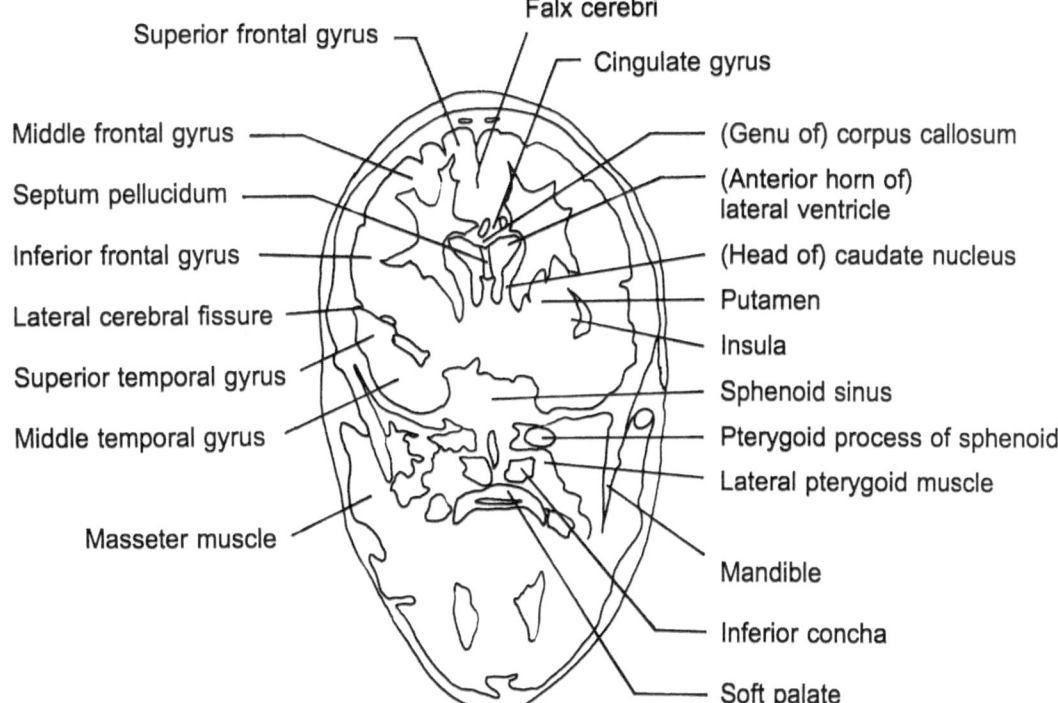

Superior frontal gyrus

Falx cerebri

Cingulate gyrus

Middle frontal gyrus

Septum pellucidum

Inferior frontal gyrus

Lateral cerebral fissure

Superior temporal gyrus

Middle temporal gyrus

Masseter muscle

(Genu of) corpus callosum

(Anterior horn of) lateral ventricle

(Head of) caudate nucleus

Putamen

Insula

Sphenoid sinus

Pterygoid process of sphenoid

Lateral pterygoid muscle

Mandible

Inferior concha

Soft palate

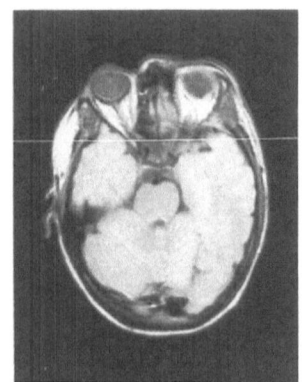

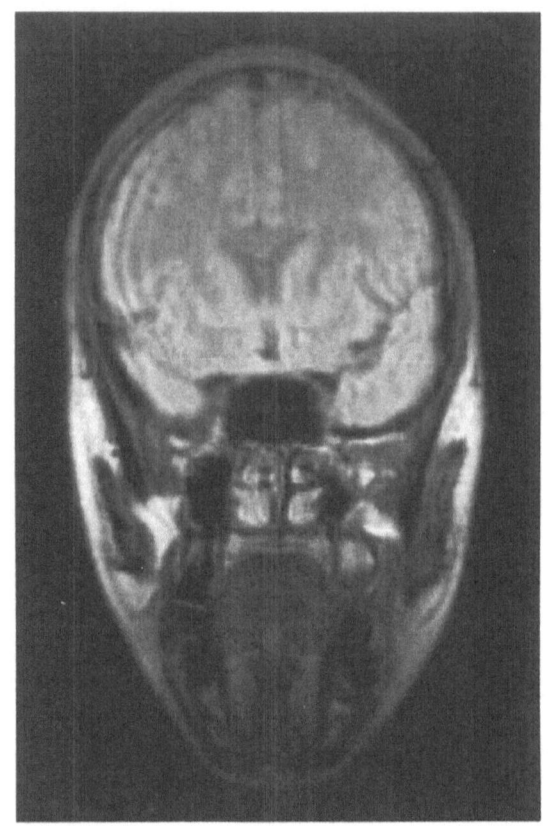

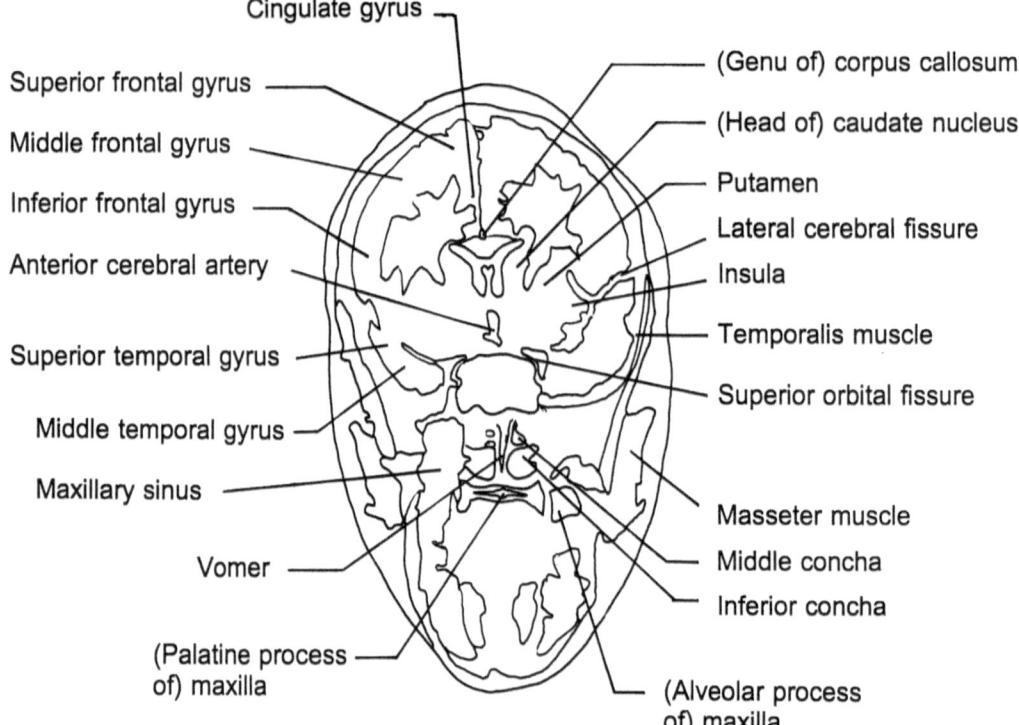

Cingulate gyrus

Superior frontal gyrus

Middle frontal gyrus

Inferior frontal gyrus

Anterior cerebral artery

Superior temporal gyrus

Middle temporal gyrus

Maxillary sinus

Vomer

(Palatine process of) maxilla

(Genu of) corpus callosum

(Head of) caudate nucleus

Putamen

Lateral cerebral fissure

Insula

Temporalis muscle

Superior orbital fissure

Masseter muscle

Middle concha

Inferior concha

(Alveolar process of) maxilla

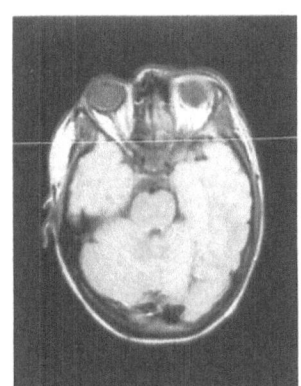

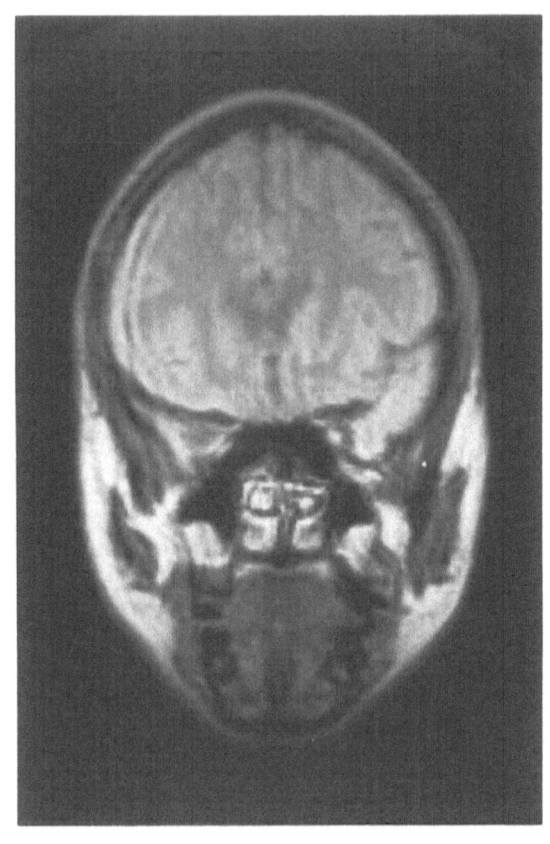

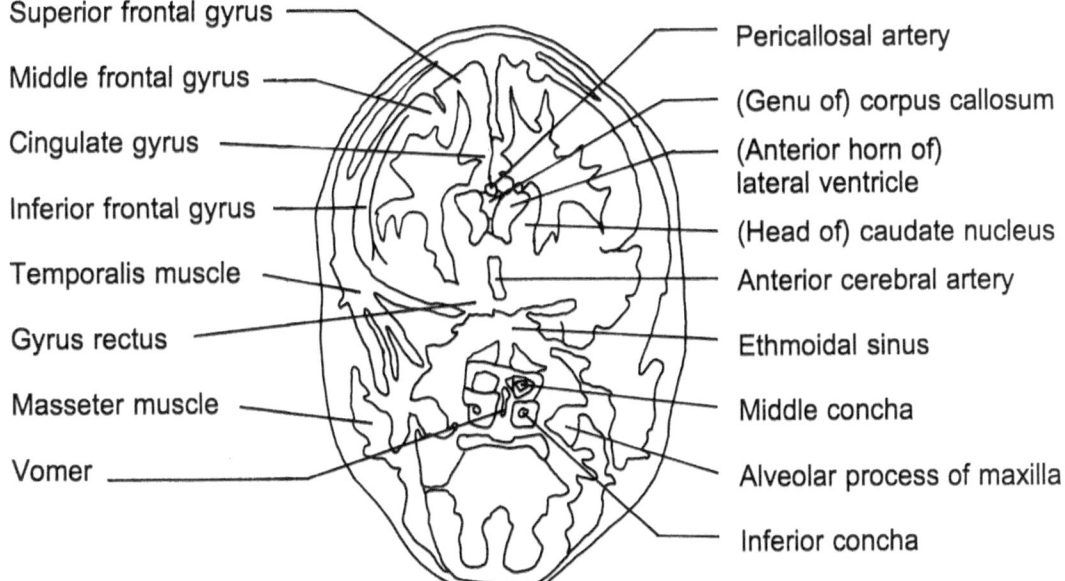

Superior frontal gyrus ⎯ Pericallosal artery

Middle frontal gyrus ⎯ (Genu of) corpus callosum

Cingulate gyrus ⎯ (Anterior horn of) lateral ventricle

Inferior frontal gyrus ⎯ (Head of) caudate nucleus

Temporalis muscle ⎯ Anterior cerebral artery

Gyrus rectus ⎯ Ethmoidal sinus

Masseter muscle ⎯ Middle concha

Vomer ⎯ Alveolar process of maxilla

Inferior concha

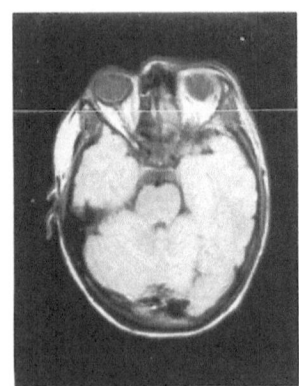

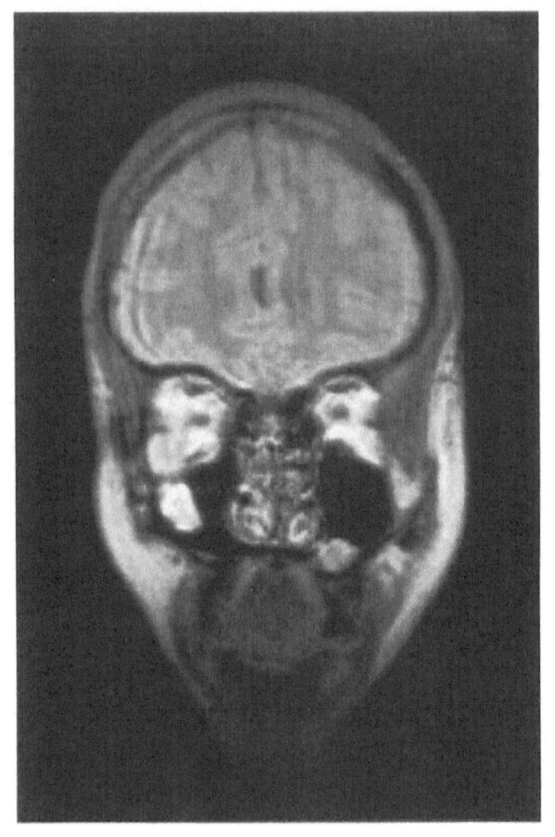

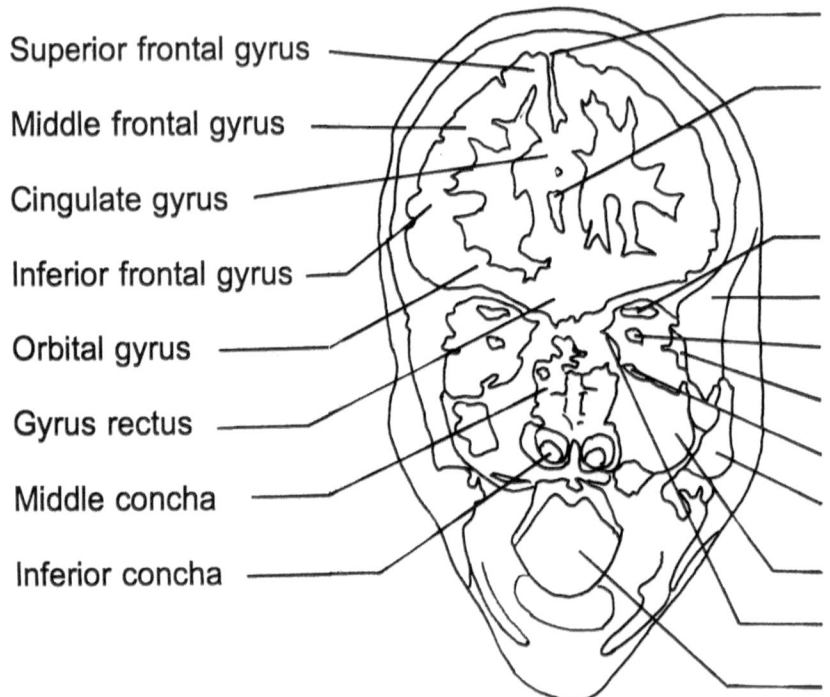

Superior frontal gyrus

Middle frontal gyrus

Cingulate gyrus

Inferior frontal gyrus

Orbital gyrus

Gyrus rectus

Middle concha

Inferior concha

Superior sagittal sinus

Pericallosal artery
(anterior cerebral artery)

Superior rectus muscle

Temporalis muscle

Optic nerve

Lateral rectus muscle

Inferior rectus muscle

Masseter muscle

Maxillary sinus

Medial rectus muscle

Tongue

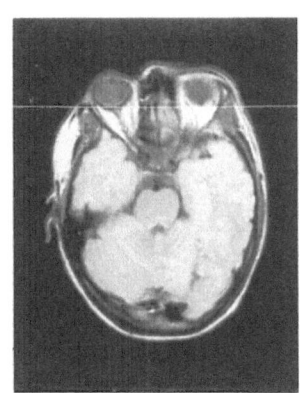

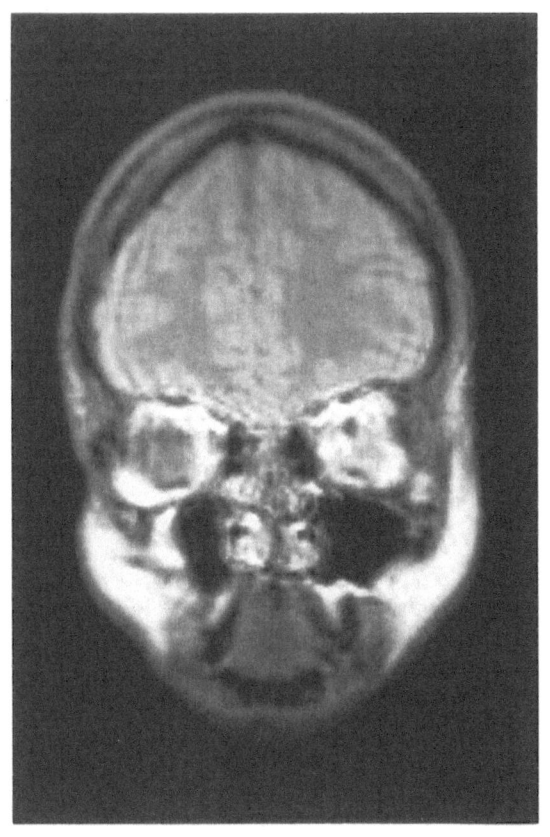

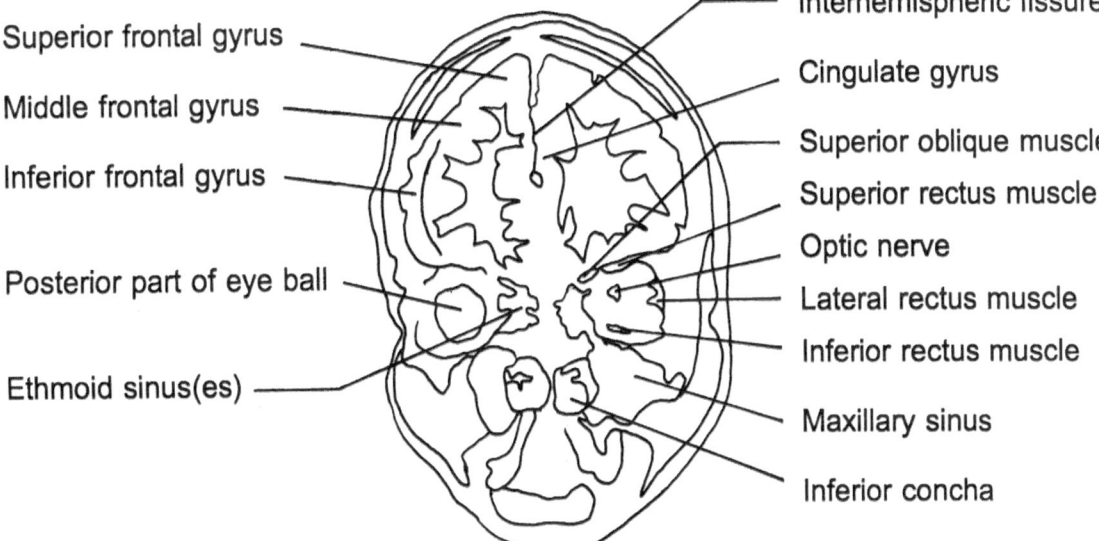

Superior frontal gyrus

Middle frontal gyrus

Inferior frontal gyrus

Posterior part of eye ball

Ethmoid sinus(es)

Interhemispheric fissure

Cingulate gyrus

Superior oblique muscle

Superior rectus muscle

Optic nerve

Lateral rectus muscle

Inferior rectus muscle

Maxillary sinus

Inferior concha

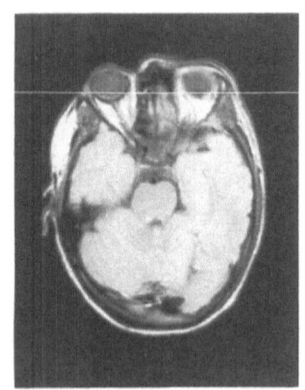

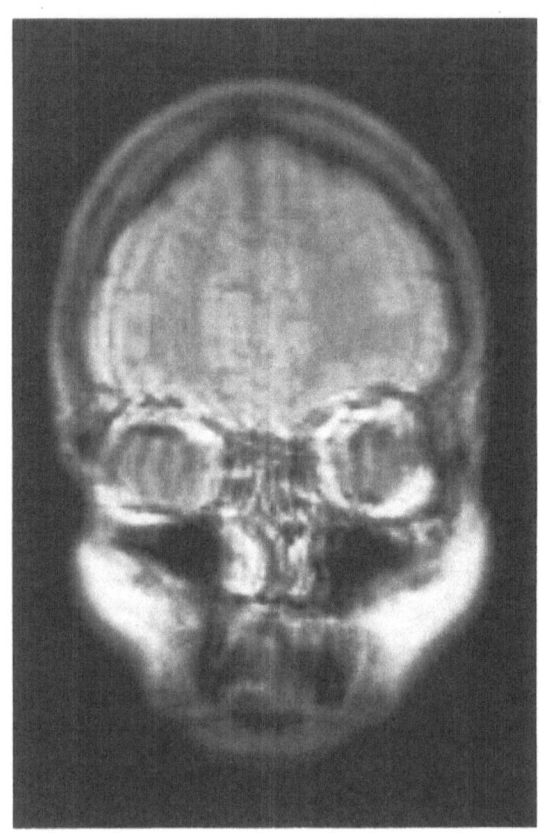

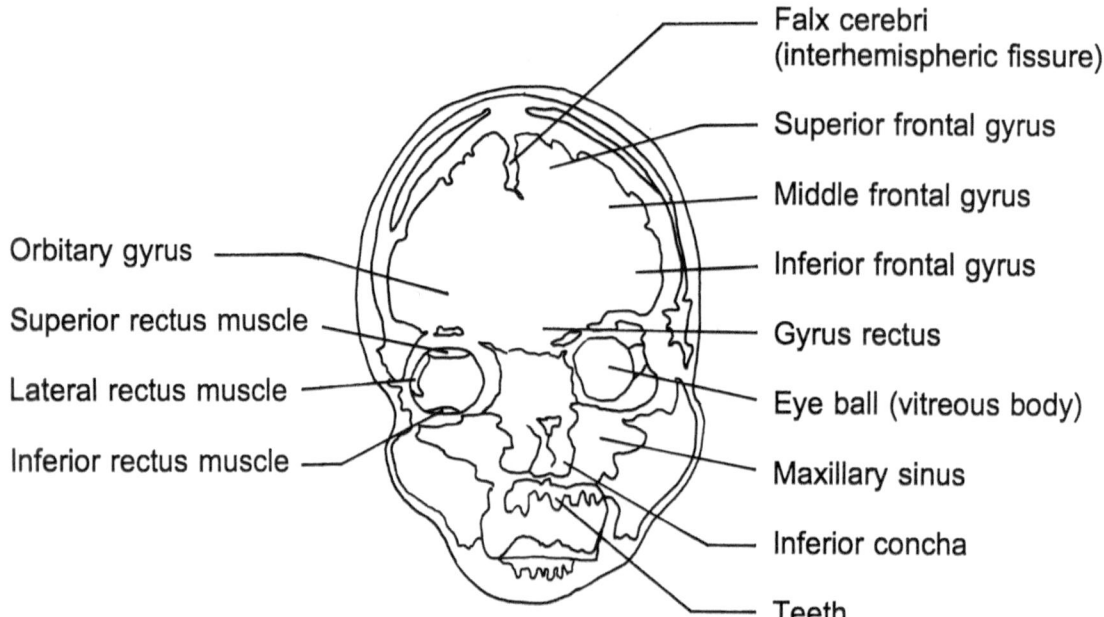

Falx cerebri
(interhemispheric fissure)

Superior frontal gyrus

Middle frontal gyrus

Inferior frontal gyrus

Orbitary gyrus

Superior rectus muscle

Gyrus rectus

Lateral rectus muscle

Eye ball (vitreous body)

Inferior rectus muscle

Maxillary sinus

Inferior concha

Teeth

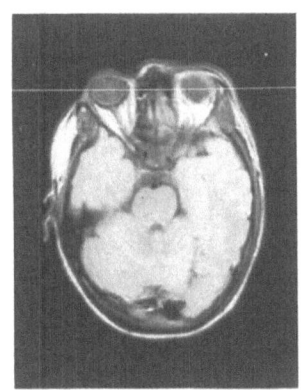

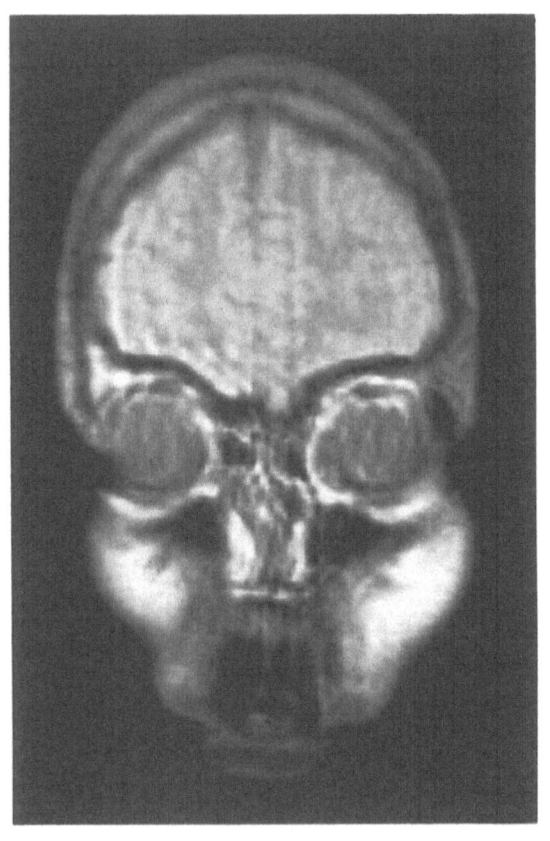

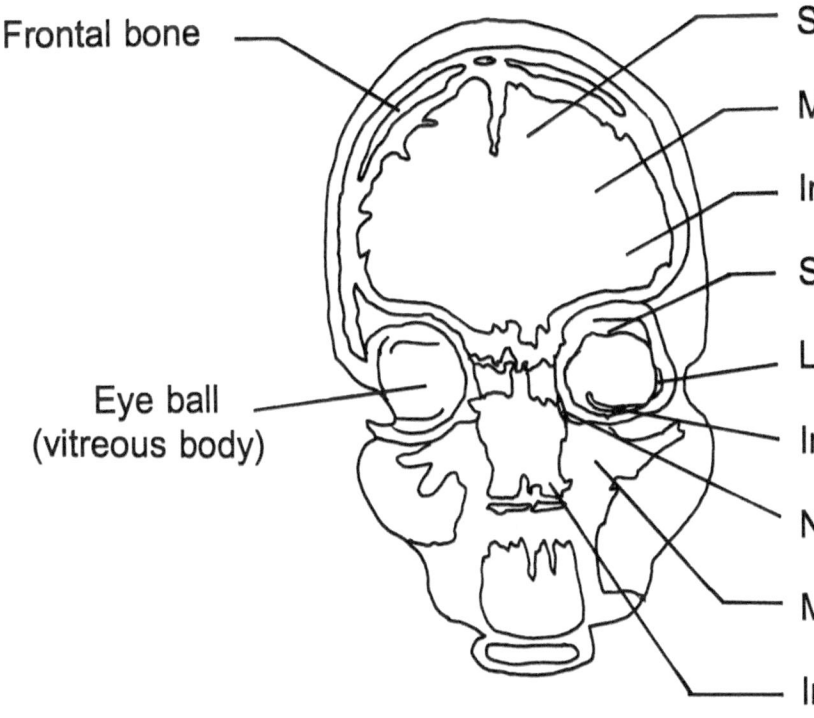

Frontal bone

Superior frontal gyrus

Middle frontal gyrus

Inferior frontal gyrus

Superior rectus muscle

Lateral rectus muscle

Eye ball
(vitreous body)

Inferior rectus muscle

Nasolacrimal duct

Maxillary sinus

Inferior concha

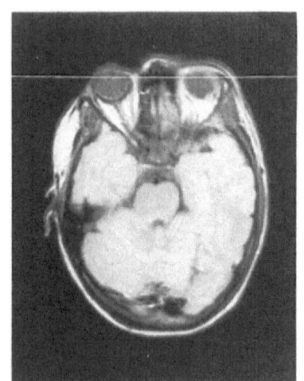

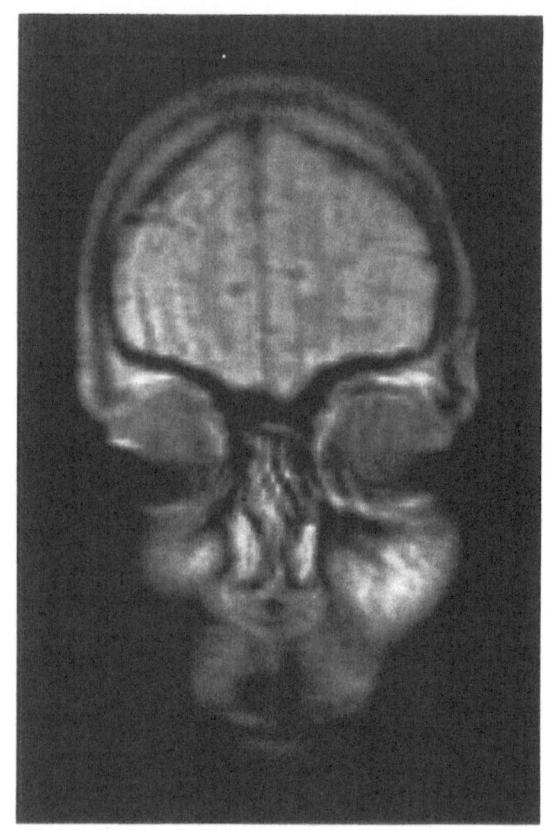

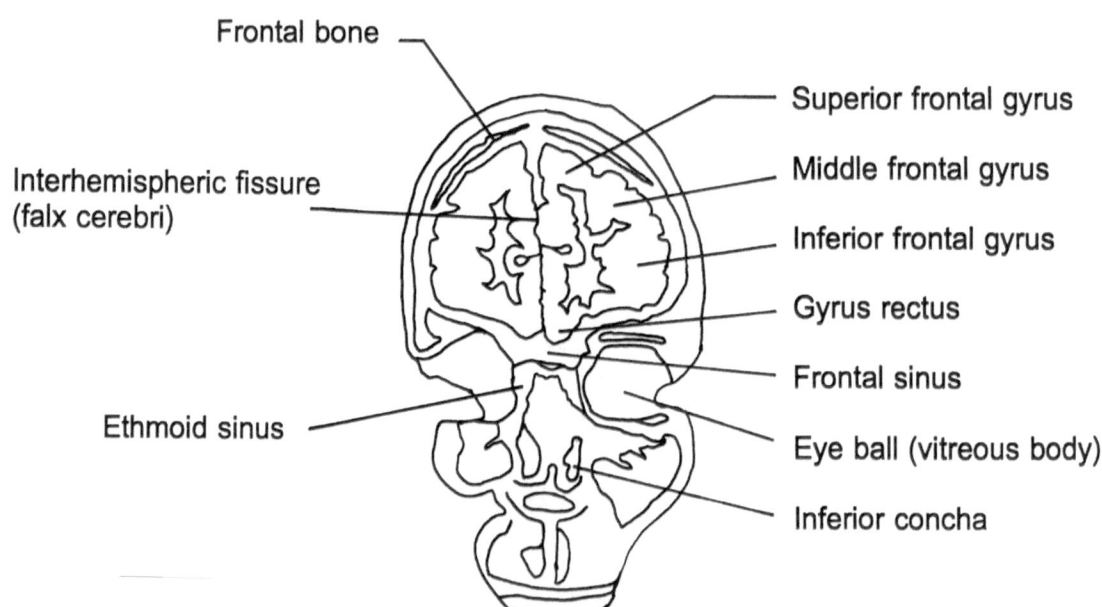

Frontal bone

Superior frontal gyrus

Middle frontal gyrus

Interhemispheric fissure
(falx cerebri)

Inferior frontal gyrus

Gyrus rectus

Frontal sinus

Ethmoid sinus

Eye ball (vitreous body)

Inferior concha

THORAX

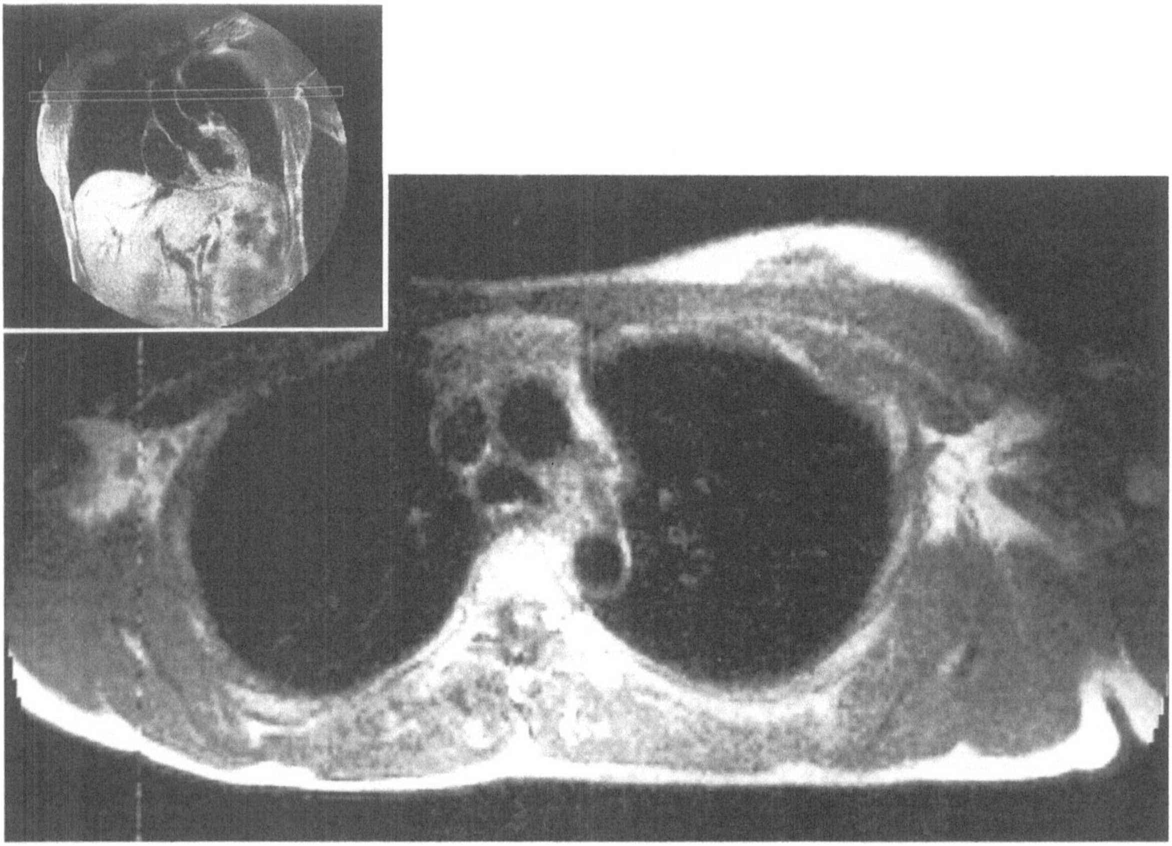

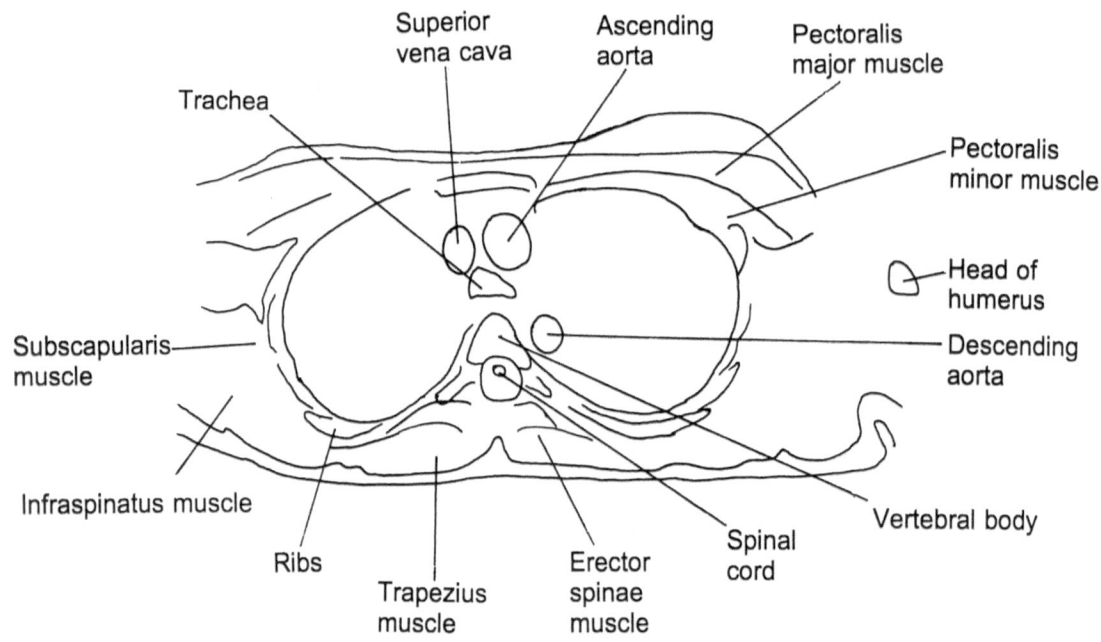

Superior
vena cava

Trachea

Ascending
aorta

Pectoralis
major muscle

Pectoralis
minor muscle

Head of
humerus

Subscapularis
muscle

Descending
aorta

Infraspinatus muscle

Ribs

Trapezius
muscle

Erector
spinae
muscle

Spinal
cord

Vertebral body

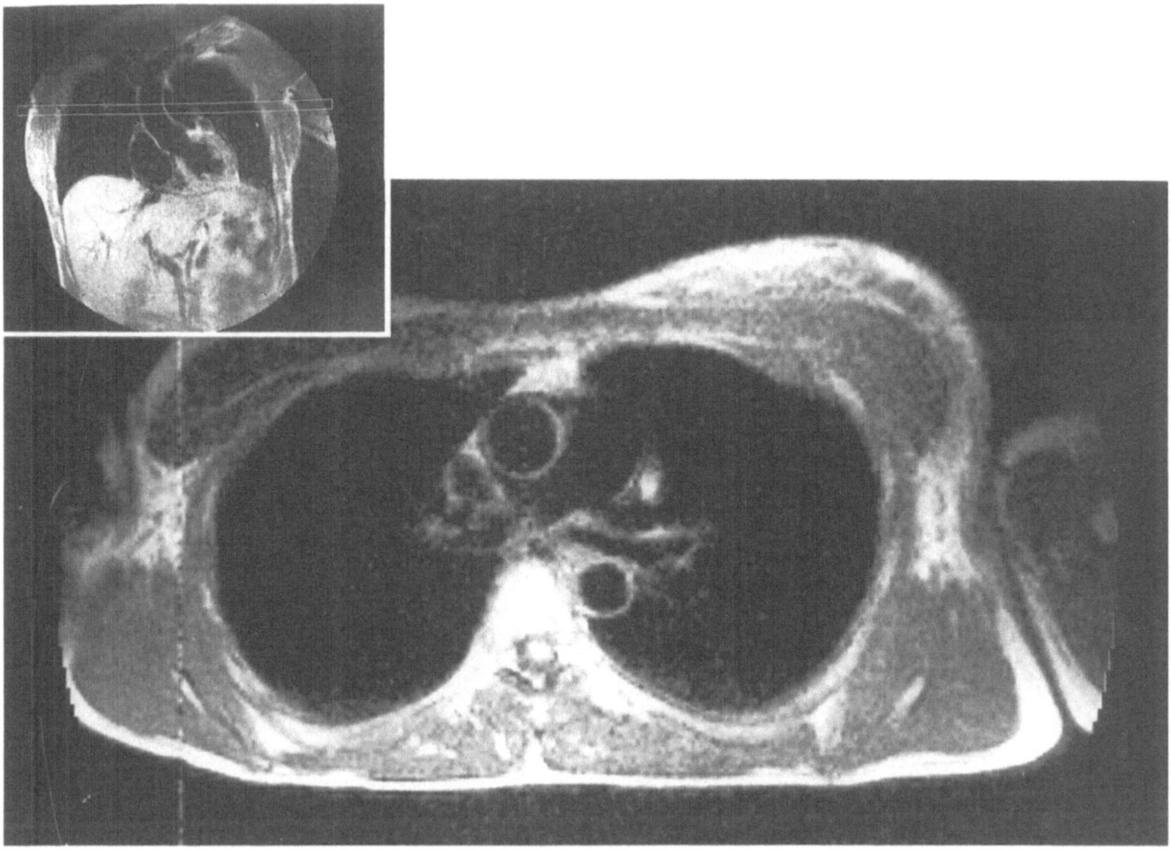

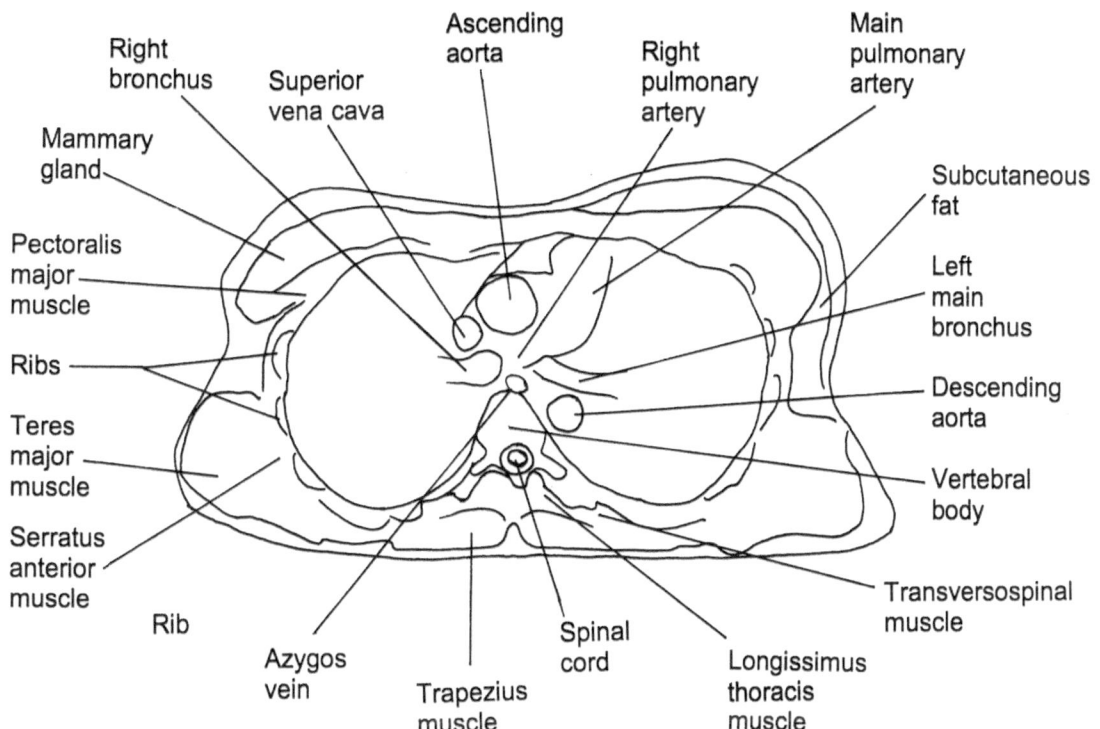

Right
bronchus

Ascending
aorta

Right
pulmonary
artery

Main
pulmonary
artery

Superior
vena cava

Mammary
gland

Subcutaneous
fat

Pectoralis
major
muscle

Left
main
bronchus

Ribs

Descending
aorta

Teres
major
muscle

Vertebral
body

Serratus
anterior
muscle

Transversospinal
muscle

Rib

Azygos
vein

Spinal
cord

Longissimus
thoracis
muscle

Trapezius
muscle

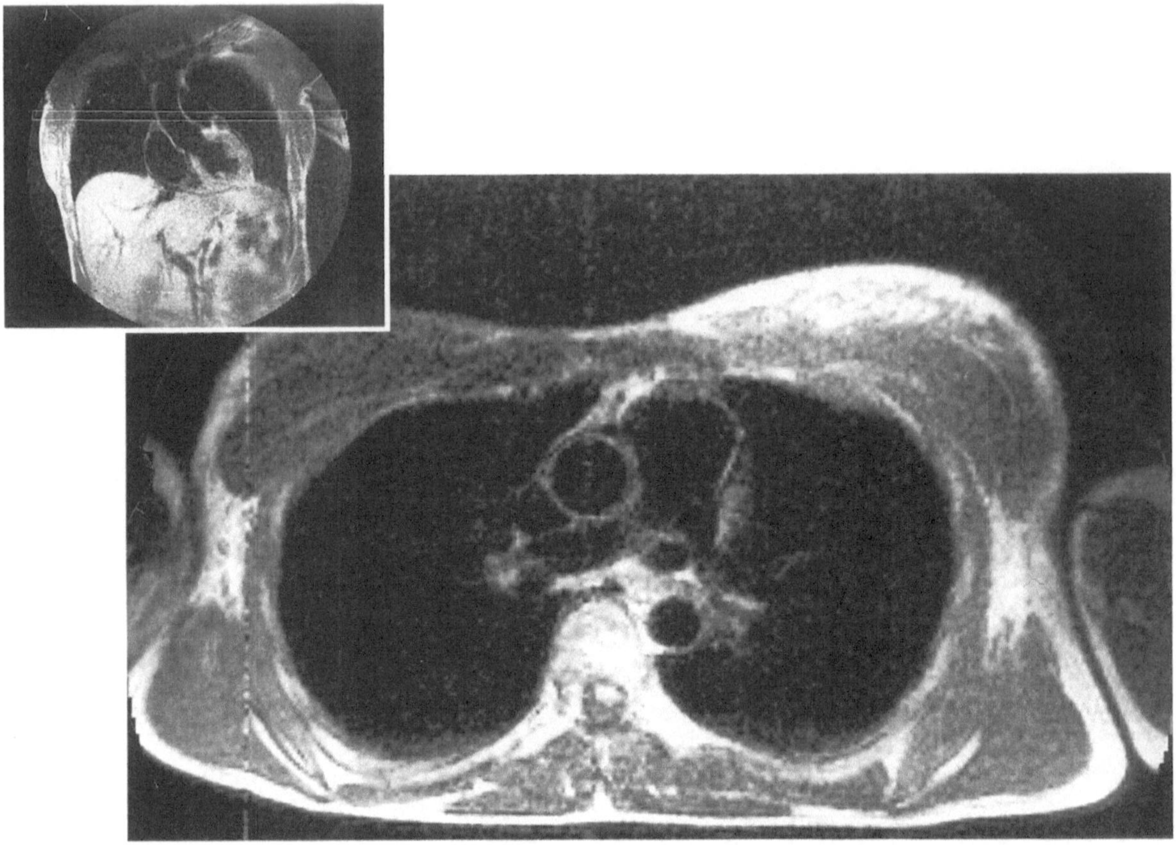

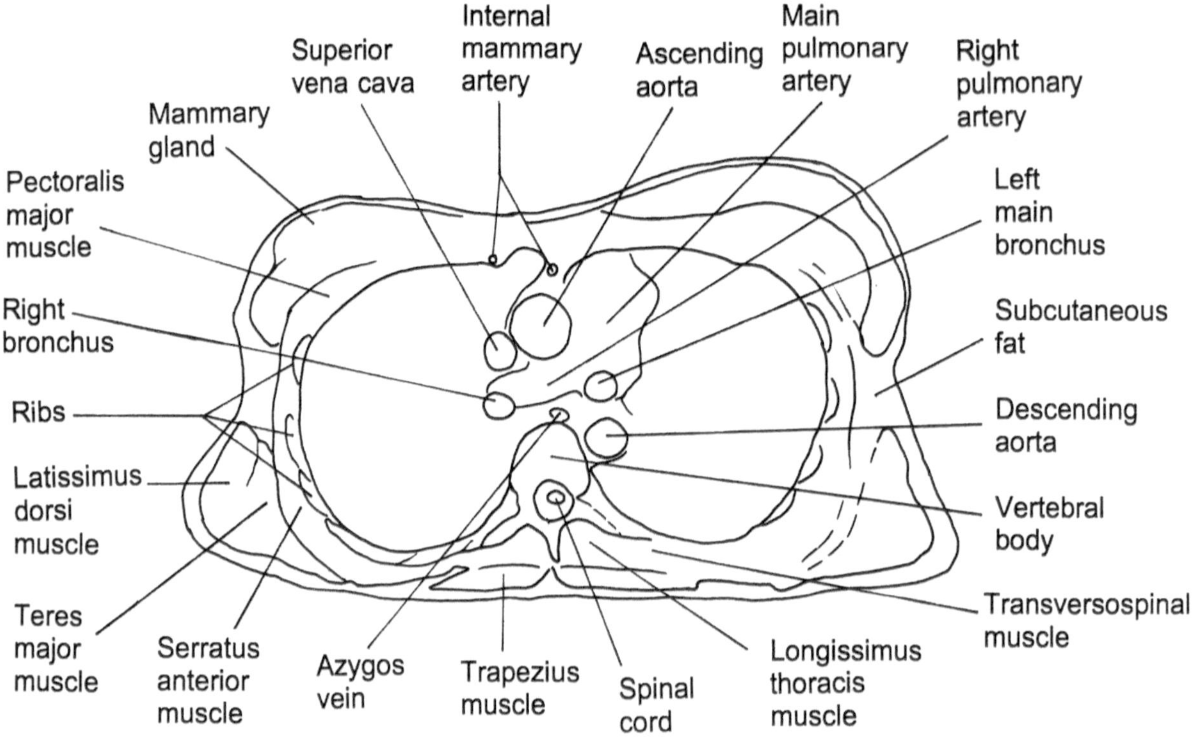

Superior vena cava

Internal mammary artery

Ascending aorta

Main pulmonary artery

Right pulmonary artery

Mammary gland

Pectoralis major muscle

Left main bronchus

Right bronchus

Subcutaneous fat

Ribs

Descending aorta

Latissimus dorsi muscle

Vertebral body

Teres major muscle

Transversospinal muscle

Serratus anterior muscle

Azygos vein

Trapezius muscle

Spinal cord

Longissimus thoracis muscle

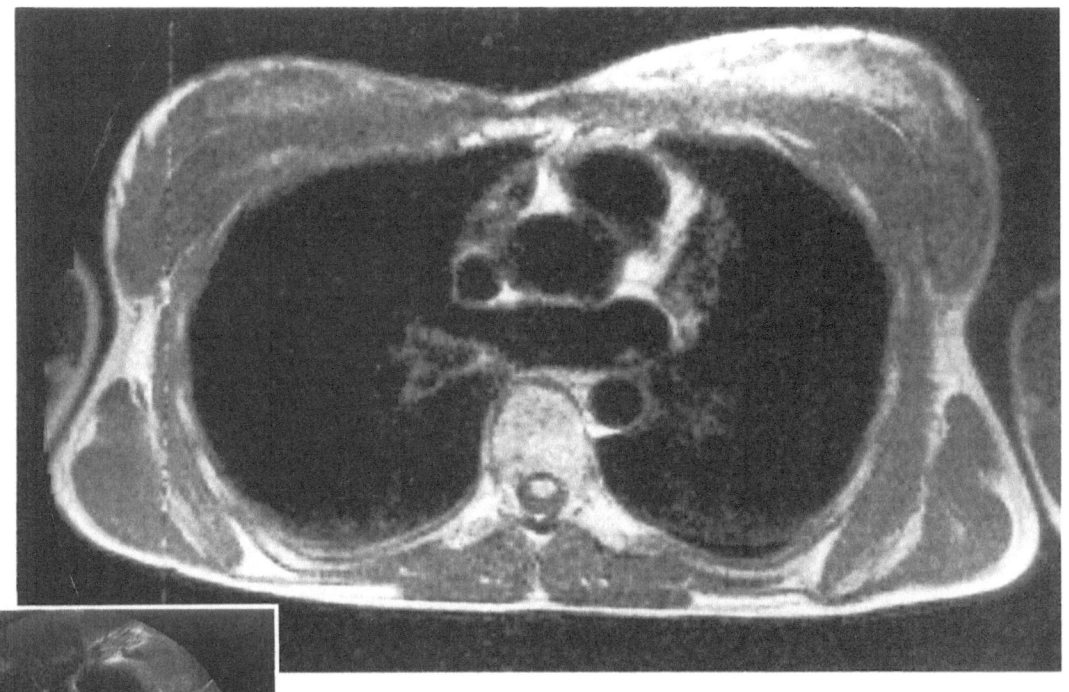

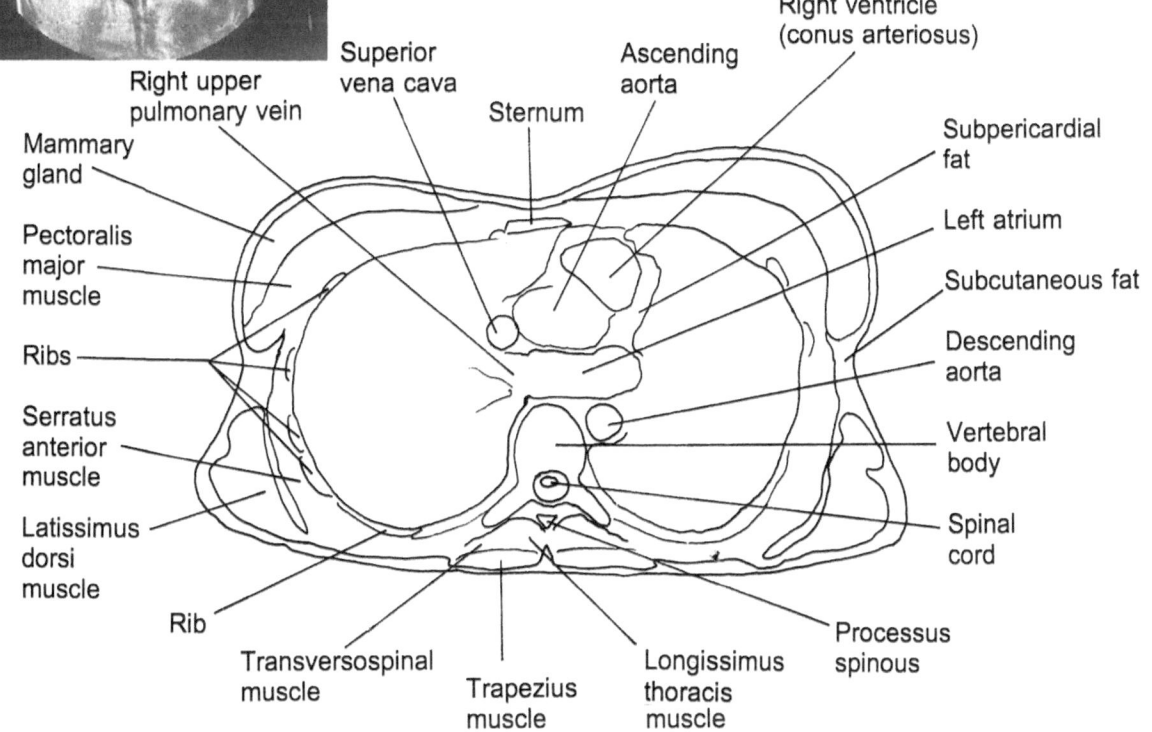

Right upper pulmonary vein

Mammary gland

Pectoralis major muscle

Ribs

Serratus anterior muscle

Latissimus dorsi muscle

Rib

Superior vena cava

Sternum

Ascending aorta

Right ventricle (conus arteriosus)

Subpericardial fat

Left atrium

Subcutaneous fat

Descending aorta

Vertebral body

Spinal cord

Processus spinous

Transversospinal muscle

Trapezius muscle

Longissimus thoracis muscle

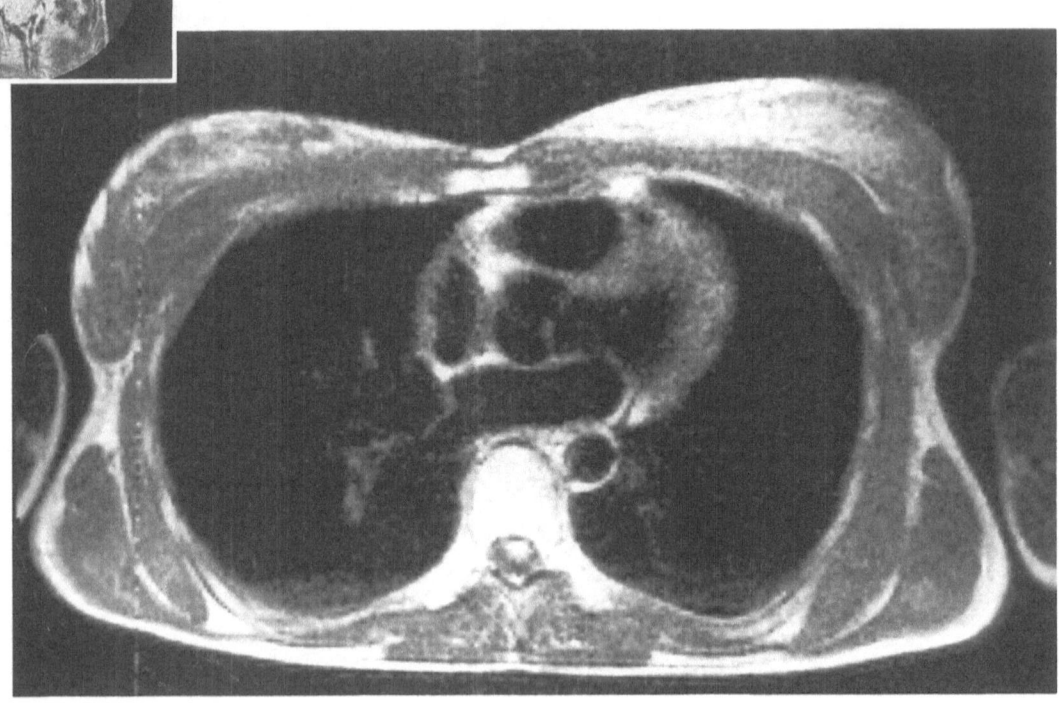

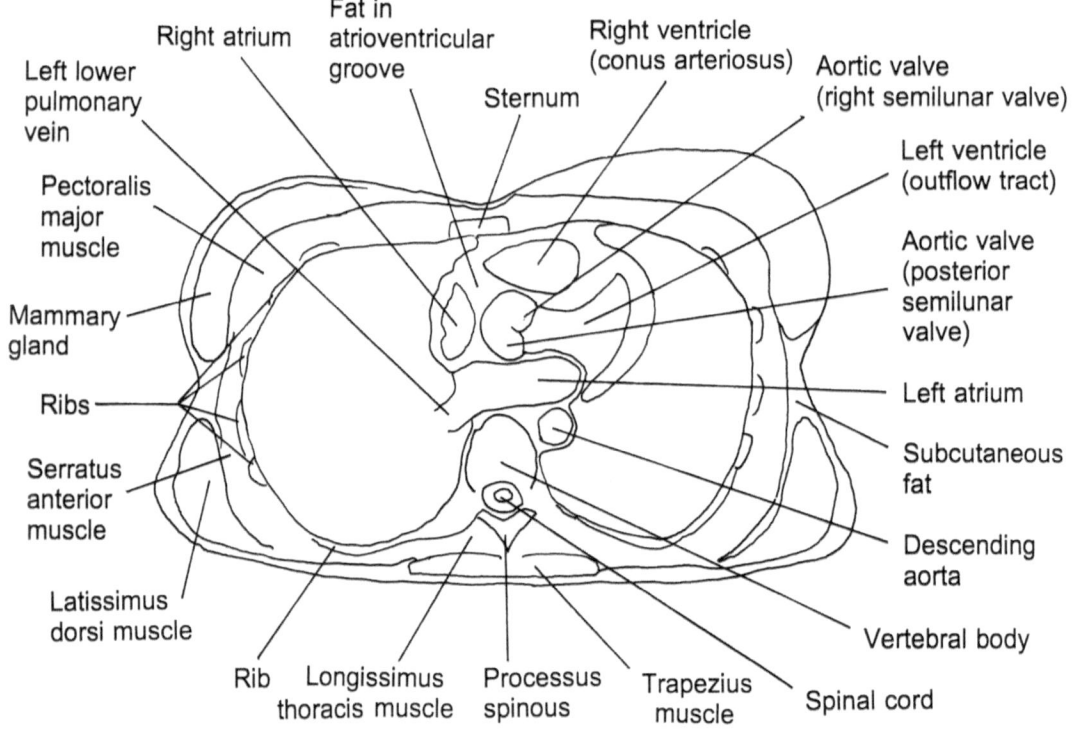

Fat in
atrioventricular
groove

Right atrium

Left lower
pulmonary
vein

Right ventricle
(conus arteriosus)

Sternum

Aortic valve
(right semilunar valve)

Pectoralis
major
muscle

Left ventricle
(outflow tract)

Aortic valve
(posterior
semilunar
valve)

Mammary
gland

Left atrium

Ribs

Subcutaneous
fat

Serratus
anterior
muscle

Descending
aorta

Latissimus
dorsi muscle

Vertebral body

Rib   Longissimus
       thoracis muscle

Processus
spinous

Trapezius
muscle

Spinal cord

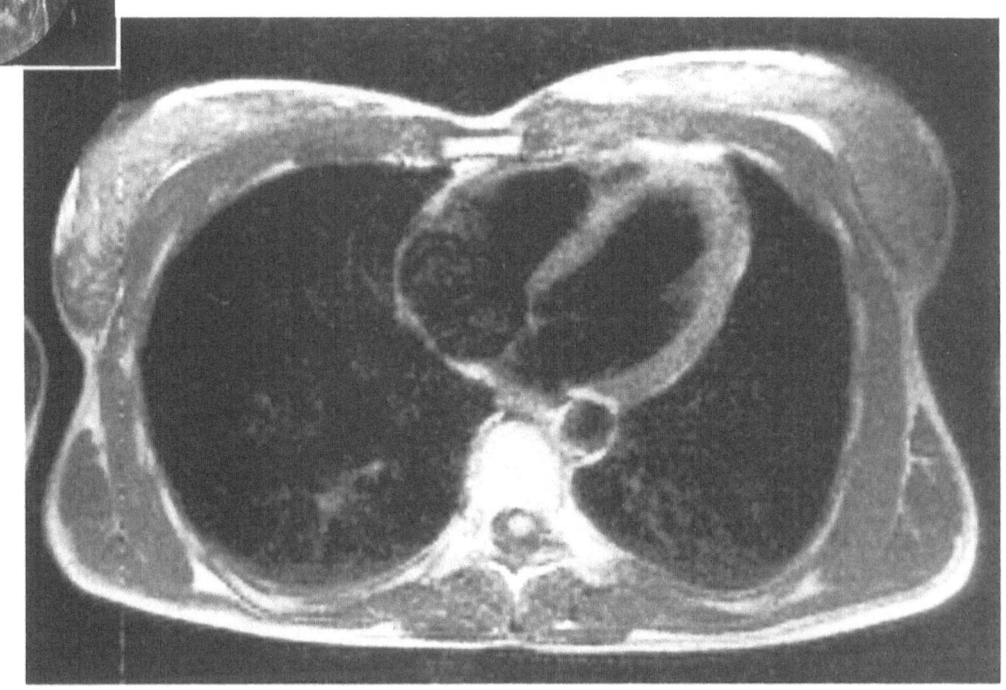

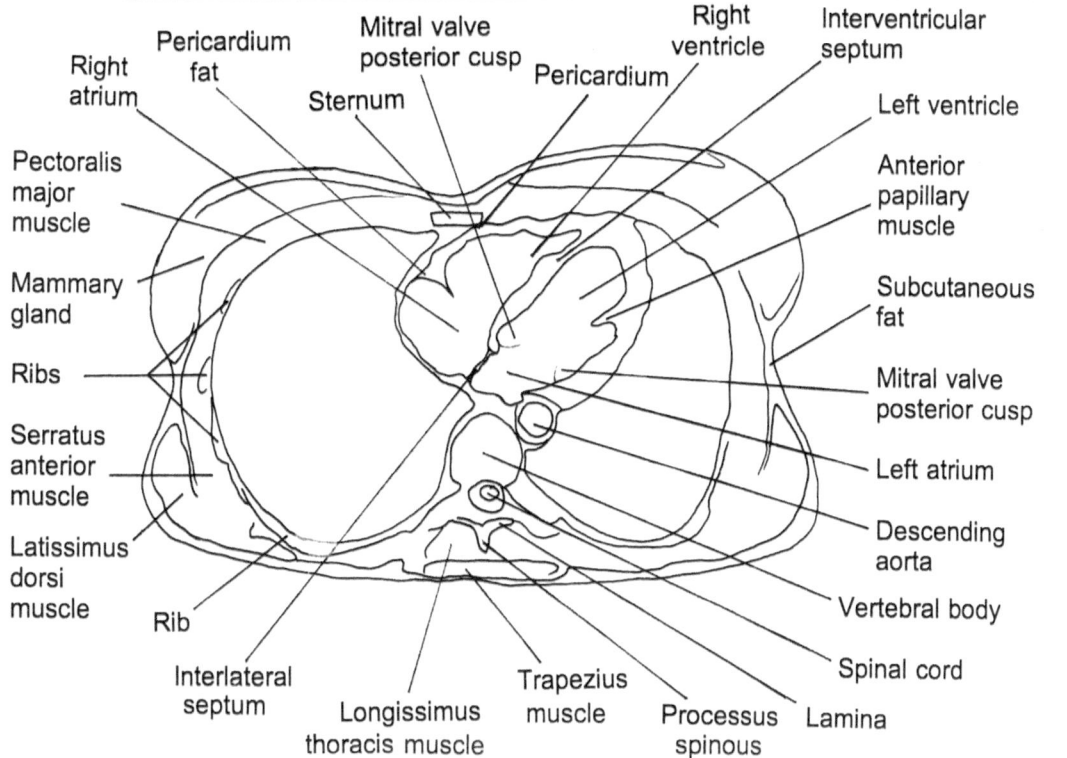

Right atrium

Pericardium fat

Mitral valve posterior cusp

Sternum

Pericardium

Right ventricle

Interventricular septum

Left ventricle

Pectoralis major muscle

Anterior papillary muscle

Mammary gland

Subcutaneous fat

Ribs

Mitral valve posterior cusp

Serratus anterior muscle

Left atrium

Latissimus dorsi muscle

Descending aorta

Rib

Vertebral body

Interlateral septum

Longissimus thoracis muscle

Trapezius muscle

Processus spinous

Lamina

Spinal cord

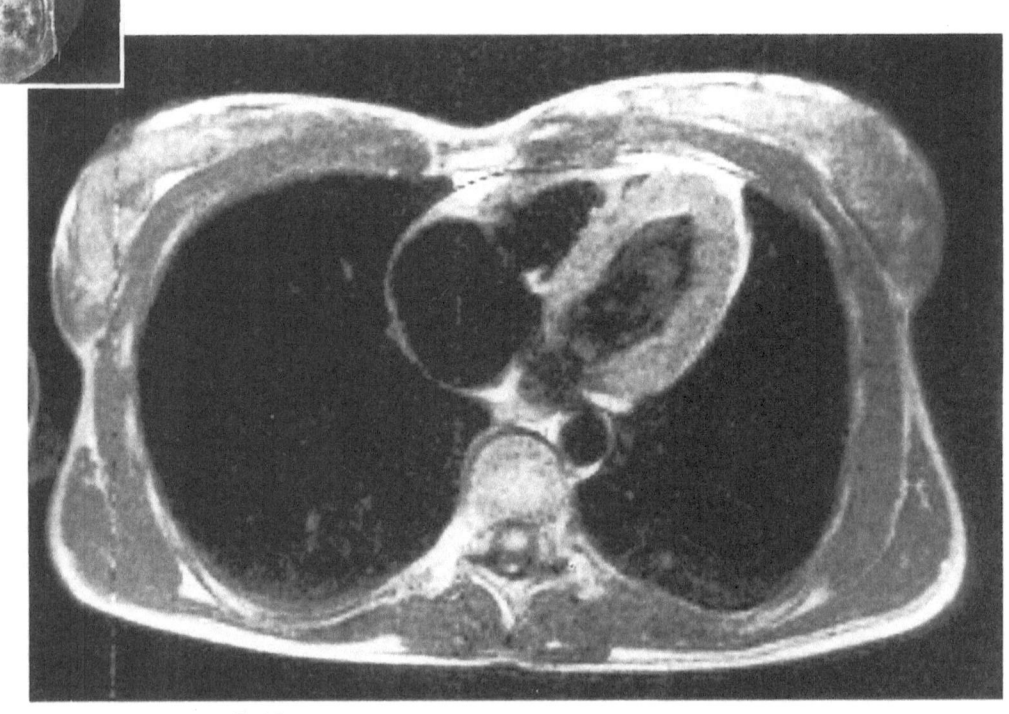

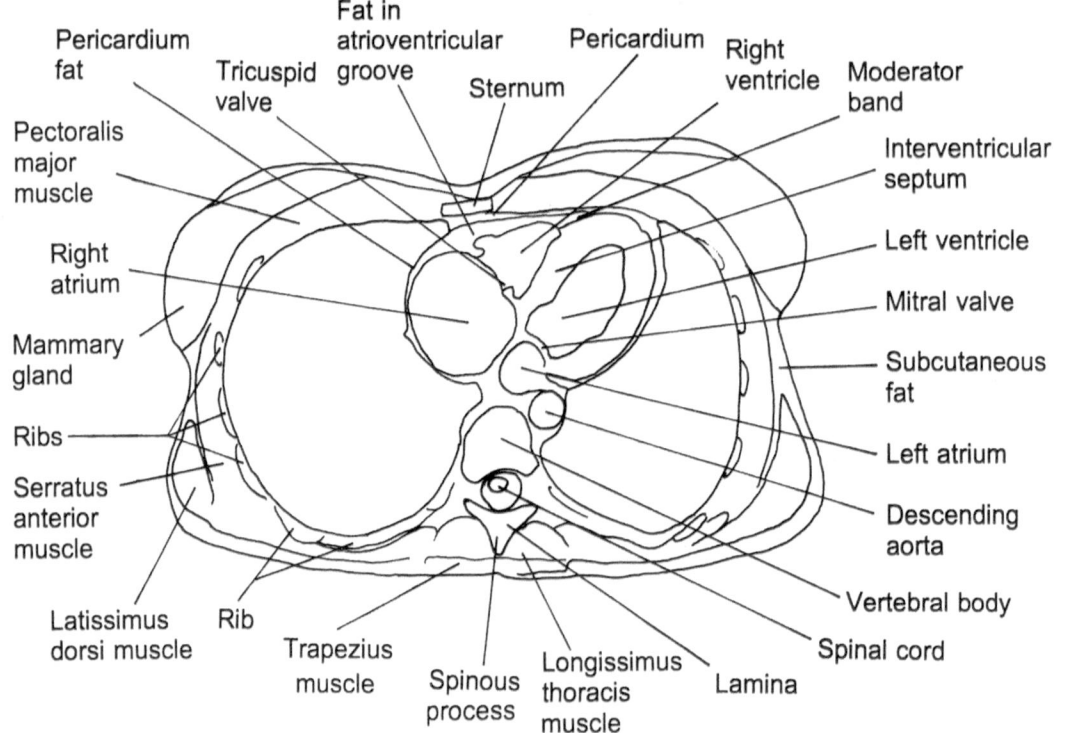

Pericardium fat

Tricuspid valve

Fat in atrioventricular groove

Sternum

Pericardium

Right ventricle

Moderator band

Pectoralis major muscle

Right atrium

Mammary gland

Ribs

Serratus anterior muscle

Latissimus dorsi muscle

Rib

Trapezius muscle

Spinous process

Longissimus thoracis muscle

Lamina

Spinal cord

Vertebral body

Descending aorta

Left atrium

Subcutaneous fat

Mitral valve

Left ventricle

Interventricular septum

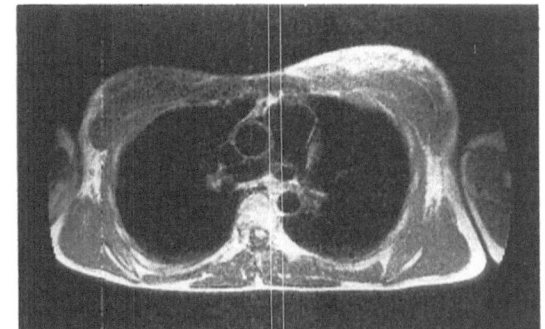
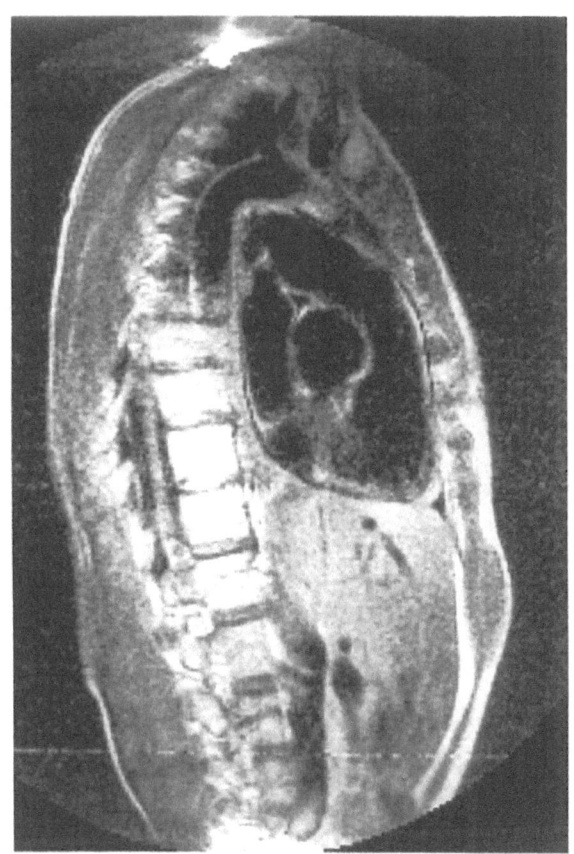

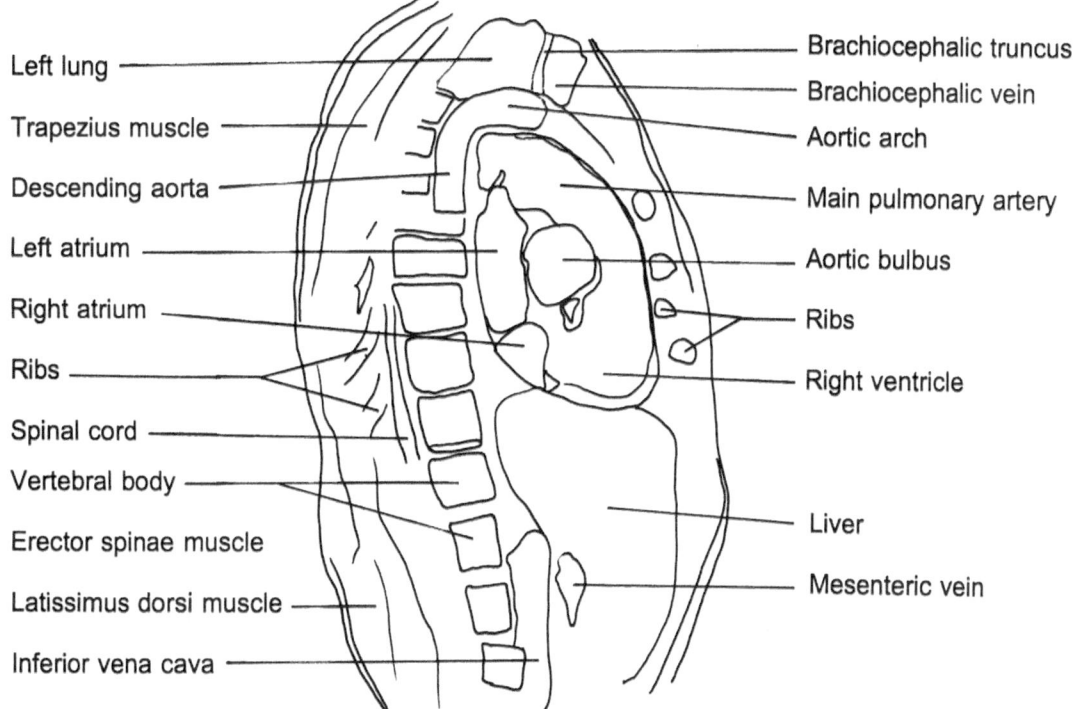

Left lung

Trapezius muscle

Descending aorta

Left atrium

Right atrium

Ribs

Spinal cord

Vertebral body

Erector spinae muscle

Latissimus dorsi muscle

Inferior vena cava

Brachiocephalic truncus

Brachiocephalic vein

Aortic arch

Main pulmonary artery

Aortic bulbus

Ribs

Right ventricle

Liver

Mesenteric vein

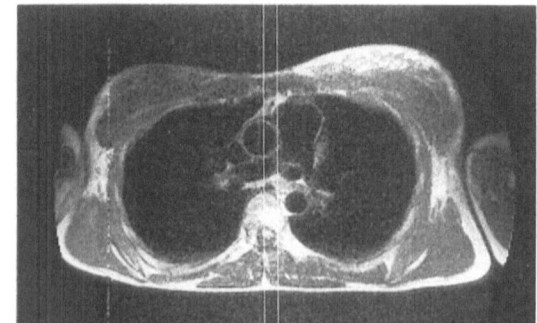

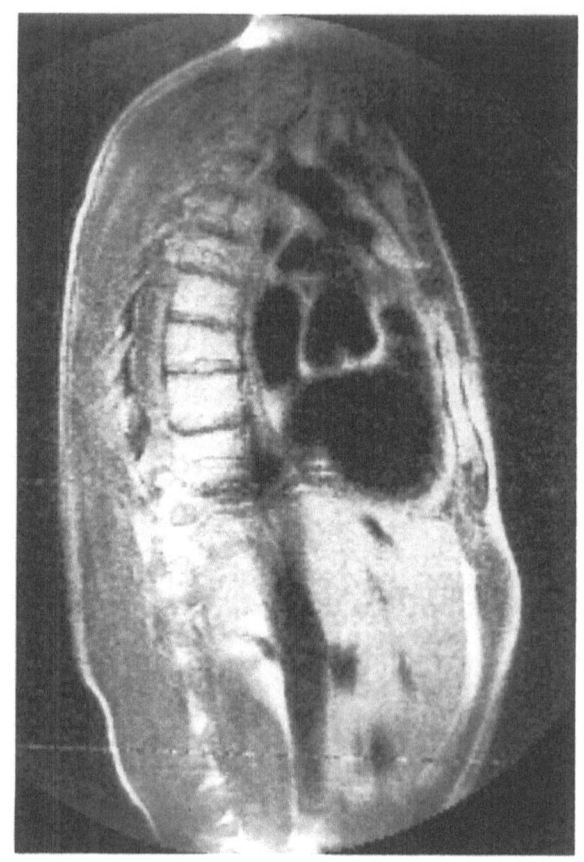

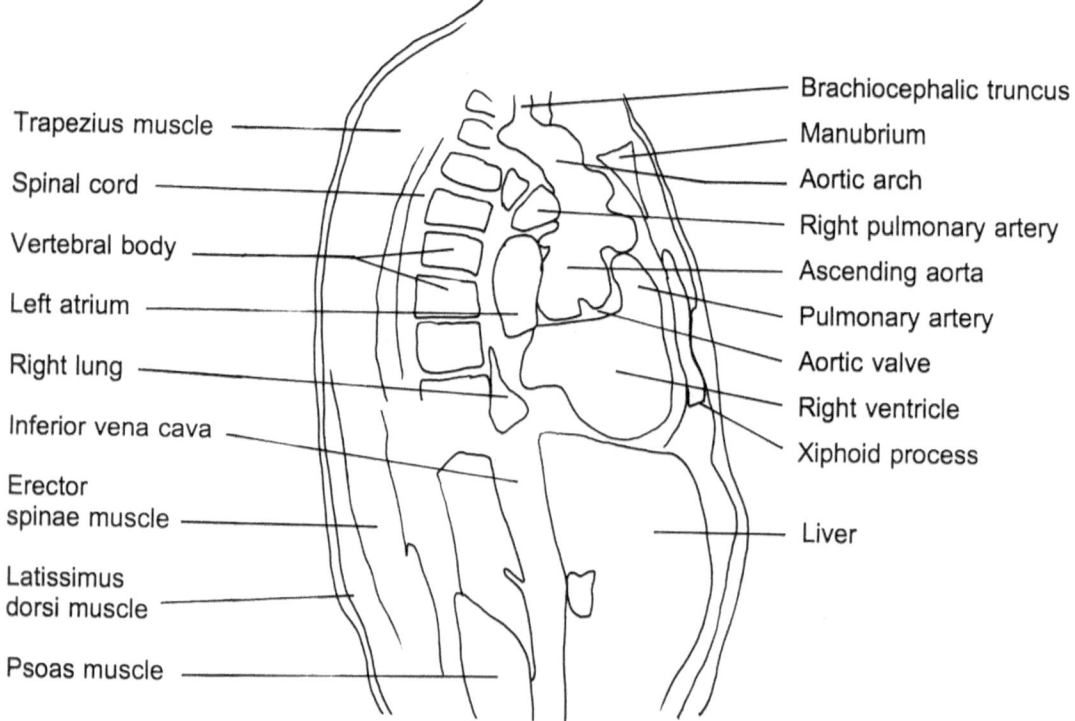

Trapezius muscle

Spinal cord

Vertebral body

Left atrium

Right lung

Inferior vena cava

Erector
spinae muscle

Latissimus
dorsi muscle

Psoas muscle

Brachiocephalic truncus

Manubrium

Aortic arch

Right pulmonary artery

Ascending aorta

Pulmonary artery

Aortic valve

Right ventricle

Xiphoid process

Liver

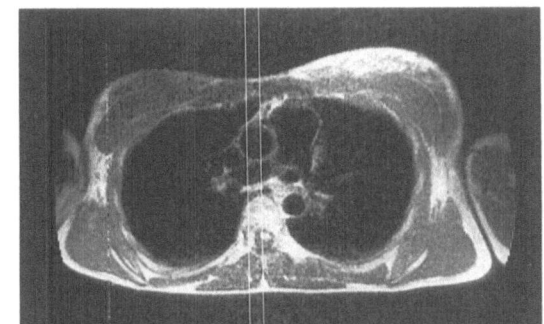

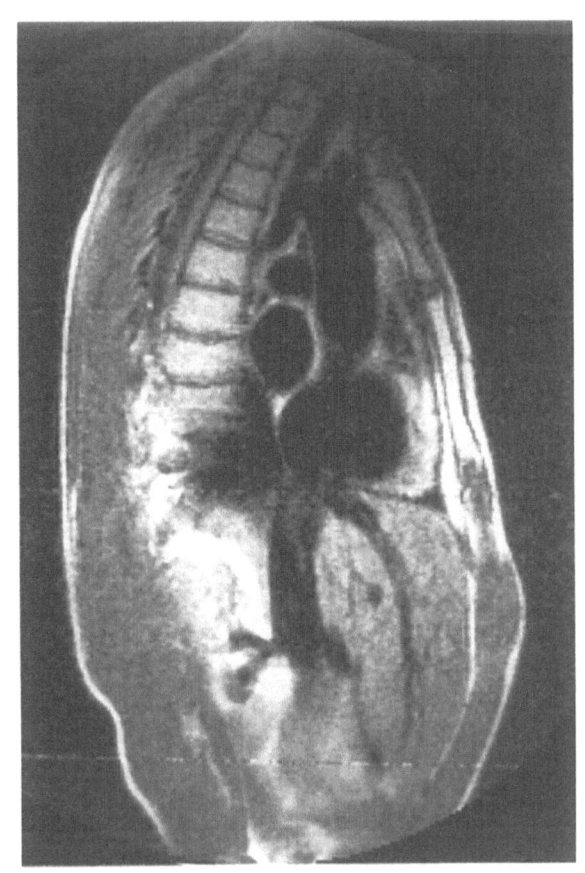

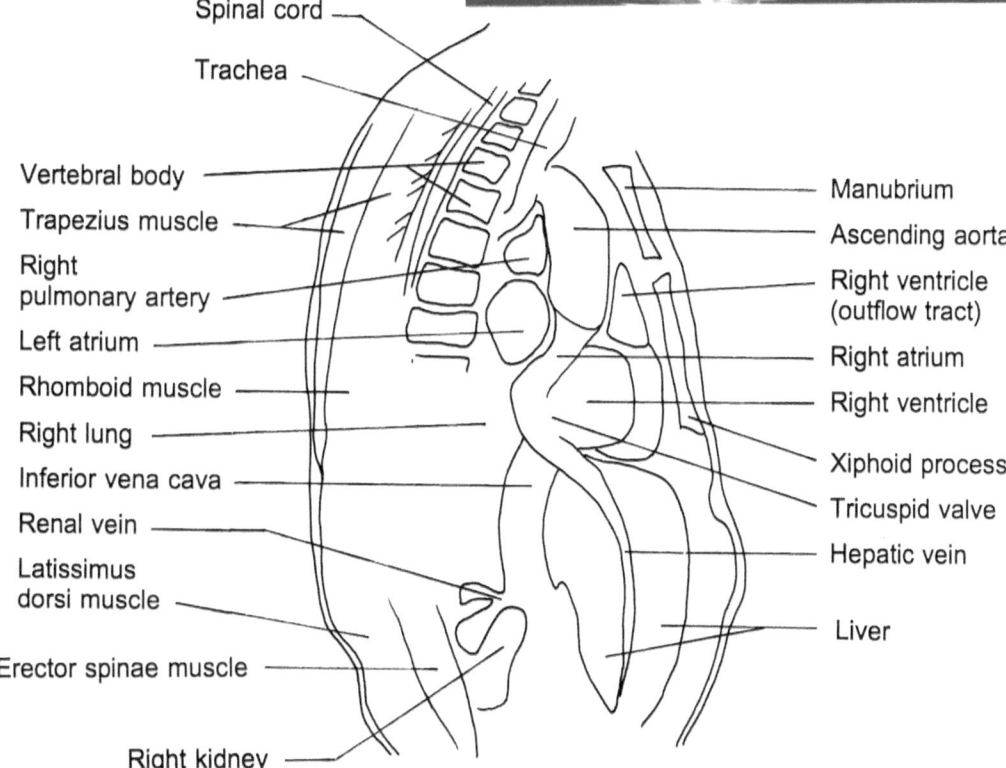

Spinal cord

Trachea

Vertebral body

Trapezius muscle

Right
pulmonary artery

Left atrium

Rhomboid muscle

Right lung

Inferior vena cava

Renal vein

Latissimus
dorsi muscle

Erector spinae muscle

Right kidney

Manubrium

Ascending aorta

Right ventricle
(outflow tract)

Right atrium

Right ventricle

Xiphoid process

Tricuspid valve

Hepatic vein

Liver

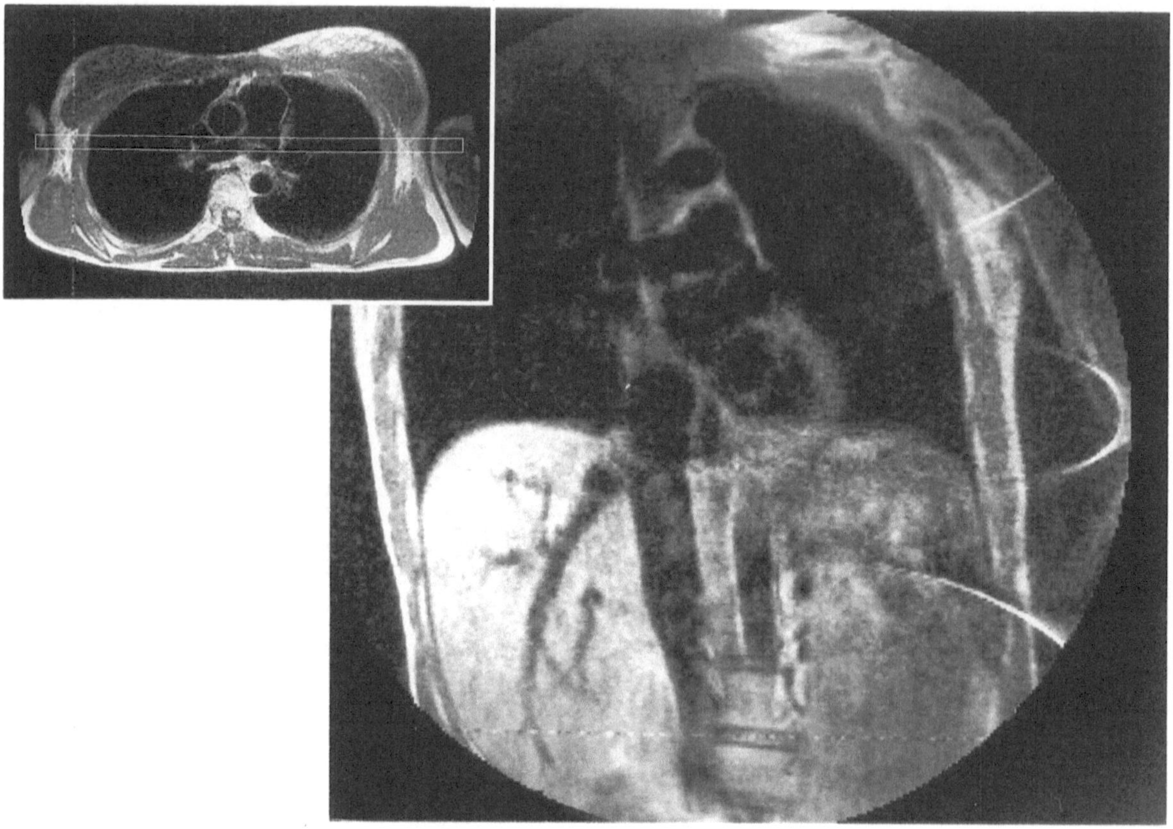

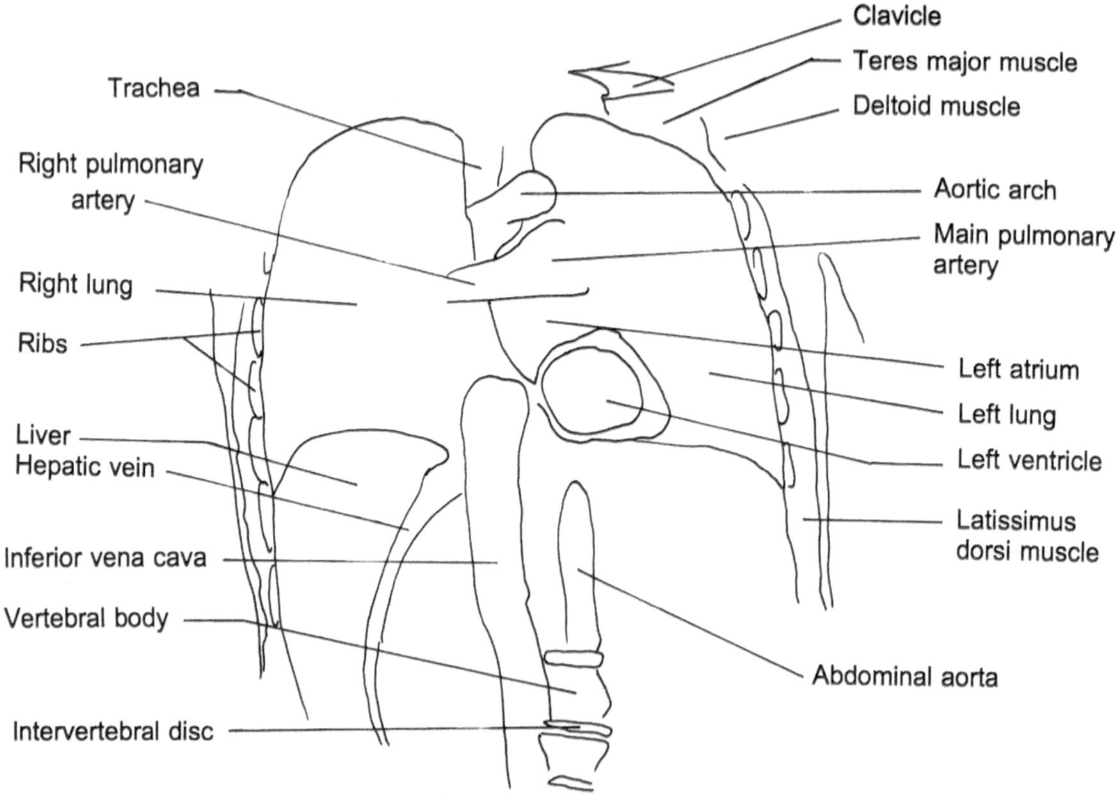

Clavicle
Teres major muscle
Deltoid muscle

Trachea

Right pulmonary
artery

Right lung

Ribs

Liver
Hepatic vein

Inferior vena cava

Vertebral body

Intervertebral disc

Aortic arch

Main pulmonary
artery

Left atrium

Left lung

Left ventricle

Latissimus
dorsi muscle

Abdominal aorta

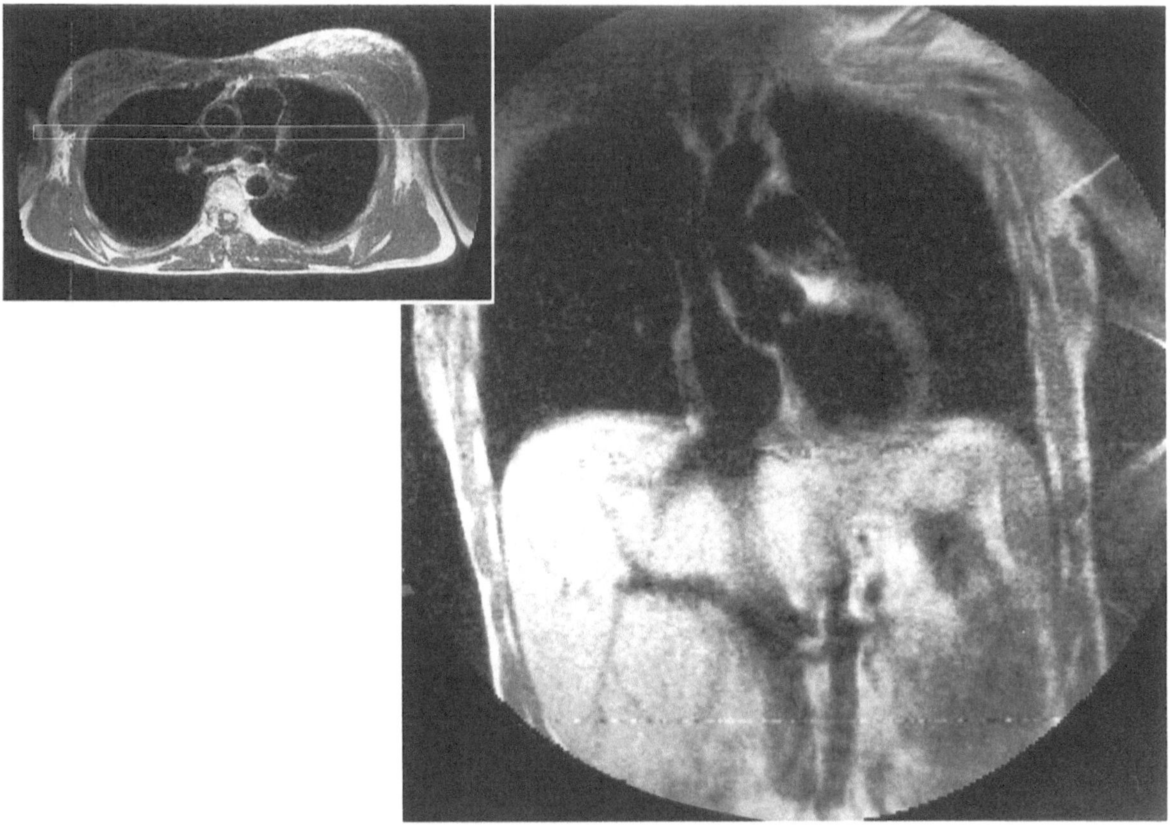

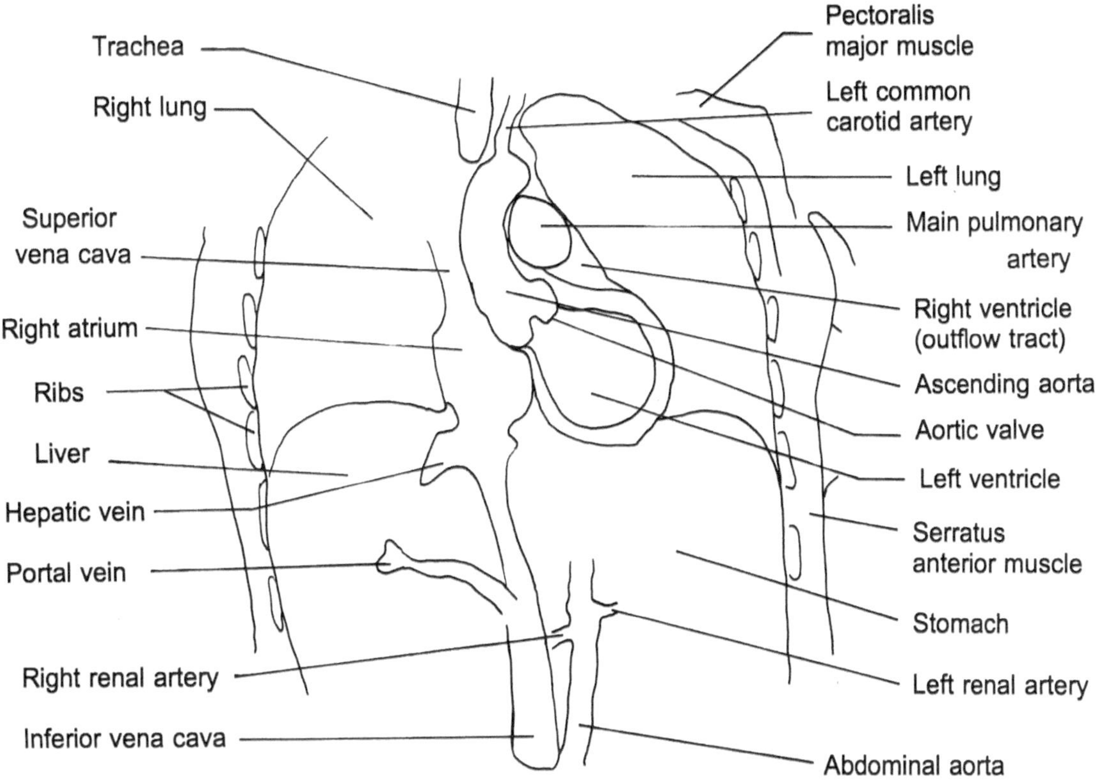

Trachea

Right lung

Superior
vena cava

Right atrium

Ribs

Liver

Hepatic vein

Portal vein

Right renal artery

Inferior vena cava

Pectoralis
major muscle

Left common
carotid artery

Left lung

Main pulmonary
artery

Right ventricle
(outflow tract)

Ascending aorta

Aortic valve

Left ventricle

Serratus
anterior muscle

Stomach

Left renal artery

Abdominal aorta

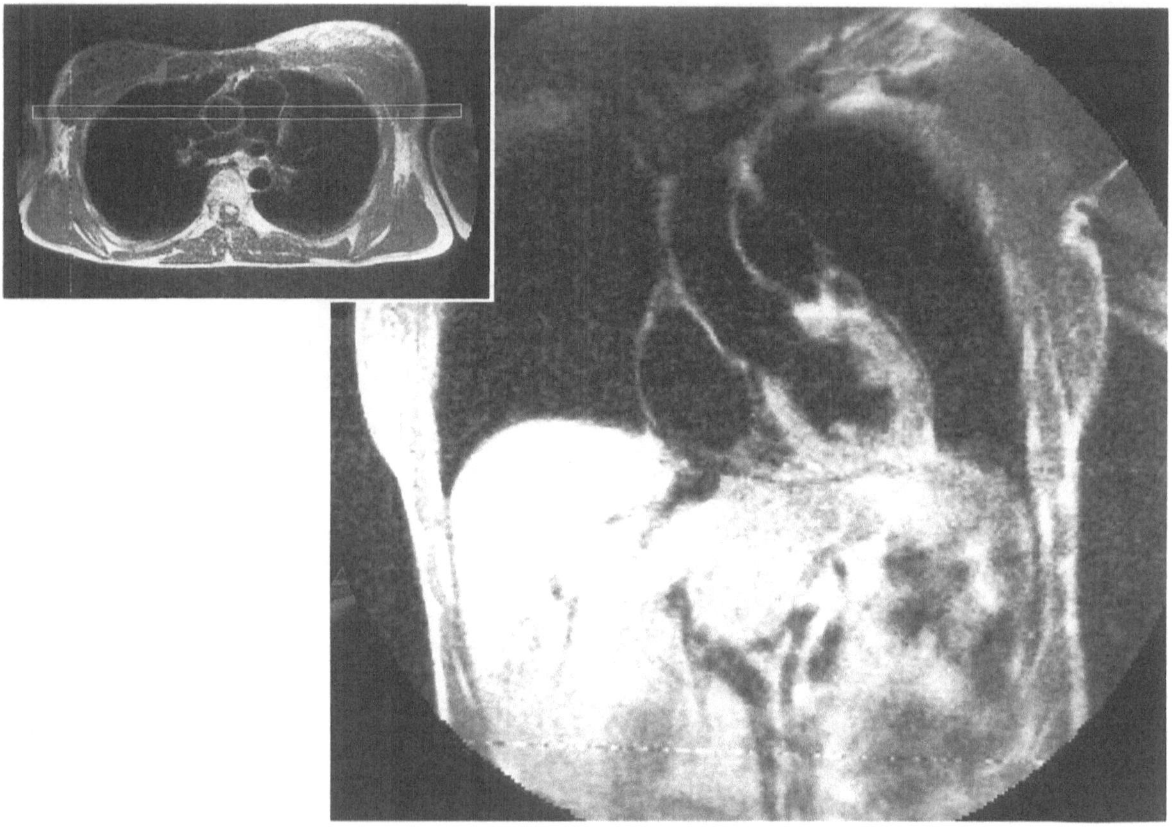

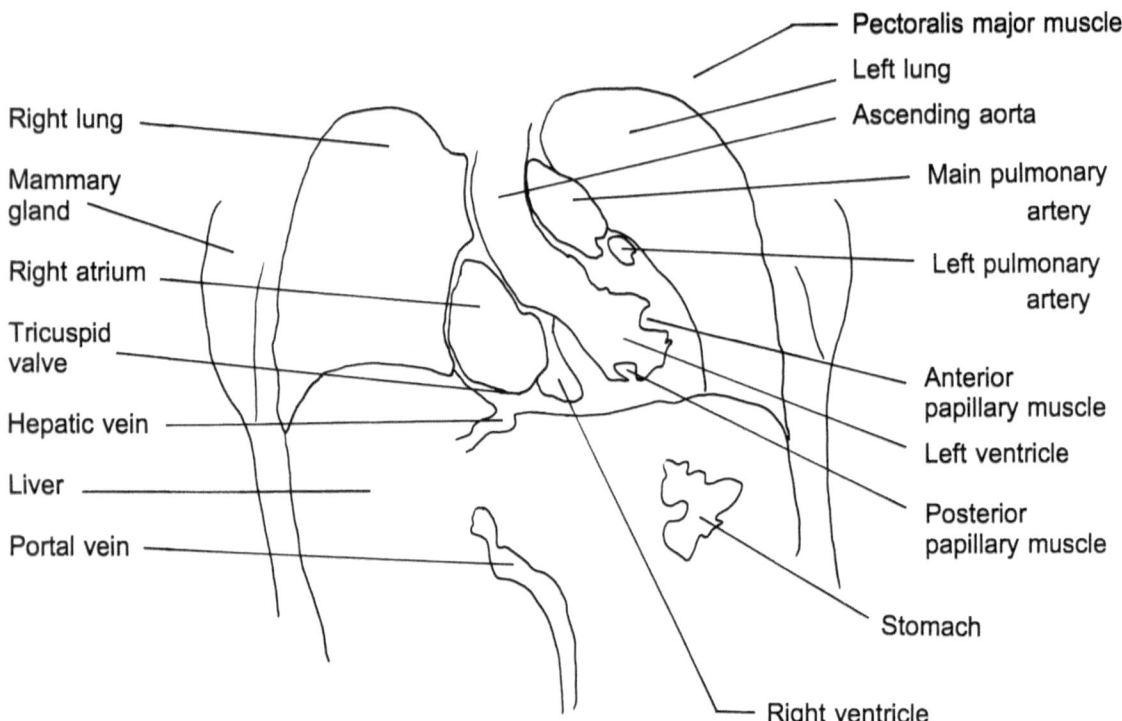

Right lung

Mammary gland

Right atrium

Tricuspid valve

Hepatic vein

Liver

Portal vein

Pectoralis major muscle

Left lung

Ascending aorta

Main pulmonary artery

Left pulmonary artery

Anterior papillary muscle

Left ventricle

Posterior papillary muscle

Stomach

Right ventricle

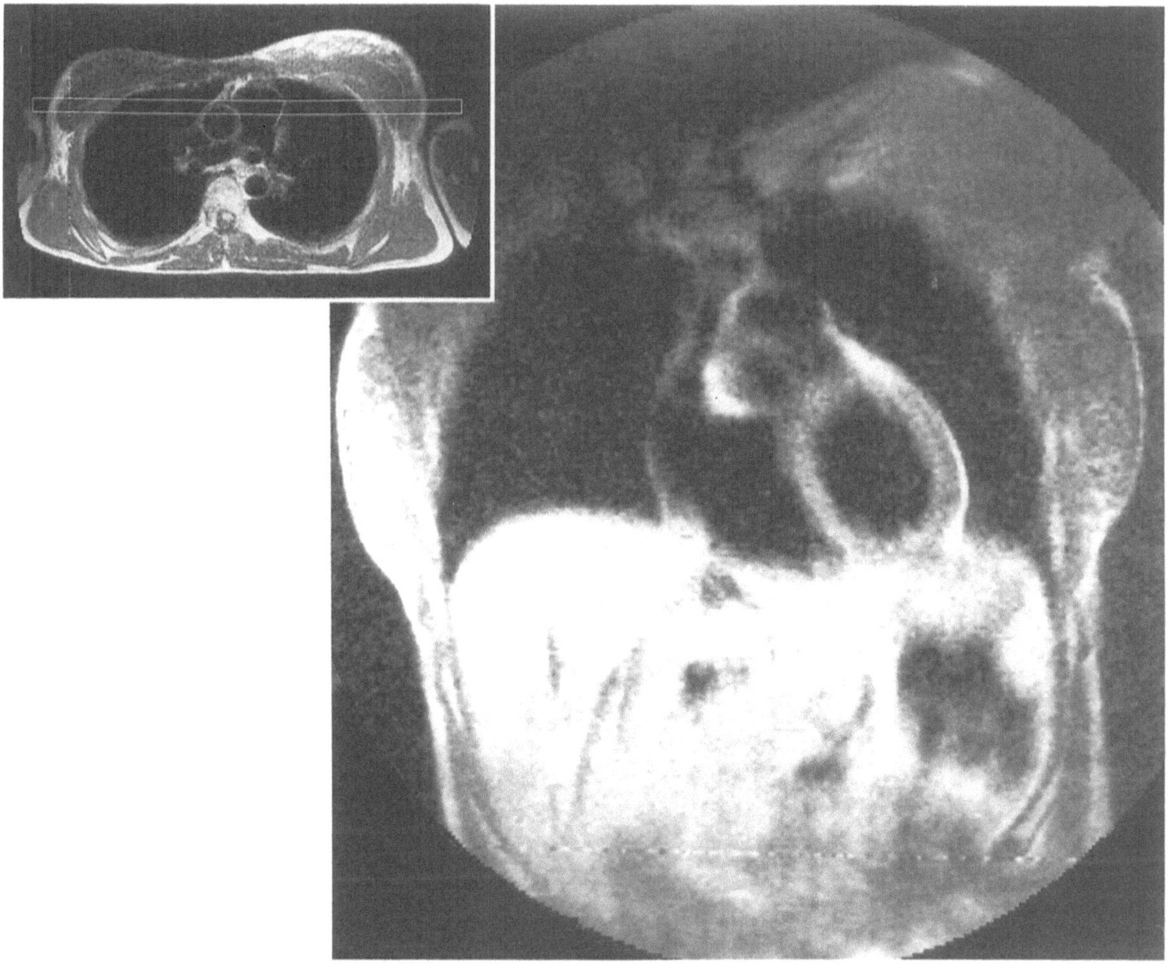

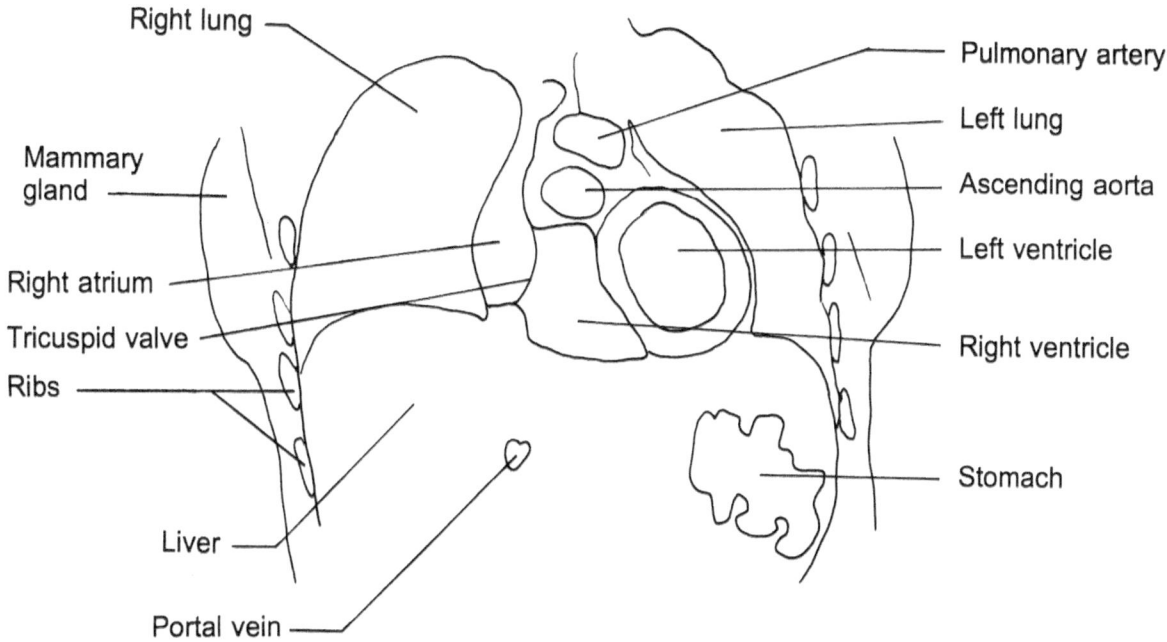

Right lung

Pulmonary artery

Left lung

Mammary gland

Ascending aorta

Left ventricle

Right atrium

Tricuspid valve

Right ventricle

Ribs

Stomach

Liver

Portal vein

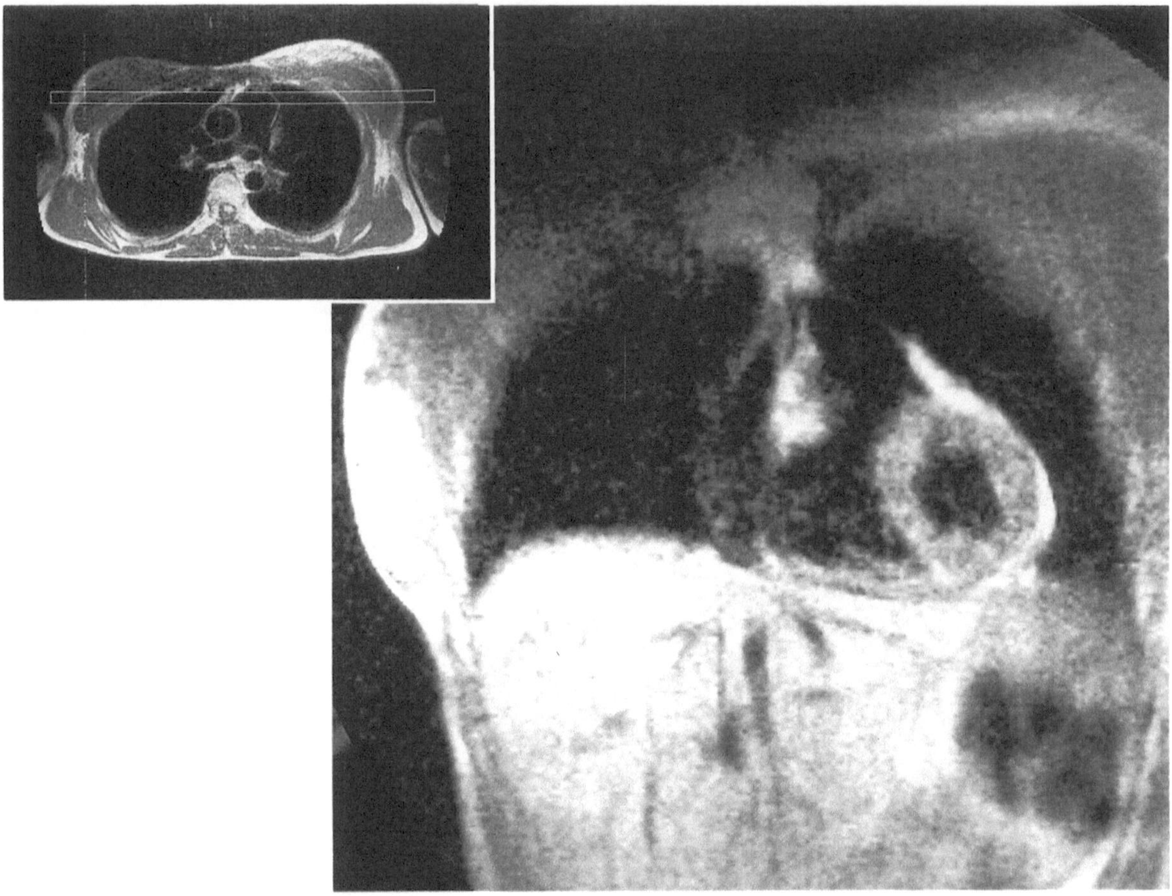

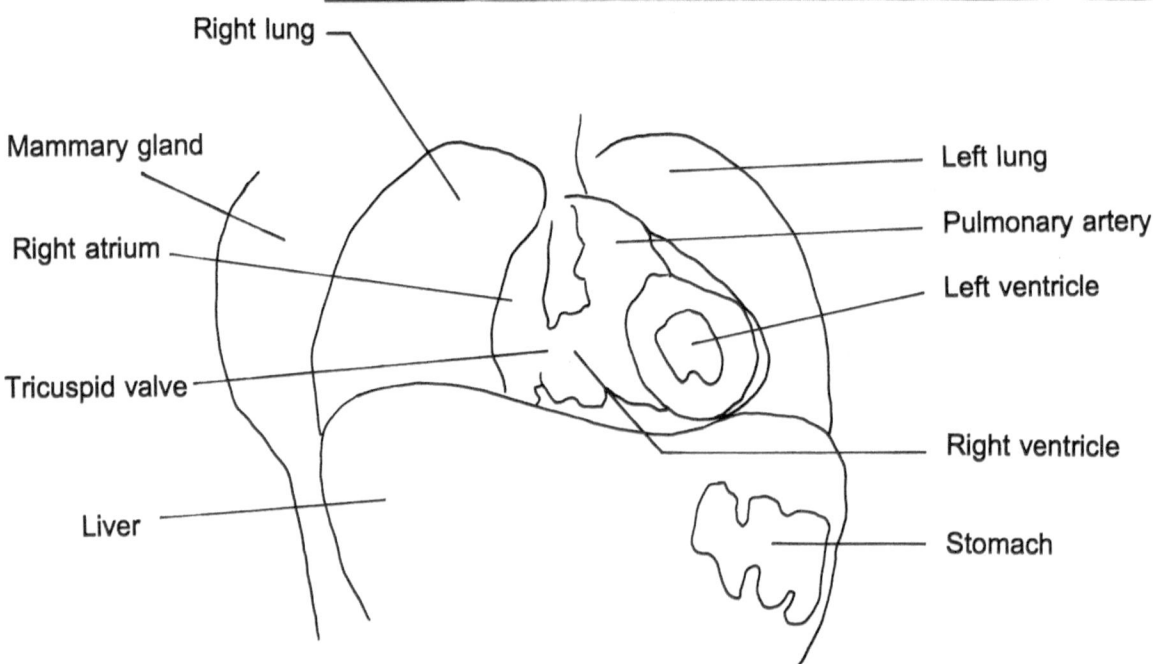

Right lung

Mammary gland

Right atrium

Tricuspid valve

Liver

Left lung

Pulmonary artery

Left ventricle

Right ventricle

Stomach

HEART

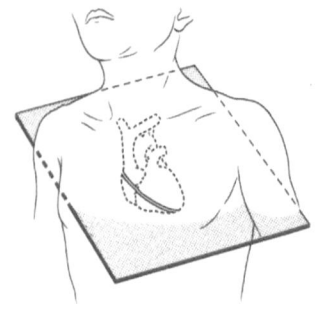

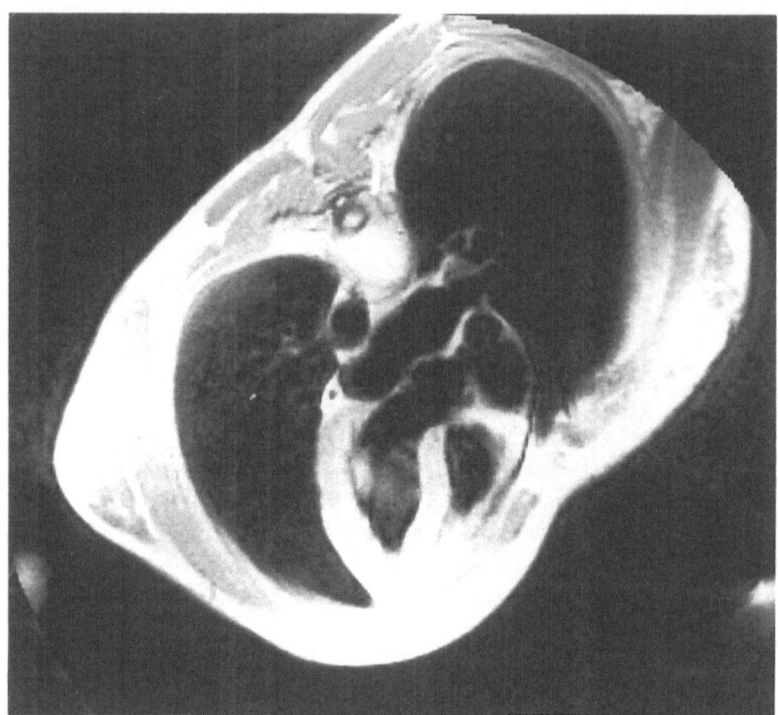

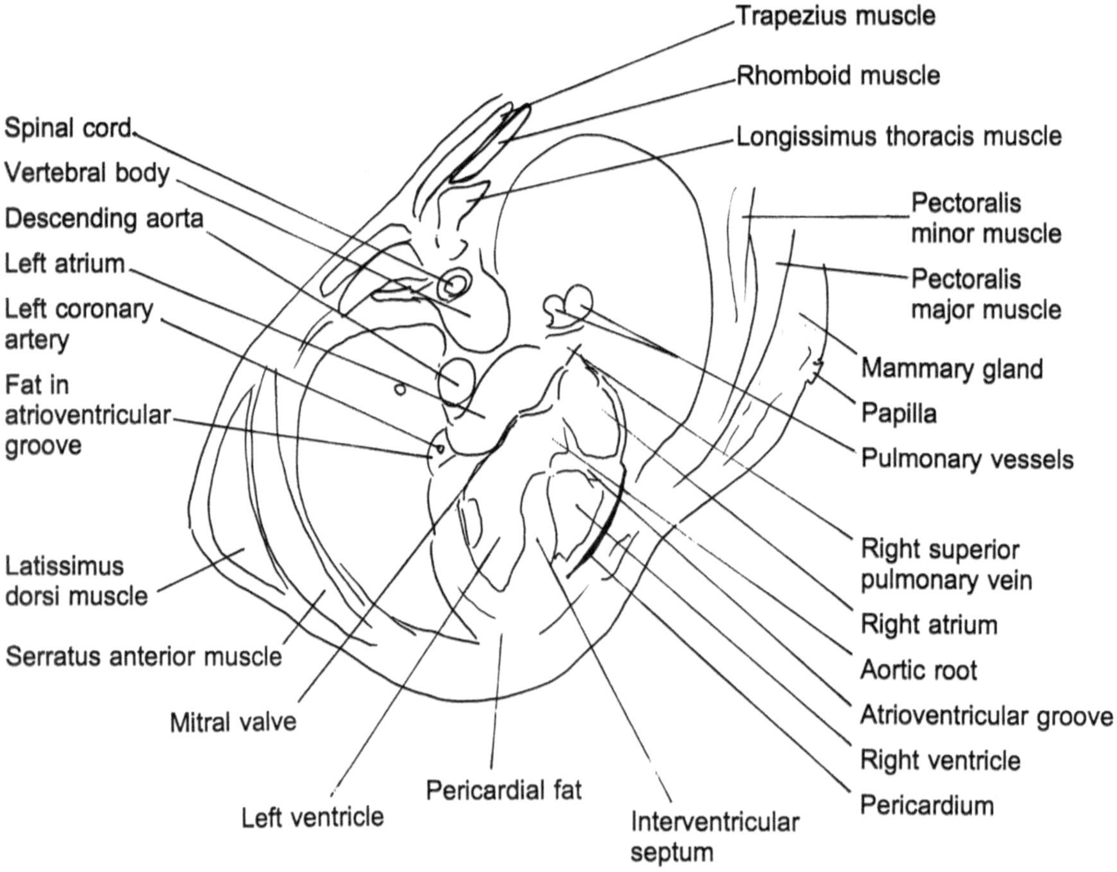

Trapezius muscle

Rhomboid muscle

Longissimus thoracis muscle

Spinal cord

Vertebral body

Descending aorta

Left atrium

Left coronary artery

Fat in atrioventricular groove

Pectoralis minor muscle

Pectoralis major muscle

Mammary gland

Papilla

Pulmonary vessels

Right superior pulmonary vein

Right atrium

Aortic root

Atrioventricular groove

Right ventricle

Pericardium

Latissimus dorsi muscle

Serratus anterior muscle

Mitral valve

Left ventricle

Pericardial fat

Interventricular septum

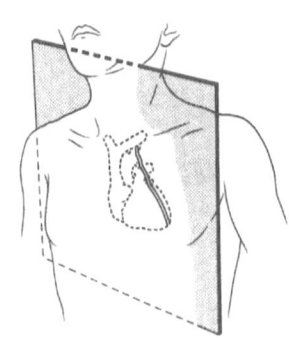

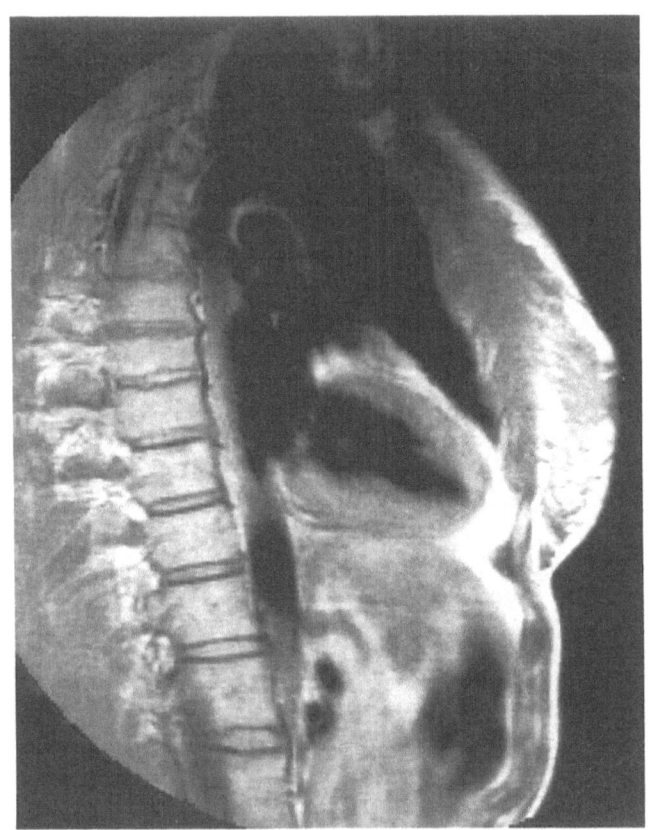

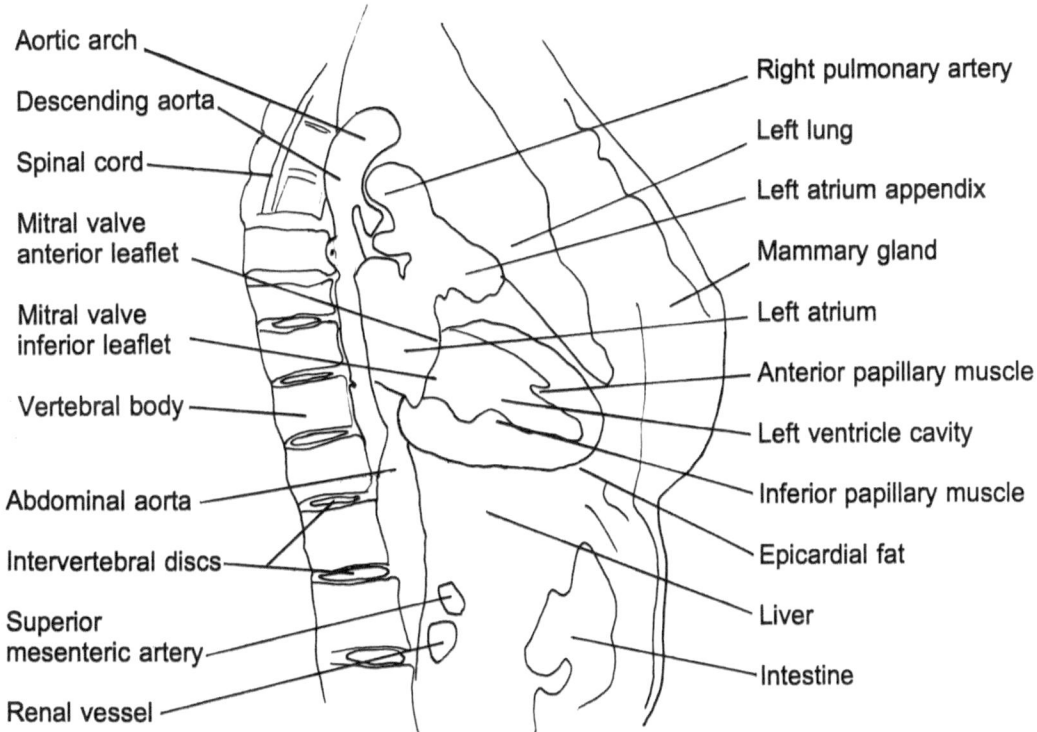

Aortic arch

Descending aorta

Spinal cord

Mitral valve anterior leaflet

Mitral valve inferior leaflet

Vertebral body

Abdominal aorta

Intervertebral discs

Superior mesenteric artery

Renal vessel

Right pulmonary artery

Left lung

Left atrium appendix

Mammary gland

Left atrium

Anterior papillary muscle

Left ventricle cavity

Inferior papillary muscle

Epicardial fat

Liver

Intestine

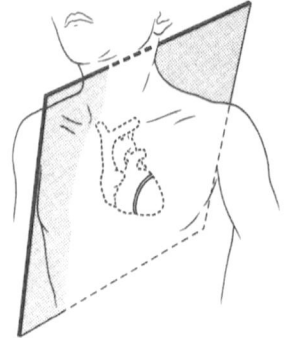

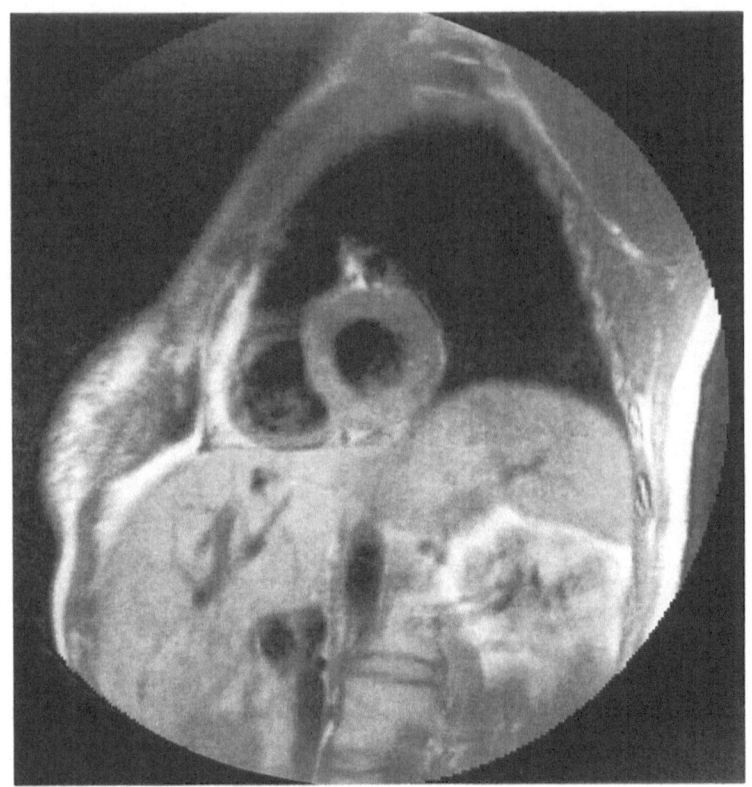

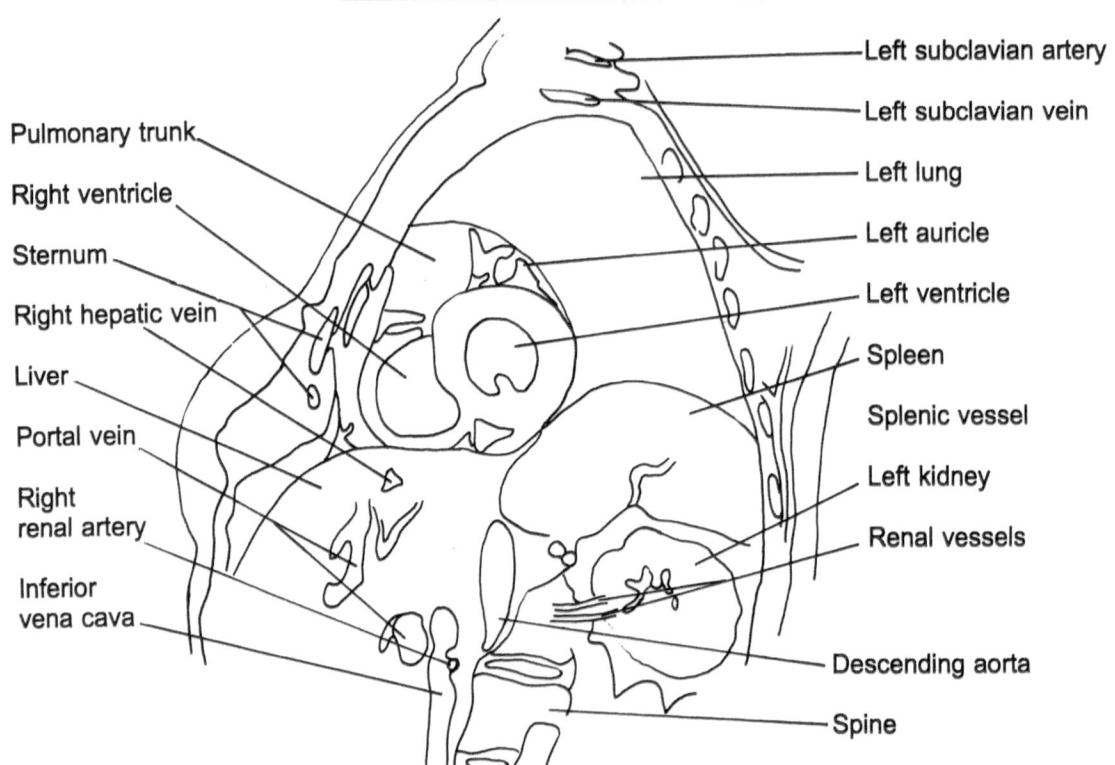

Pulmonary trunk

Right ventricle

Sternum

Right hepatic vein

Liver

Portal vein

Right renal artery

Inferior vena cava

Left subclavian artery

Left subclavian vein

Left lung

Left auricle

Left ventricle

Spleen

Splenic vessel

Left kidney

Renal vessels

Descending aorta

Spine

BREAST

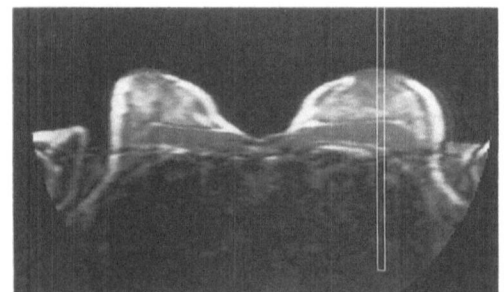

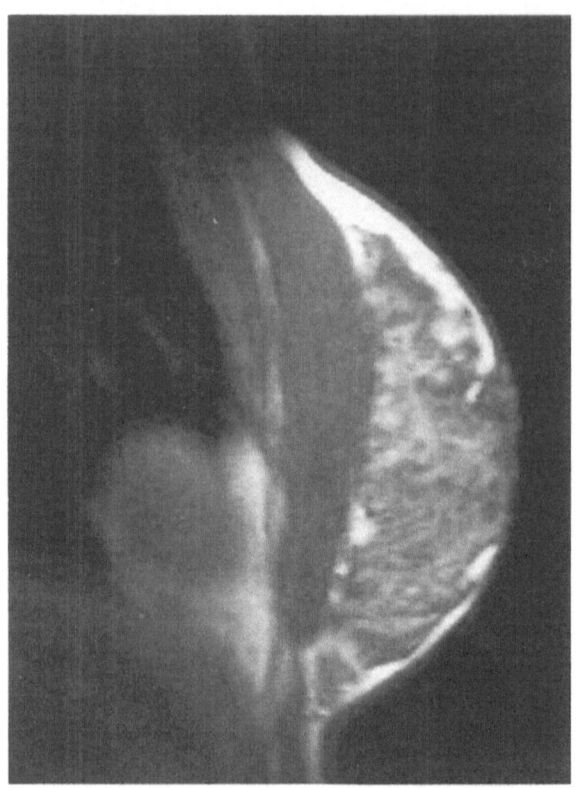

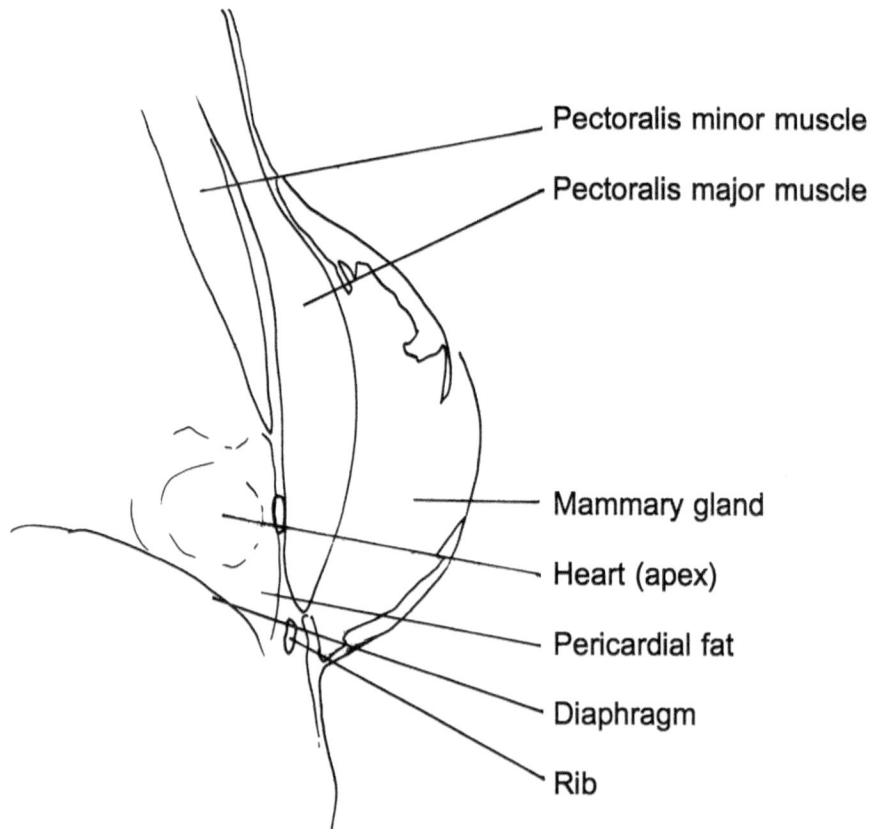

Pectoralis minor muscle

Pectoralis major muscle

Mammary gland

Heart (apex)

Pericardial fat

Diaphragm

Rib

ABDOMEN

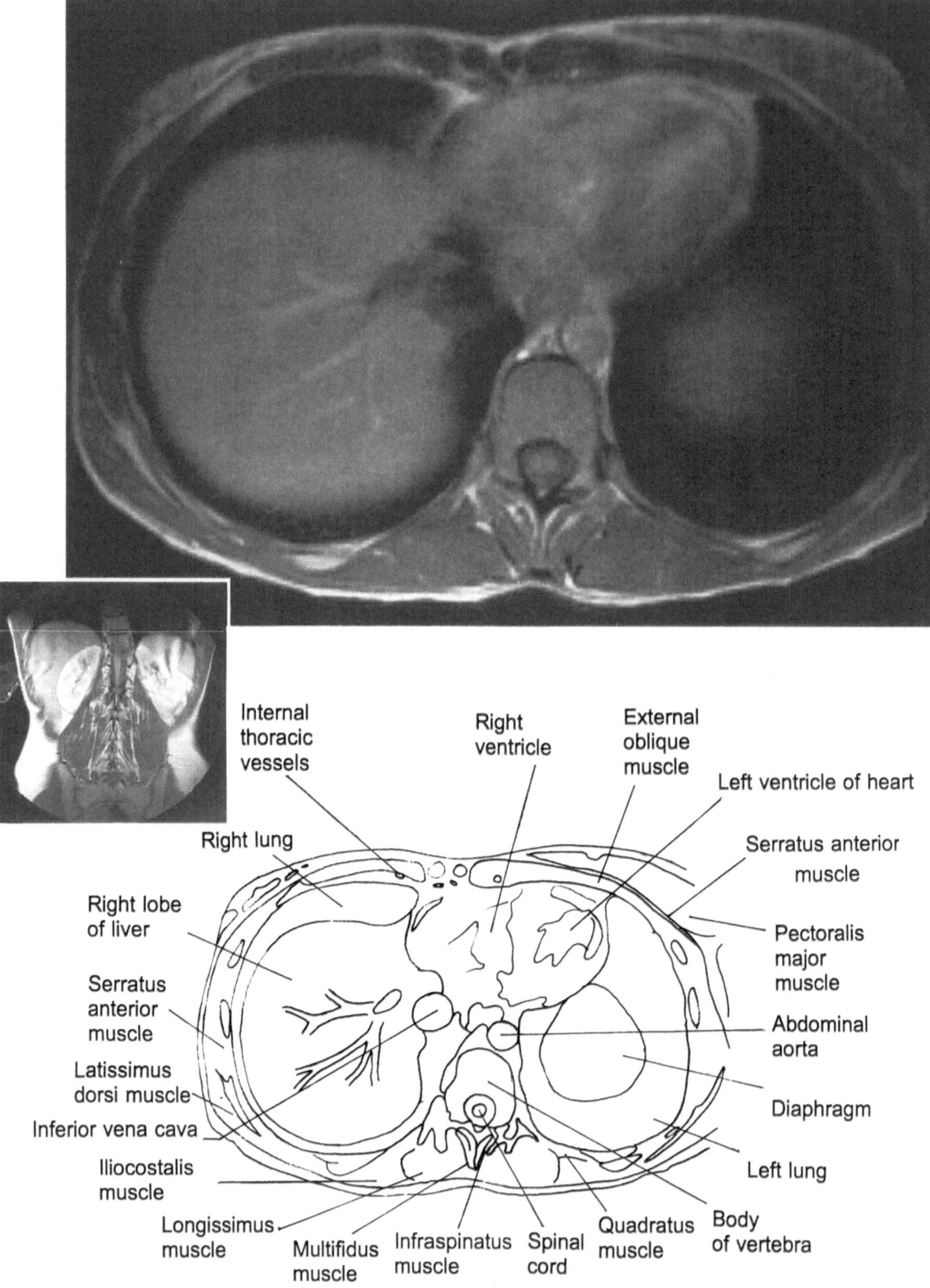

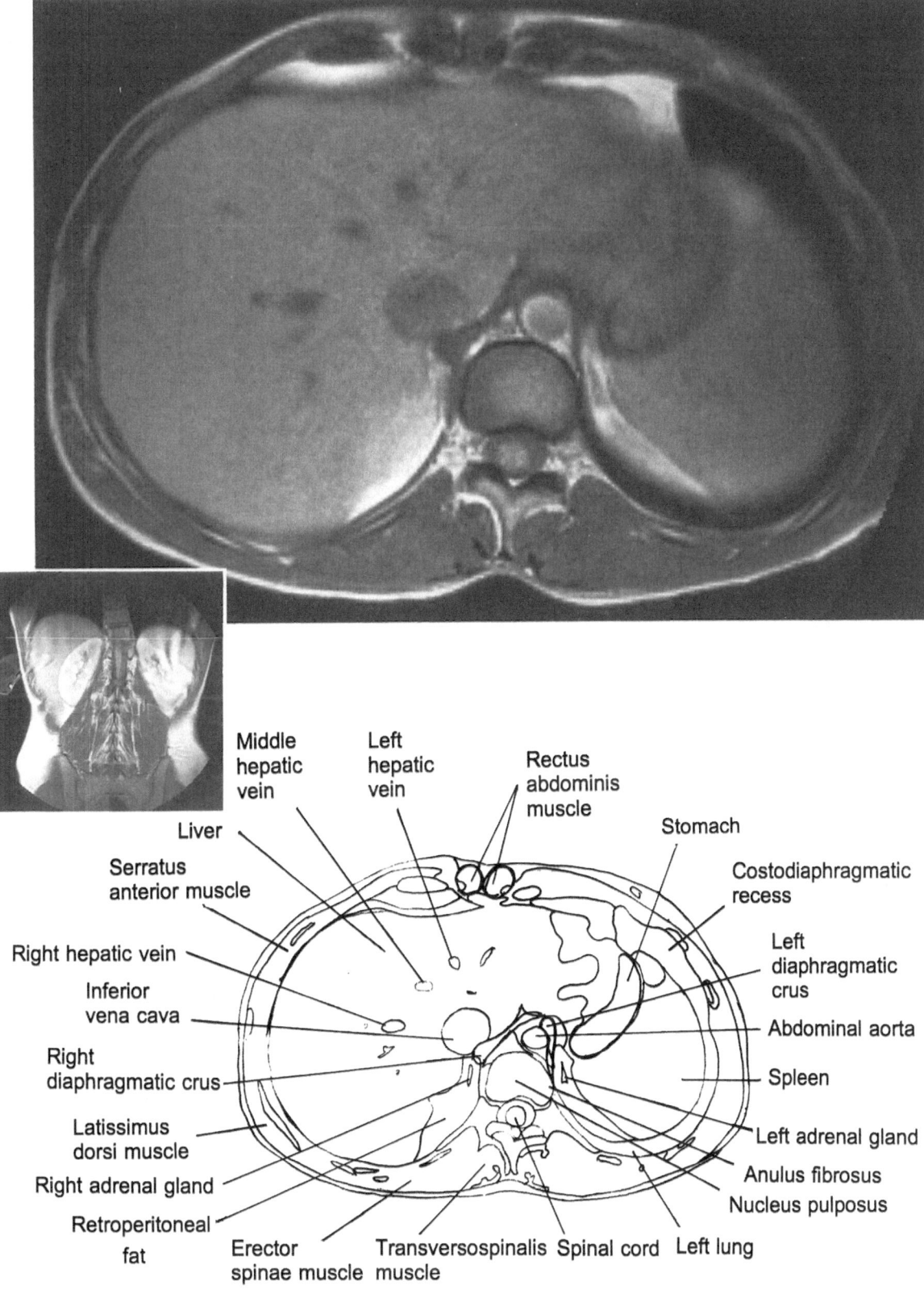

Middle hepatic vein

Left hepatic vein

Rectus abdominis muscle

Stomach

Costodiaphragmatic recess

Liver

Serratus anterior muscle

Left diaphragmatic crus

Right hepatic vein

Inferior vena cava

Abdominal aorta

Spleen

Right diaphragmatic crus

Latissimus dorsi muscle

Left adrenal gland

Right adrenal gland

Anulus fibrosus

Nucleus pulposus

Retroperitoneal fat

Erector spinae muscle

Transversospinalis muscle

Spinal cord

Left lung

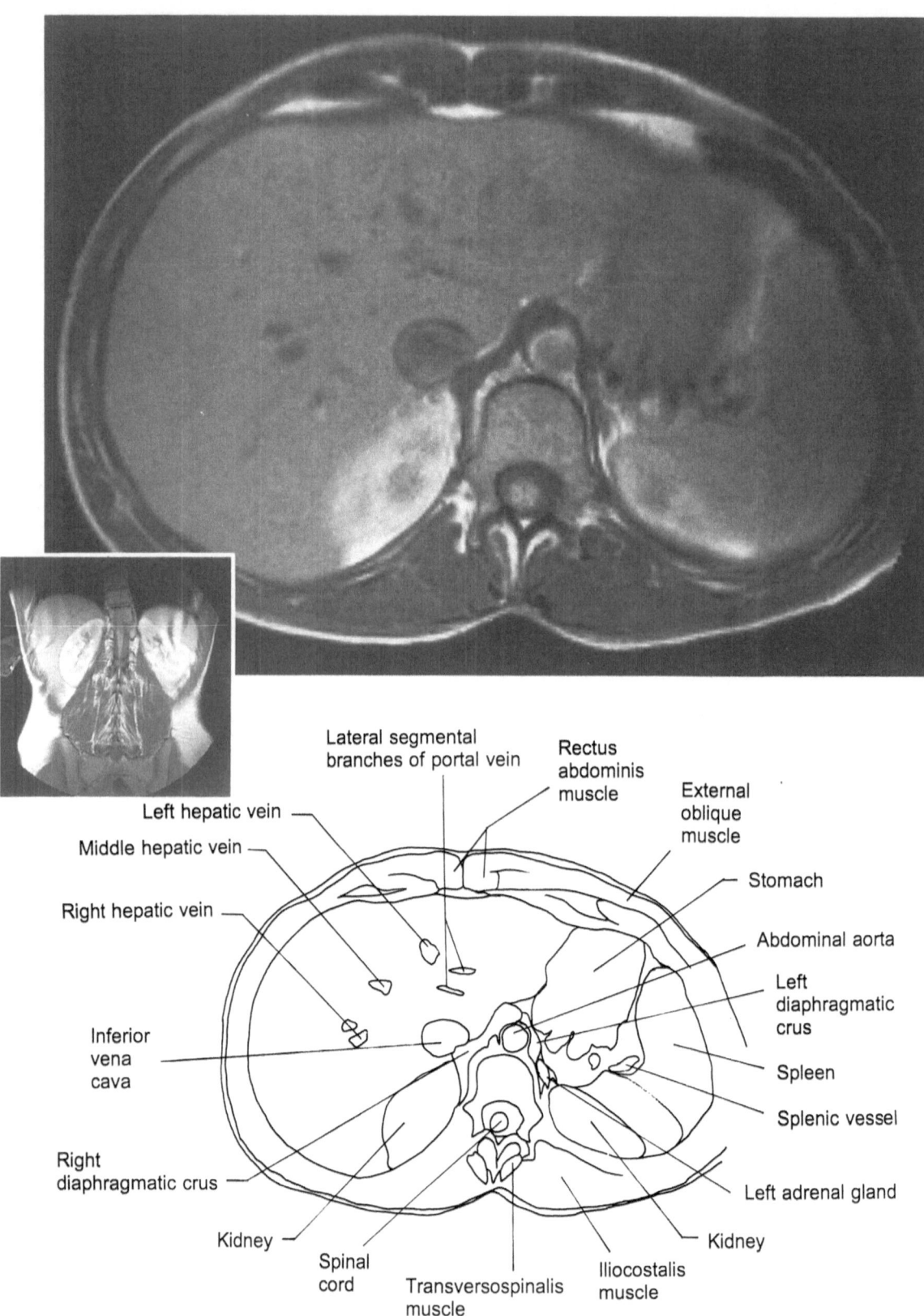

Lateral segmental branches of portal vein

Rectus abdominis muscle

External oblique muscle

Left hepatic vein

Middle hepatic vein

Right hepatic vein

Stomach

Abdominal aorta

Left diaphragmatic crus

Inferior vena cava

Spleen

Splenic vessel

Right diaphragmatic crus

Left adrenal gland

Kidney

Kidney

Spinal cord

Transversospinalis muscle

Iliocostalis muscle

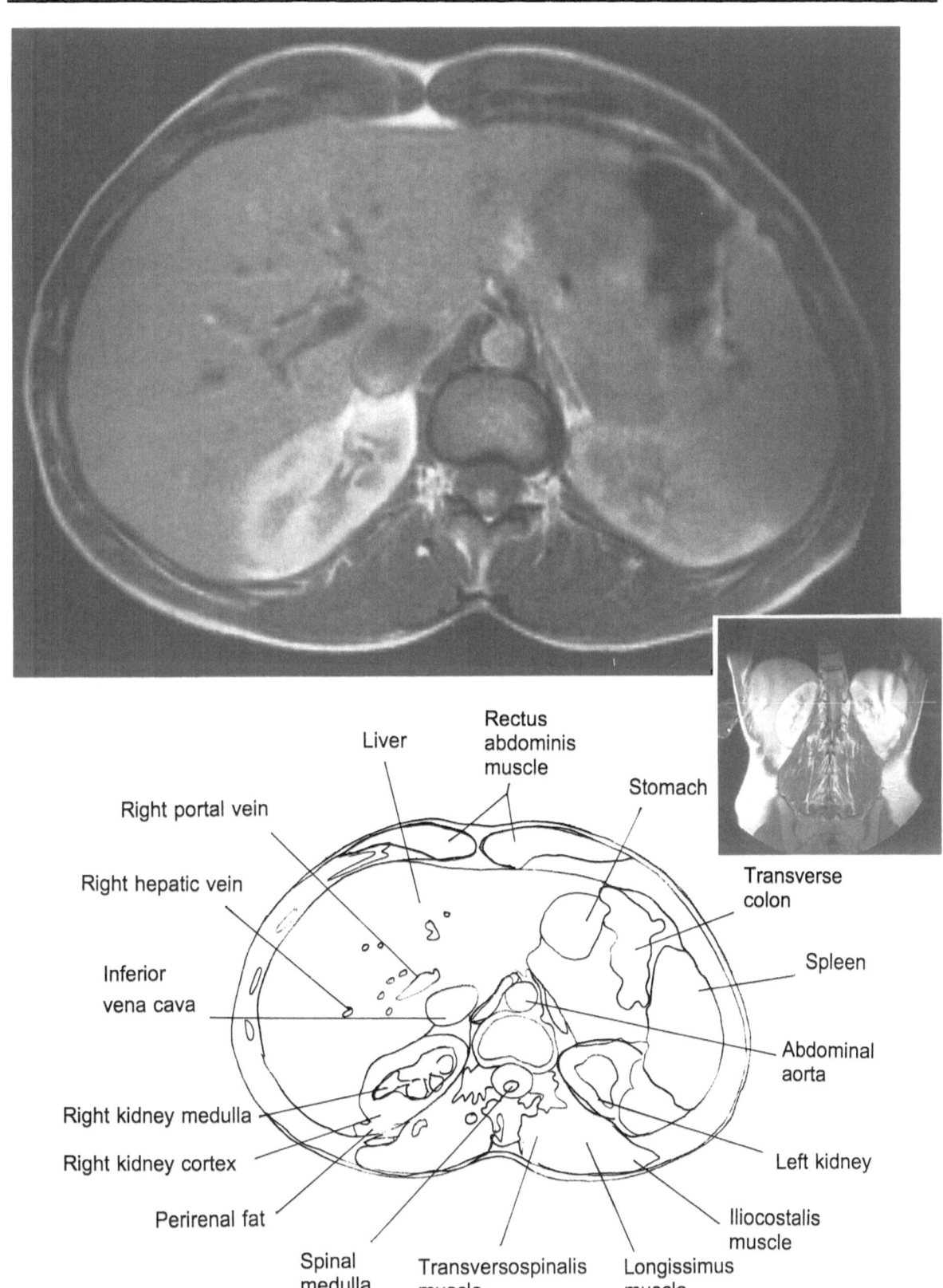

Liver

Rectus abdominis muscle

Right portal vein

Stomach

Right hepatic vein

Transverse colon

Inferior vena cava

Spleen

Abdominal aorta

Right kidney medulla

Right kidney cortex

Left kidney

Perirenal fat

Iliocostalis muscle

Spinal medulla

Transversospinalis muscle

Longissimus muscle

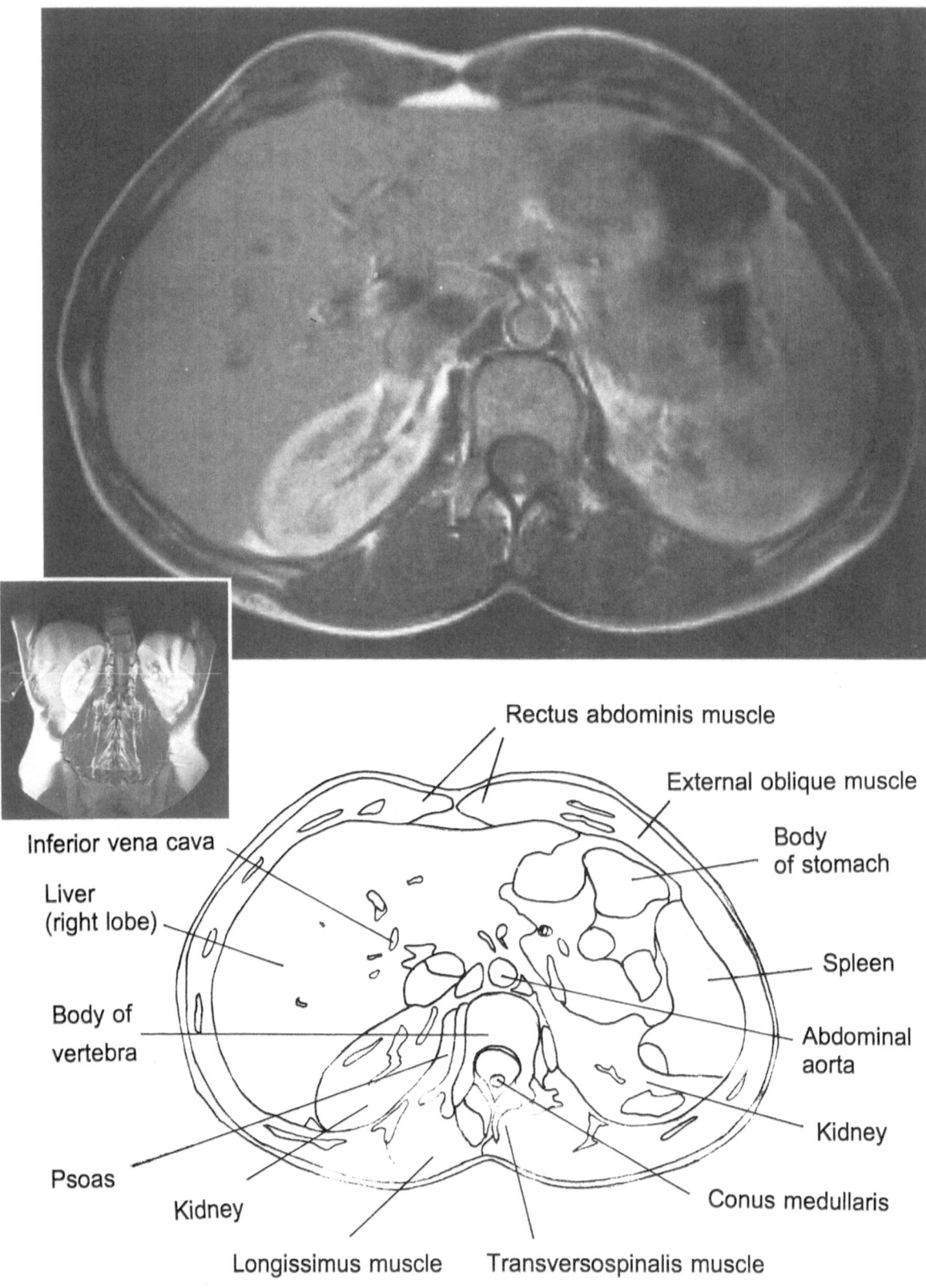

Rectus abdominis muscle

External oblique muscle

Body
of stomach

Spleen

Inferior vena cava

Liver
(right lobe)

Abdominal
aorta

Body of
vertebra

Kidney

Psoas

Conus medullaris

Kidney

Longissimus muscle   Transversospinalis muscle

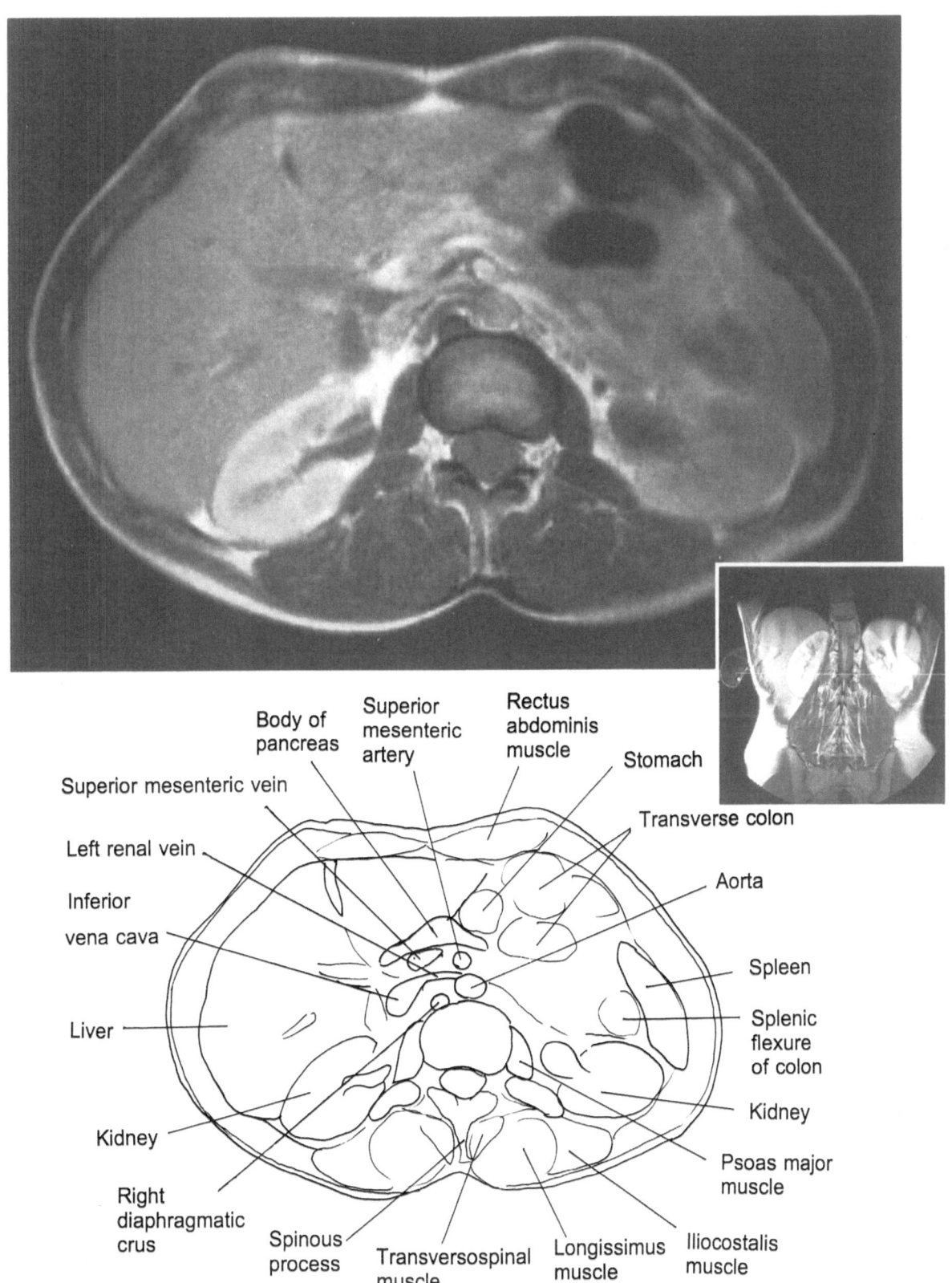

Body of pancreas

Superior mesenteric artery

Rectus abdominis muscle

Stomach

Superior mesenteric vein

Transverse colon

Left renal vein

Aorta

Inferior vena cava

Spleen

Splenic flexure of colon

Liver

Kidney

Kidney

Psoas major muscle

Right diaphragmatic crus

Spinous process

Transversospinal muscle

Longissimus muscle

Iliocostalis muscle

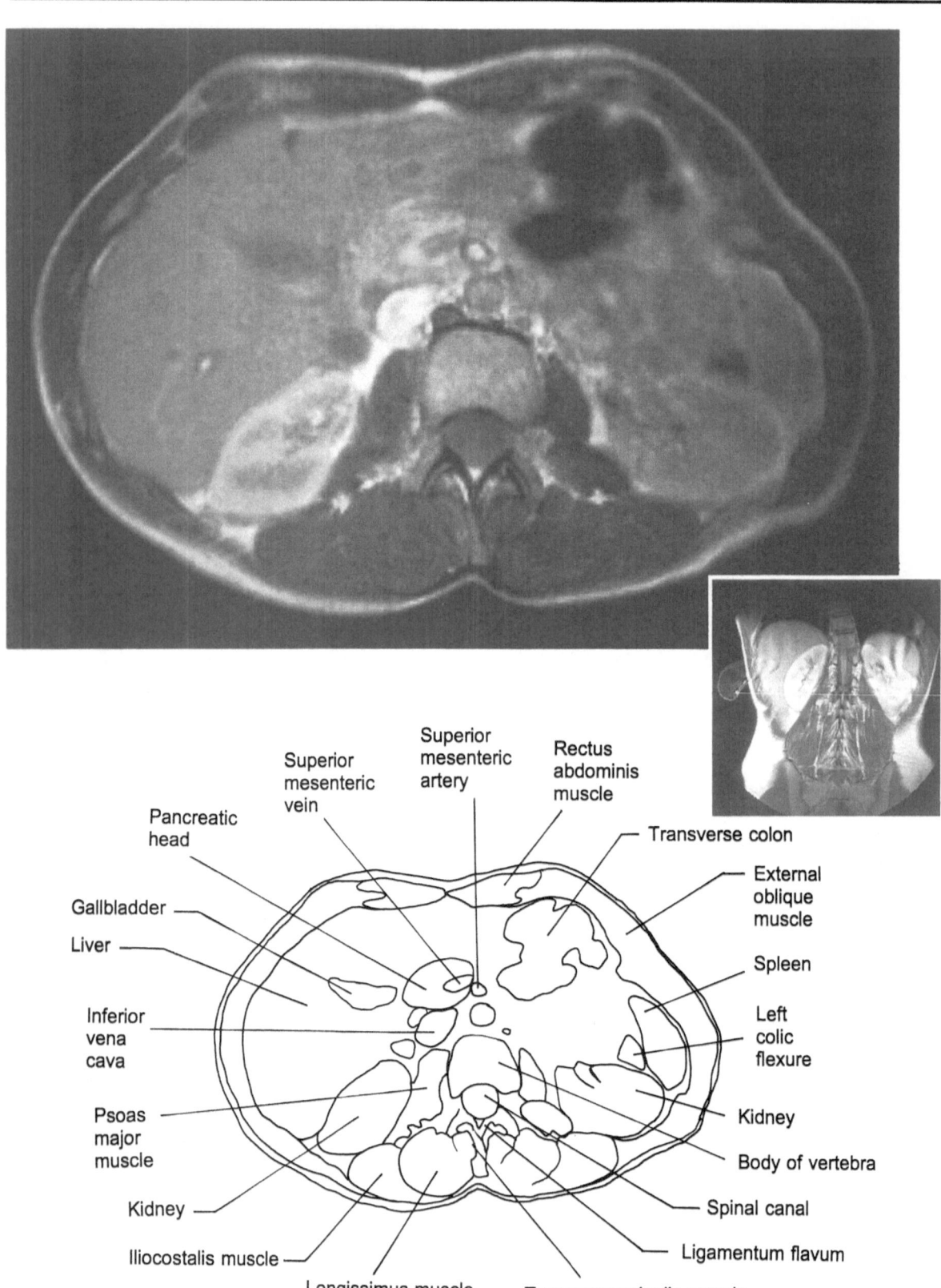

Superior mesenteric vein

Superior mesenteric artery

Rectus abdominis muscle

Pancreatic head

Transverse colon

External oblique muscle

Gallbladder

Liver

Spleen

Inferior vena cava

Left colic flexure

Psoas major muscle

Kidney

Body of vertebra

Kidney

Spinal canal

Iliocostalis muscle

Ligamentum flavum

Longissimus muscle

Transversospinalis muscle

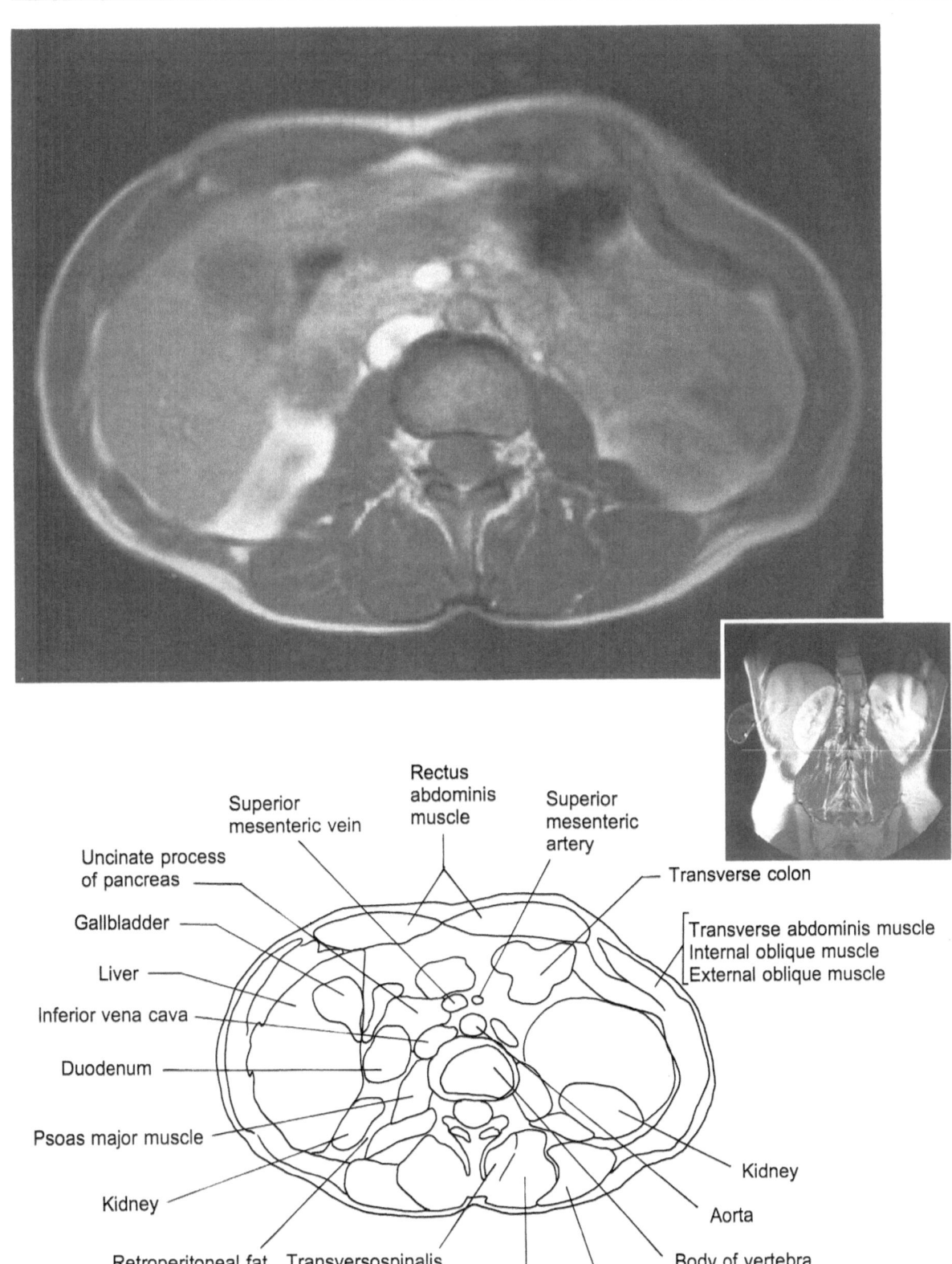

Superior
mesenteric vein

Rectus
abdominis
muscle

Superior
mesenteric
artery

Uncinate process
of pancreas

Transverse colon

Gallbladder

Transverse abdominis muscle
Internal oblique muscle
External oblique muscle

Liver

Inferior vena cava

Duodenum

Psoas major muscle

Kidney

Kidney

Aorta

Retroperitoneal fat

Transversospinalis
muscle

Body of vertebra

Longissimus
muscle

Iliocostalis muscle

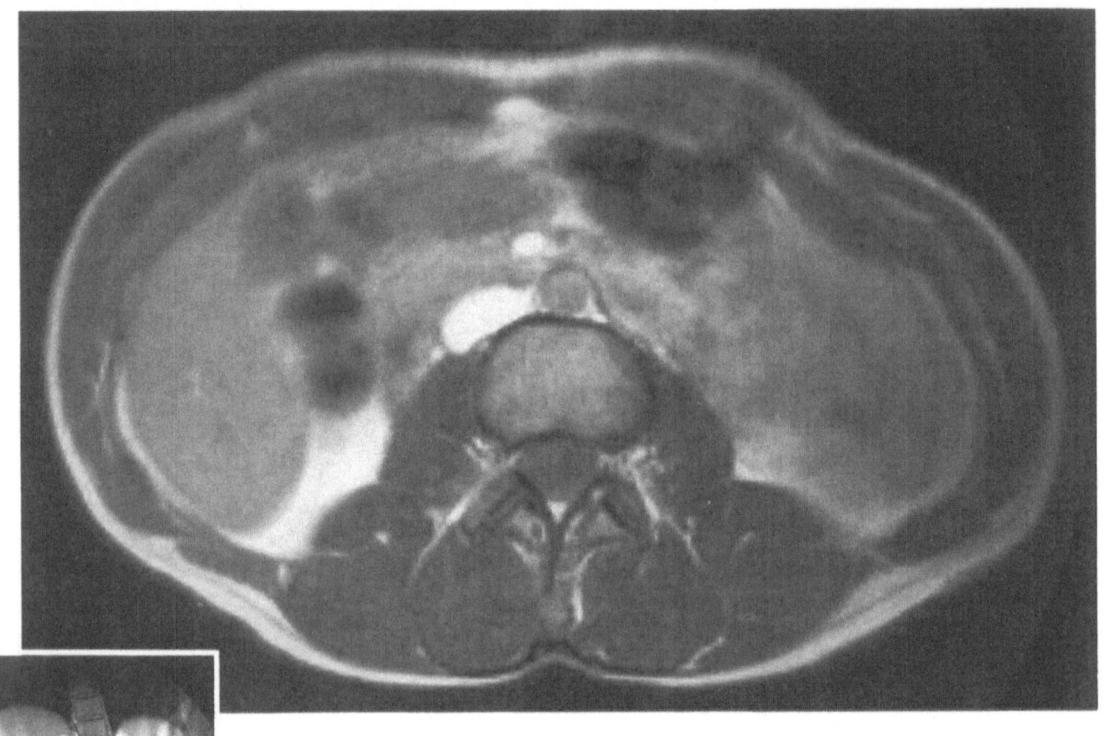

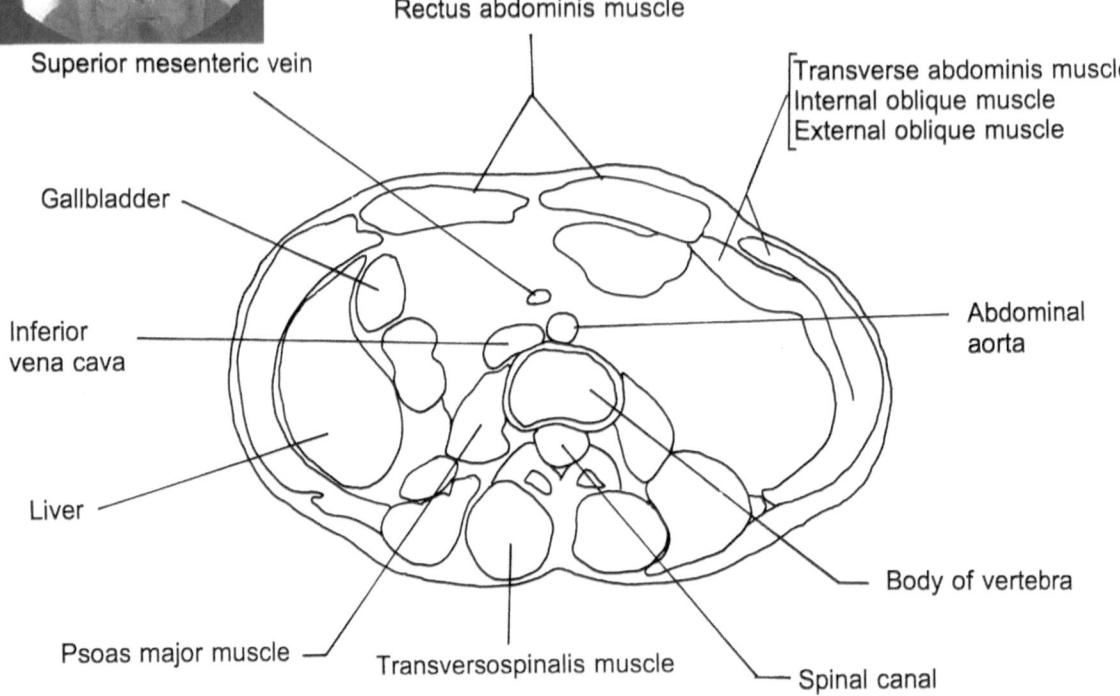

Rectus abdominis muscle

Superior mesenteric vein

Transverse abdominis muscle
Internal oblique muscle
External oblique muscle

Gallbladder

Abdominal
aorta

Inferior
vena cava

Liver

Body of vertebra

Psoas major muscle

Transversospinalis muscle

Spinal canal

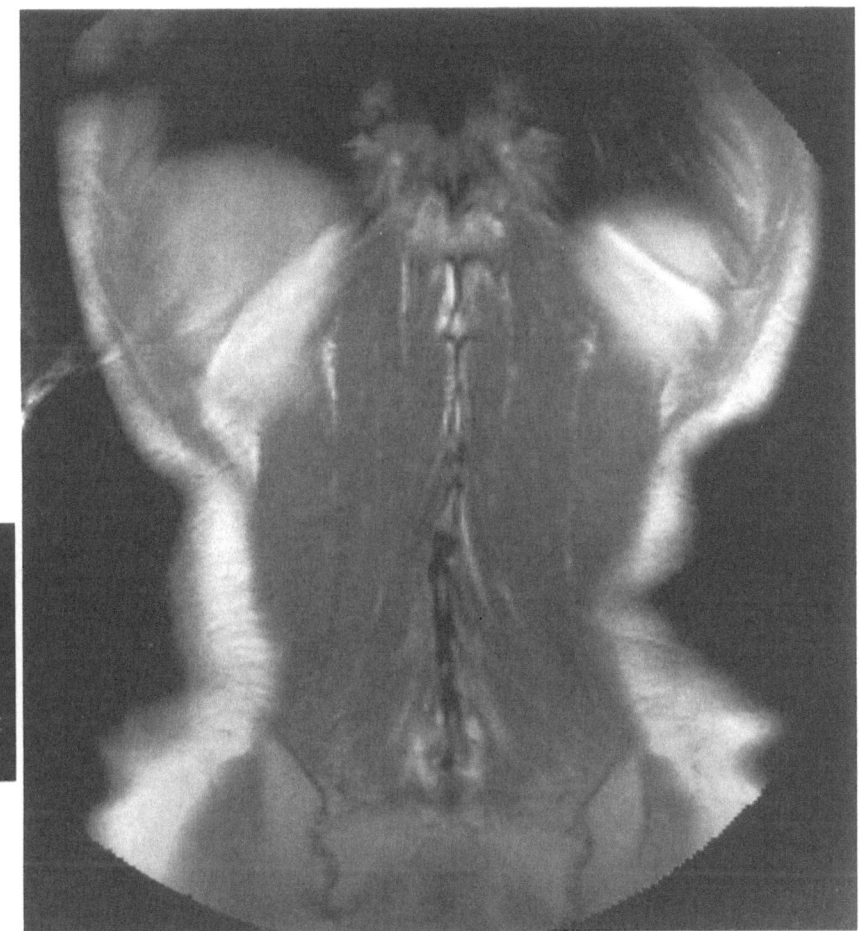

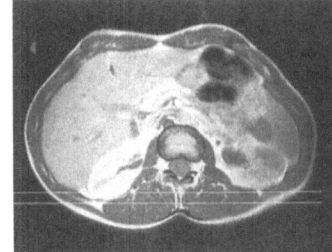

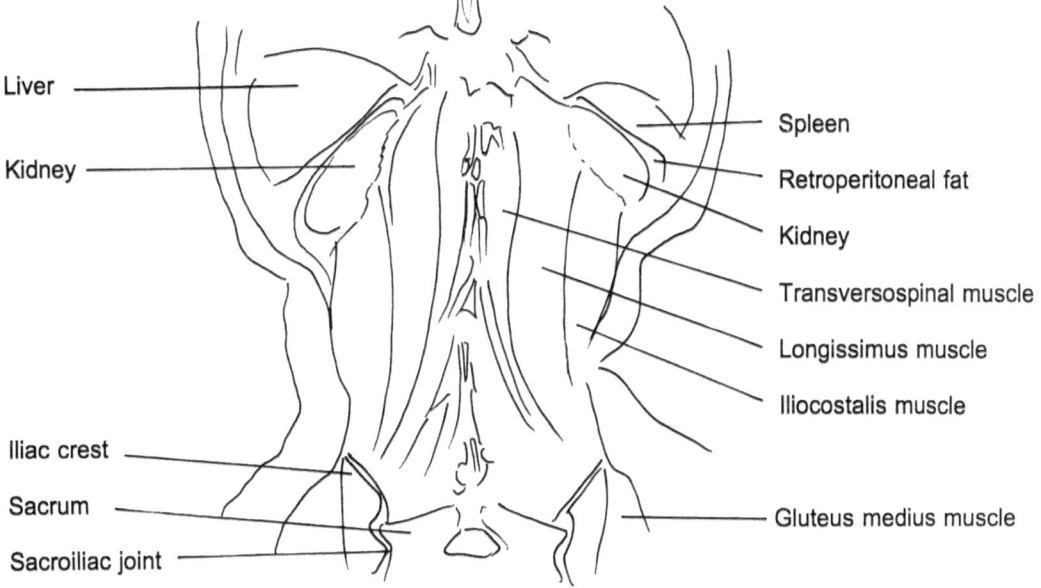

Liver

Kidney

Spleen

Retroperitoneal fat

Kidney

Transversospinal muscle

Longissimus muscle

Iliocostalis muscle

Iliac crest

Sacrum

Sacroiliac joint

Gluteus medius muscle

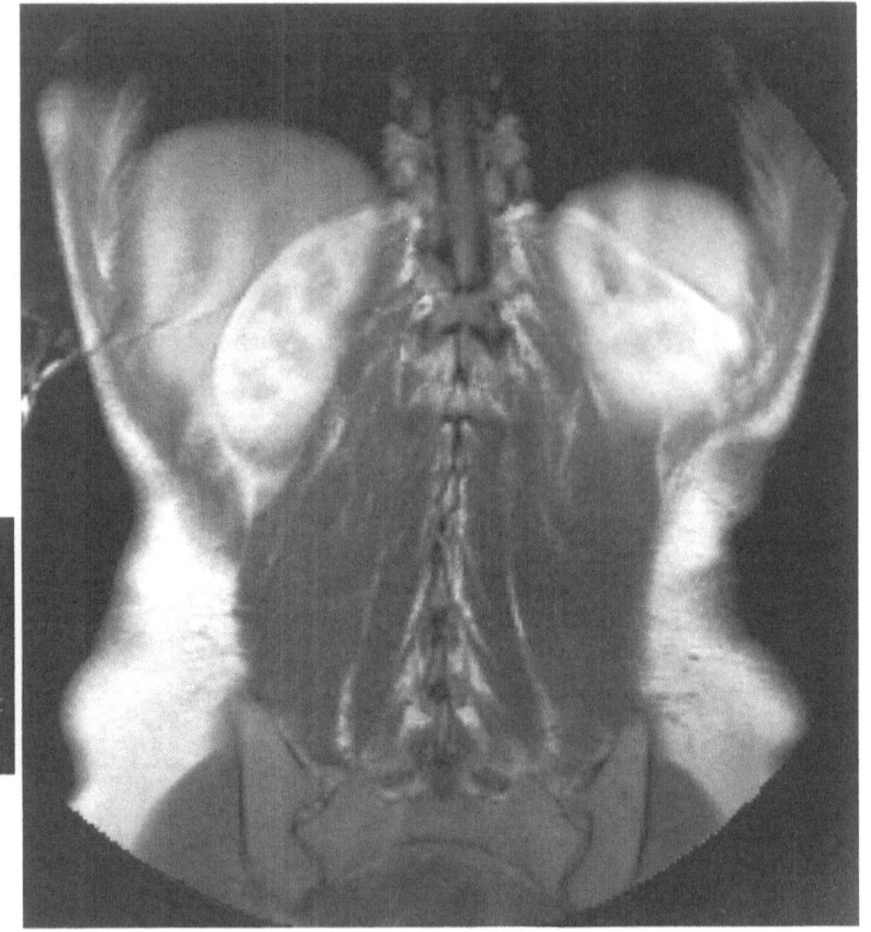

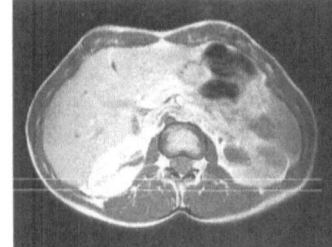

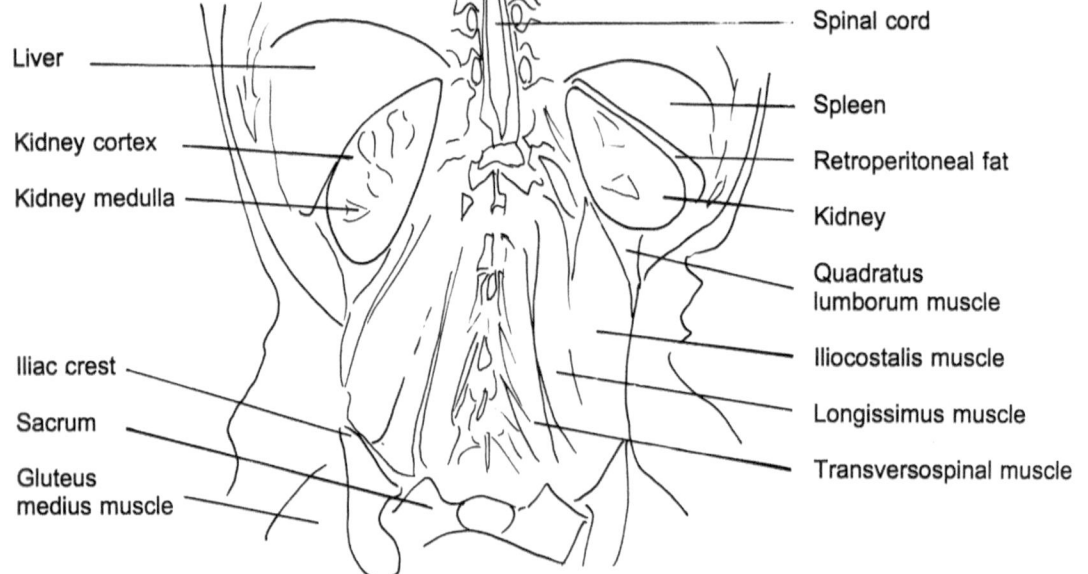

Liver

Kidney cortex

Kidney medulla

Iliac crest

Sacrum

Gluteus
medius muscle

Spinal cord

Spleen

Retroperitoneal fat

Kidney

Quadratus
lumborum muscle

Iliocostalis muscle

Longissimus muscle

Transversospinal muscle

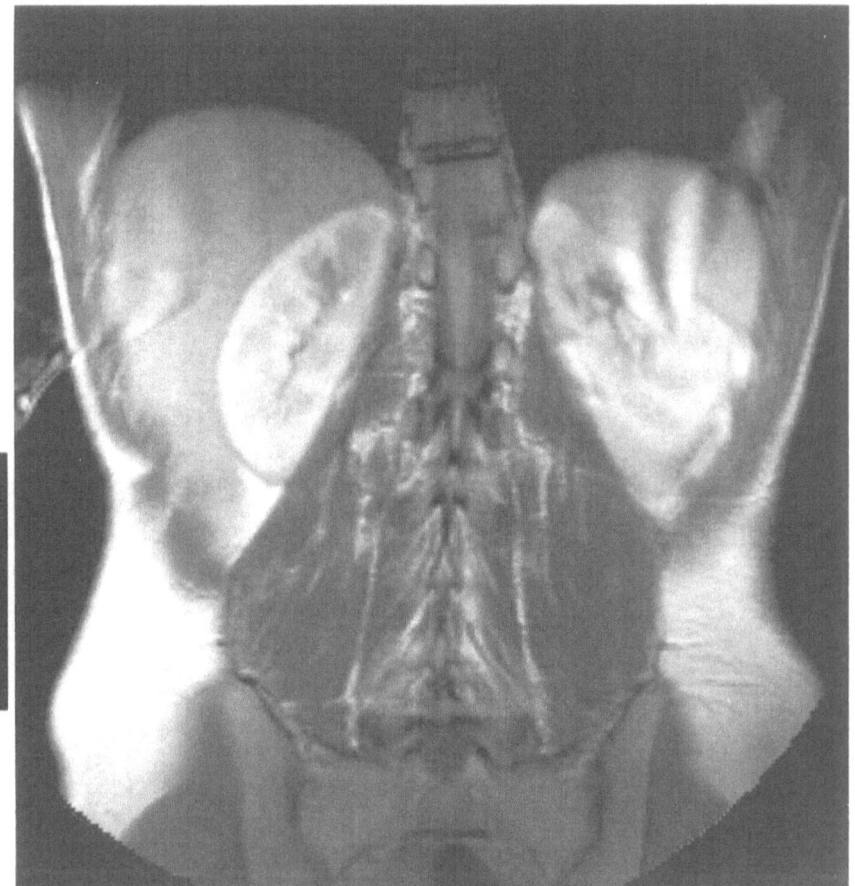

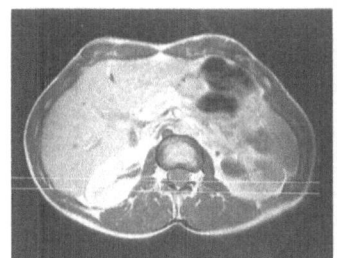

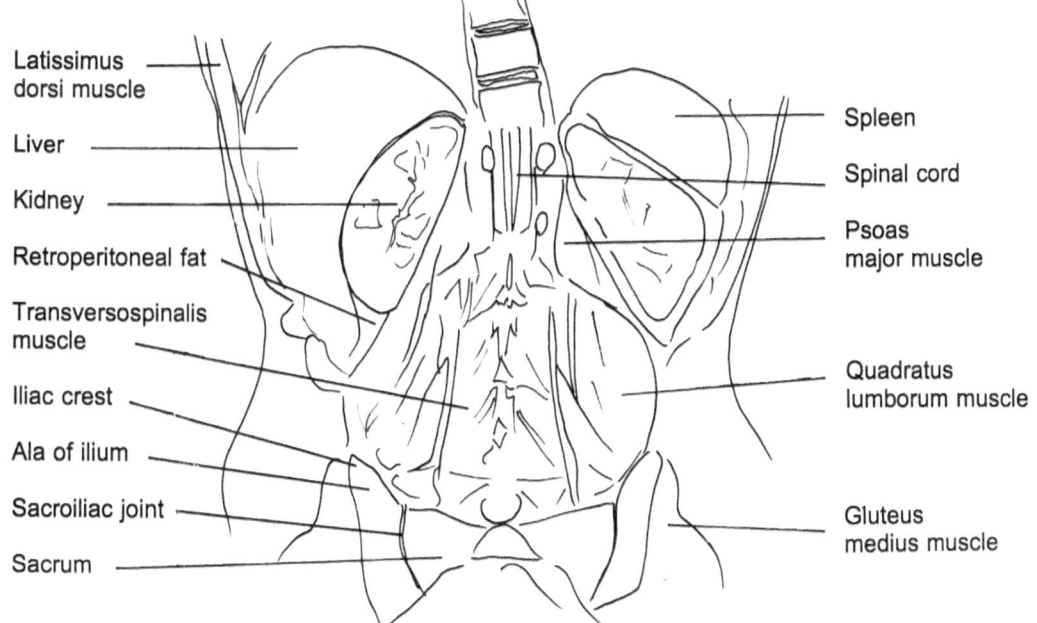

Latissimus dorsi muscle

Liver

Kidney

Retroperitoneal fat

Transversospinalis muscle

Iliac crest

Ala of ilium

Sacroiliac joint

Sacrum

Spleen

Spinal cord

Psoas major muscle

Quadratus lumborum muscle

Gluteus medius muscle

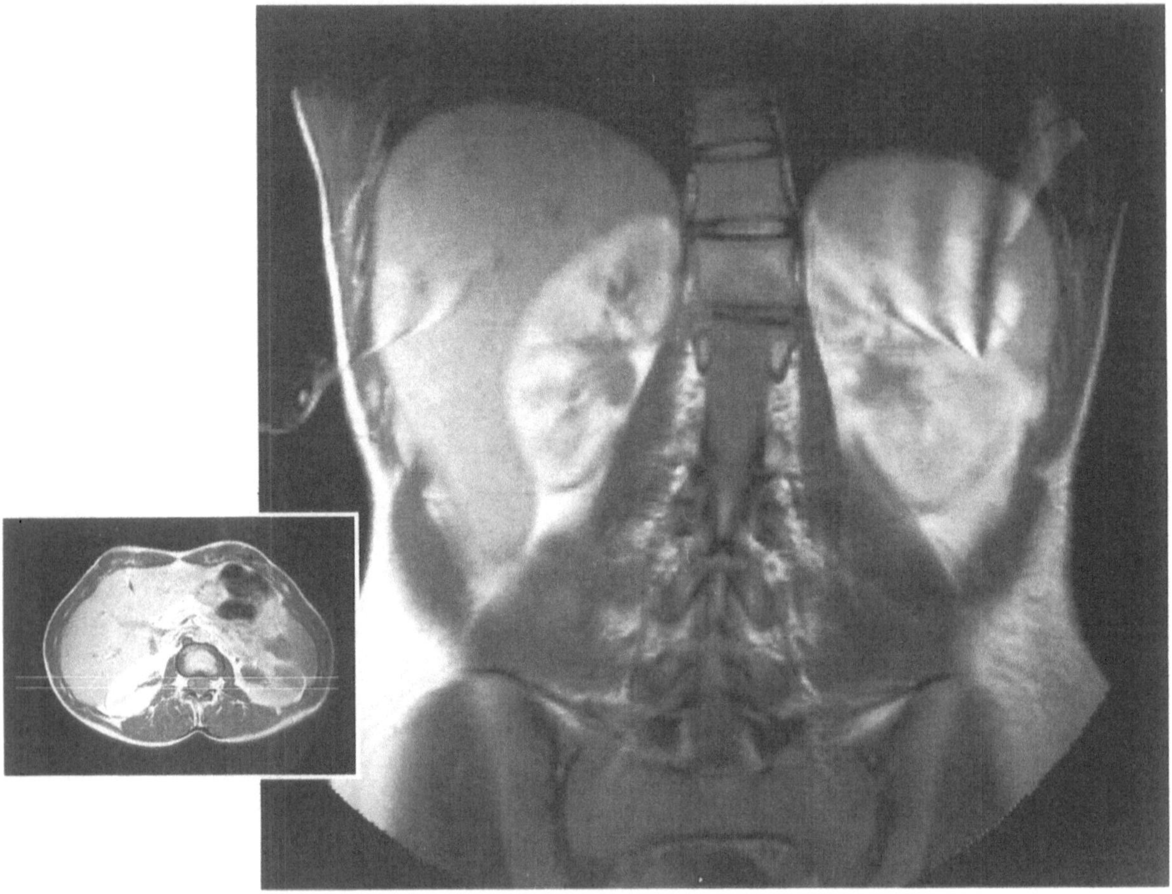

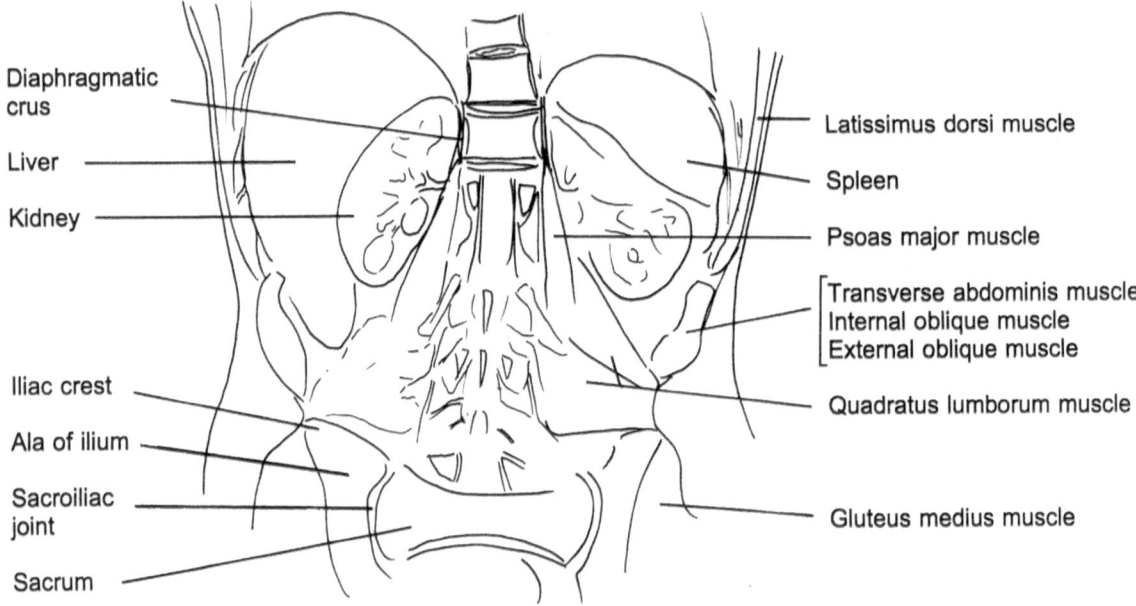

Diaphragmatic crus

Liver

Kidney

Iliac crest

Ala of ilium

Sacroiliac joint

Sacrum

Latissimus dorsi muscle

Spleen

Psoas major muscle

Transverse abdominis muscle
Internal oblique muscle
External oblique muscle

Quadratus lumborum muscle

Gluteus medius muscle

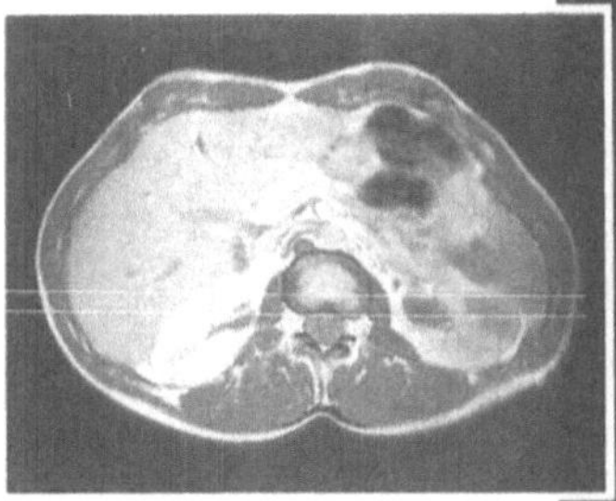

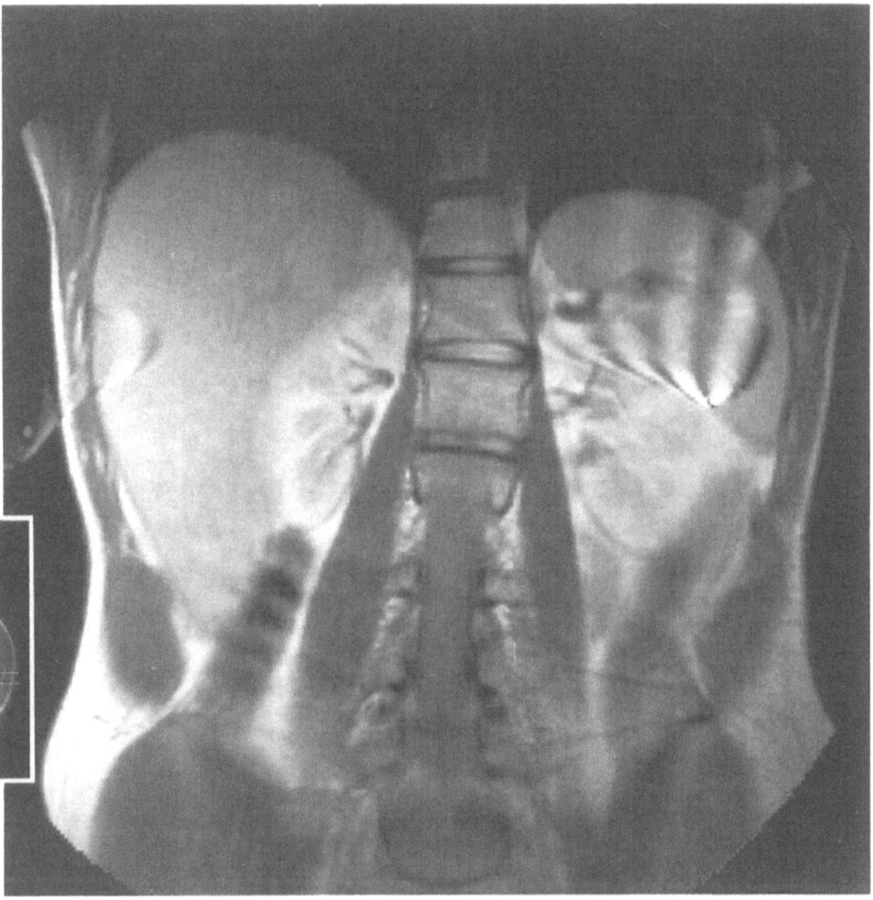

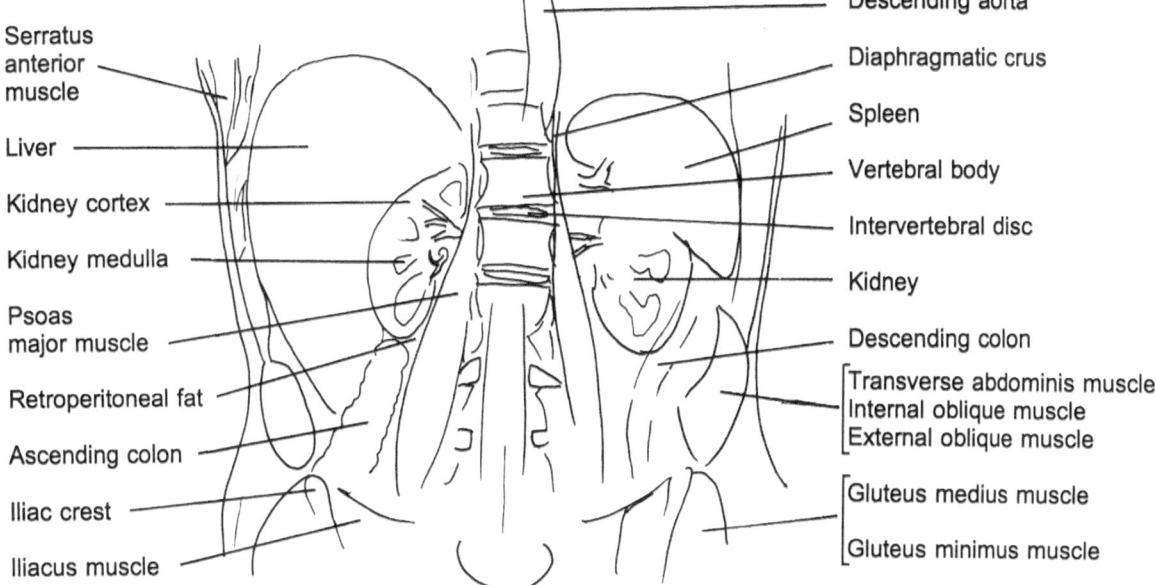

Descending aorta

Serratus anterior muscle

Diaphragmatic crus

Liver

Spleen

Kidney cortex

Vertebral body

Kidney medulla

Intervertebral disc

Psoas major muscle

Kidney

Retroperitoneal fat

Descending colon

Ascending colon

Transverse abdominis muscle
Internal oblique muscle
External oblique muscle

Iliac crest

Gluteus medius muscle

Iliacus muscle

Gluteus minimus muscle

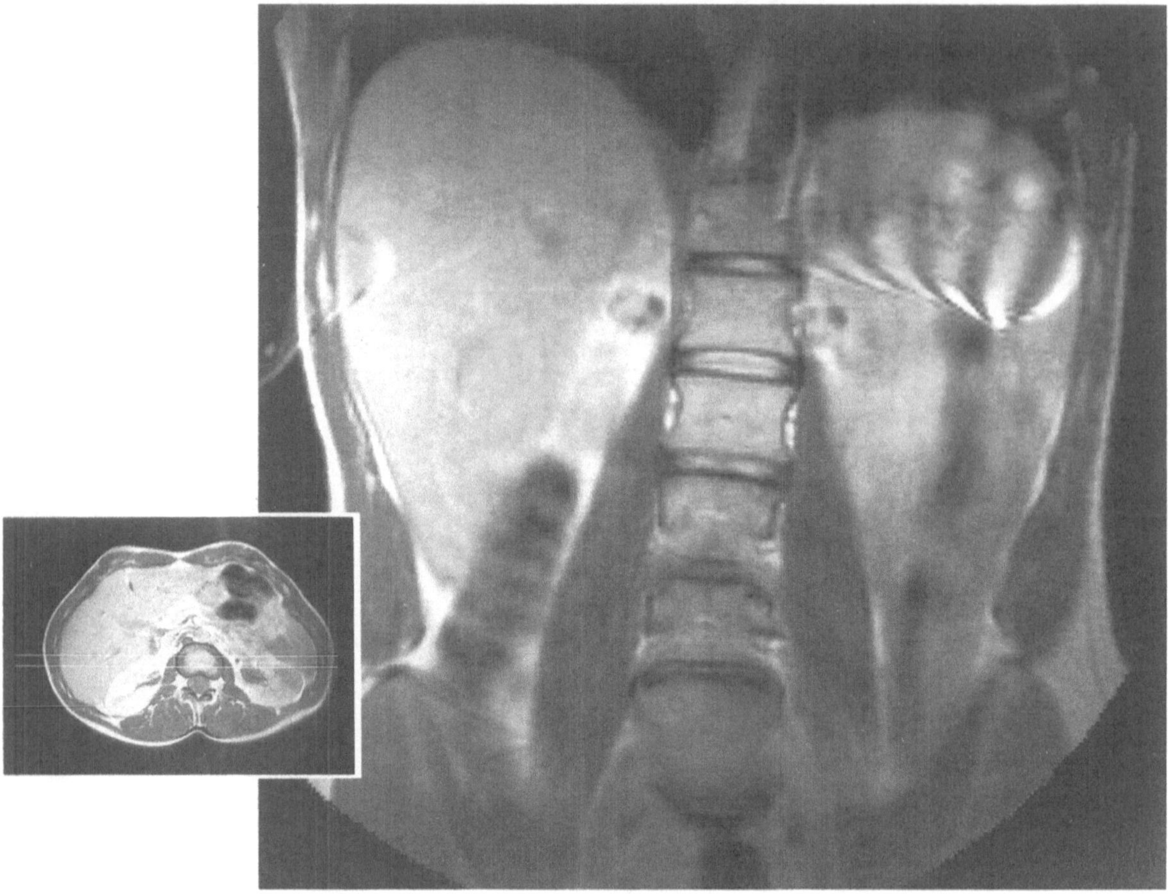

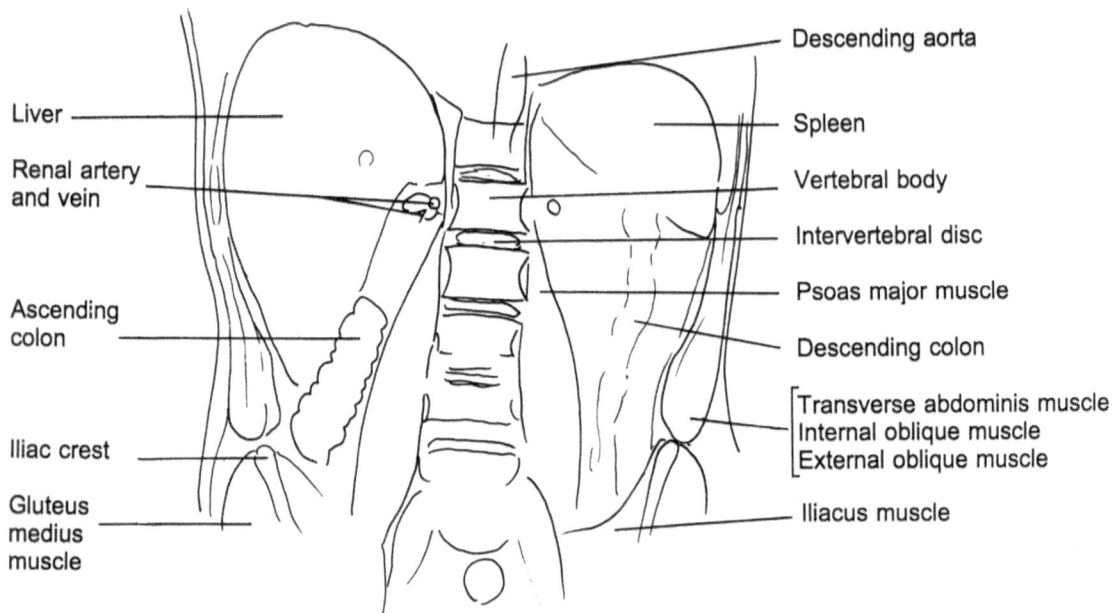

Liver

Renal artery
and vein

Ascending
colon

Iliac crest

Gluteus
medius
muscle

Descending aorta

Spleen

Vertebral body

Intervertebral disc

Psoas major muscle

Descending colon

Transverse abdominis muscle
Internal oblique muscle
External oblique muscle

Iliacus muscle

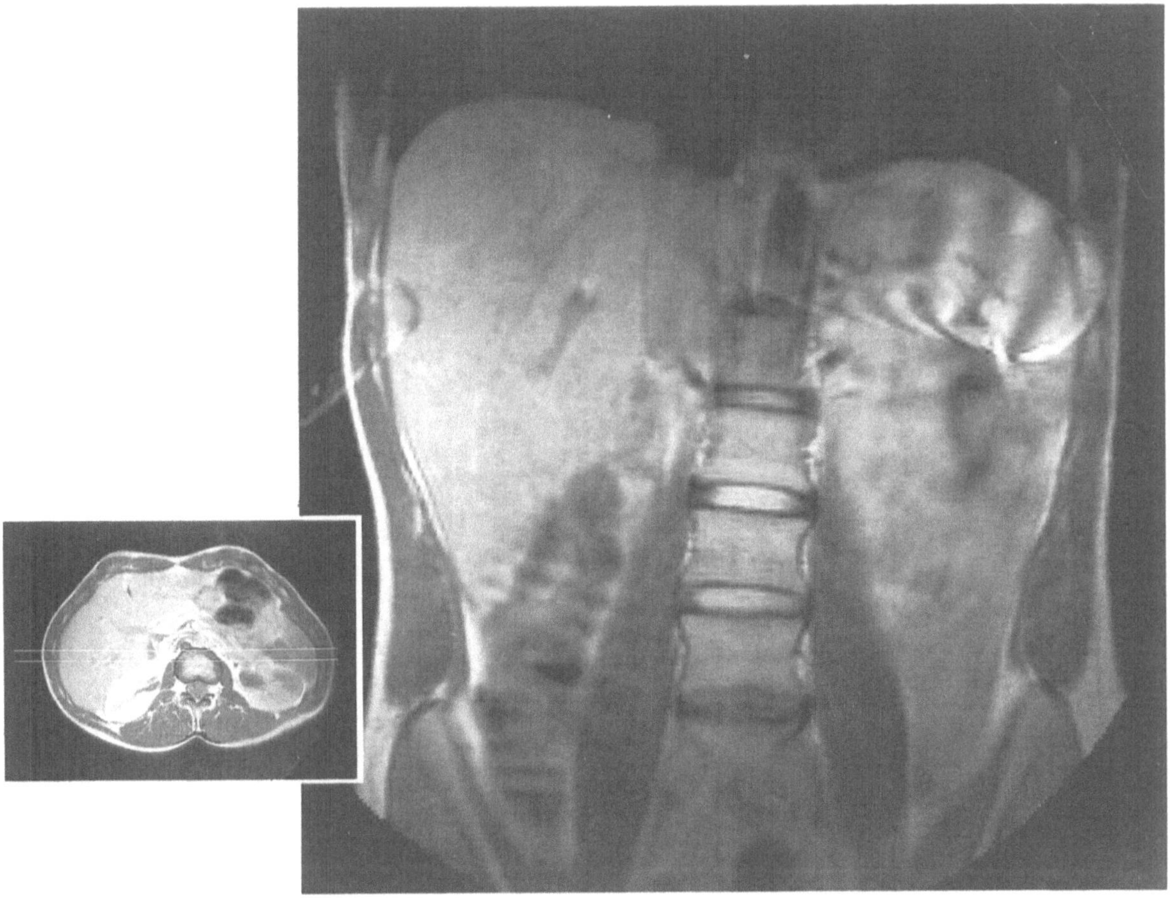

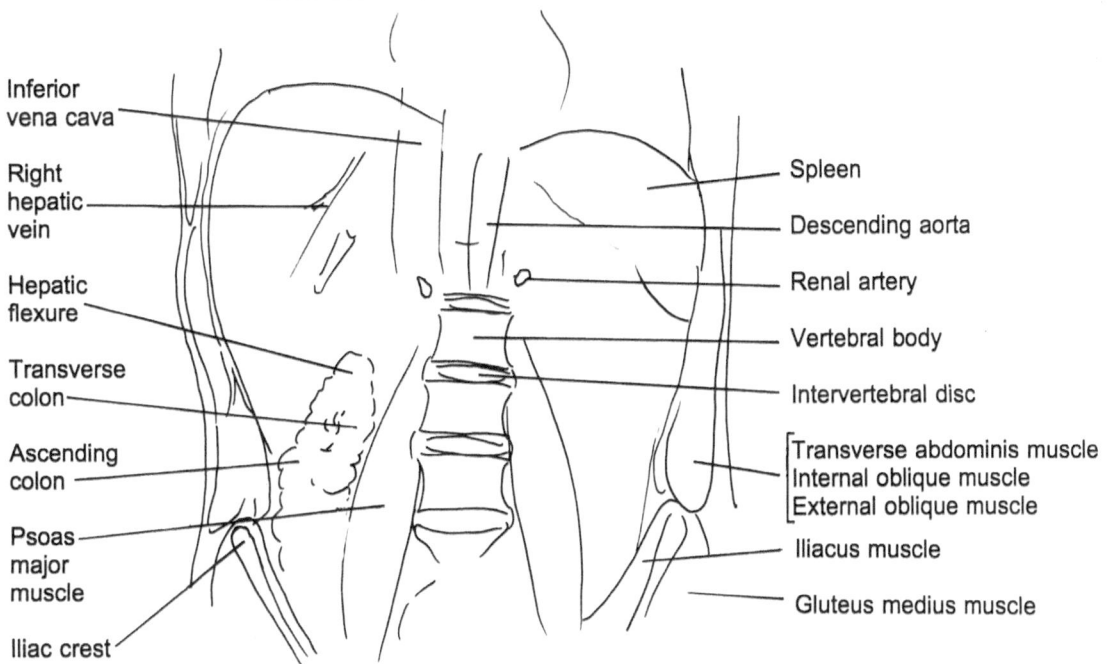

Inferior vena cava

Right hepatic vein

Hepatic flexure

Transverse colon

Ascending colon

Psoas major muscle

Iliac crest

Spleen

Descending aorta

Renal artery

Vertebral body

Intervertebral disc

Transverse abdominis muscle
Internal oblique muscle
External oblique muscle

Iliacus muscle

Gluteus medius muscle

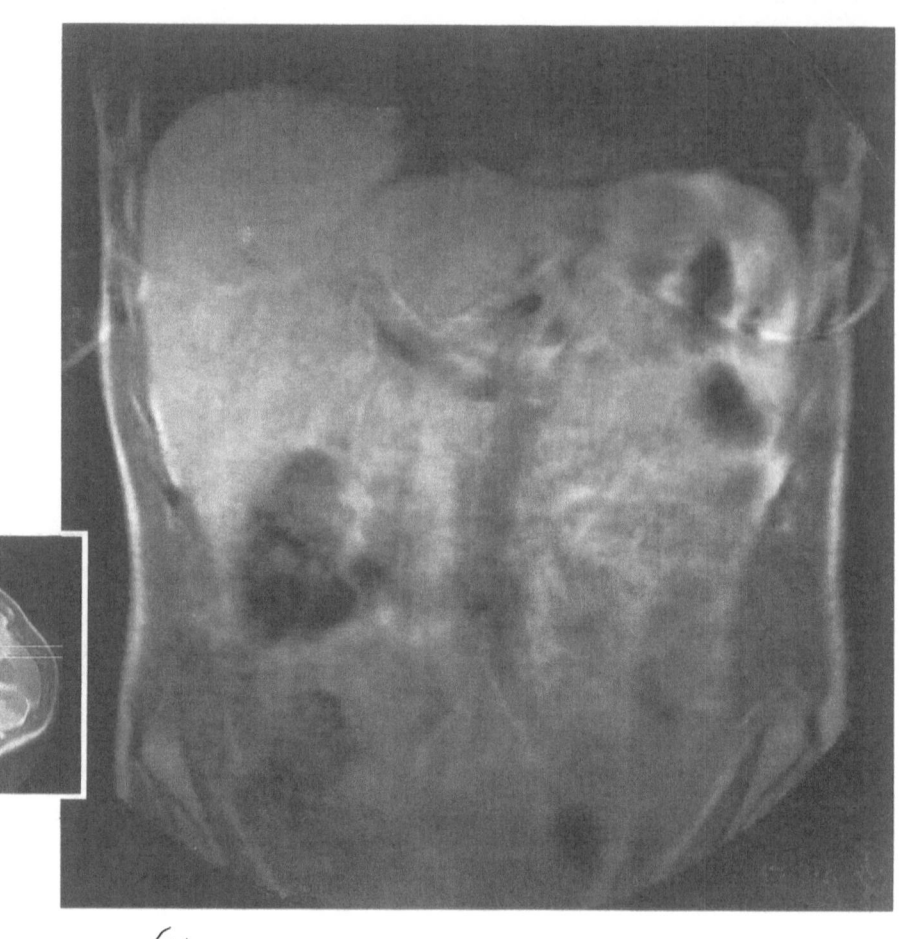

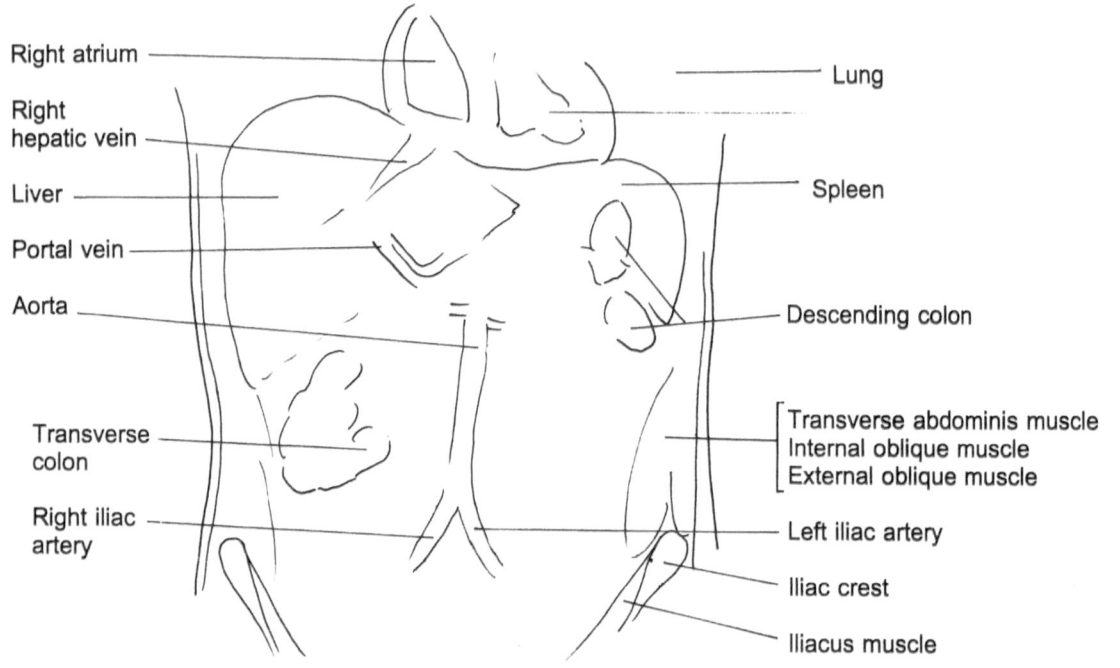

Right atrium

Right
hepatic vein

Liver

Portal vein

Aorta

Transverse
colon

Right iliac
artery

Lung

Spleen

Descending colon

Transverse abdominis muscle
Internal oblique muscle
External oblique muscle

Left iliac artery

Iliac crest

Iliacus muscle

ELBOW

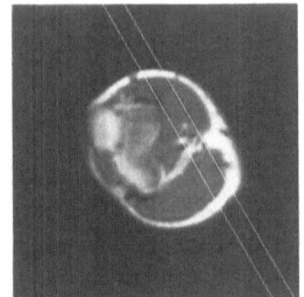

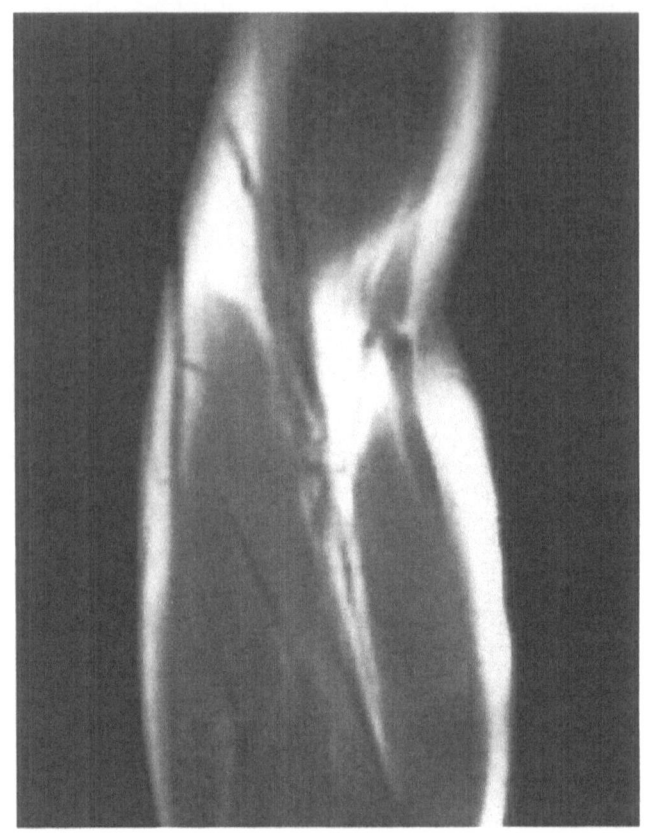

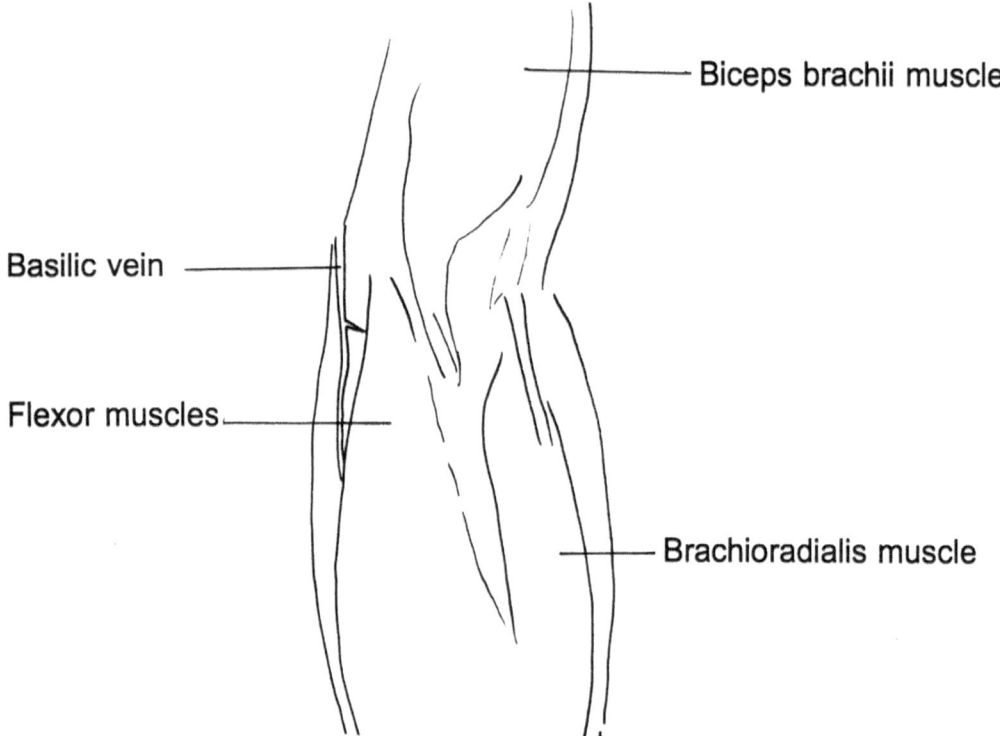

Biceps brachii muscle

Basilic vein

Flexor muscles

Brachioradialis muscle

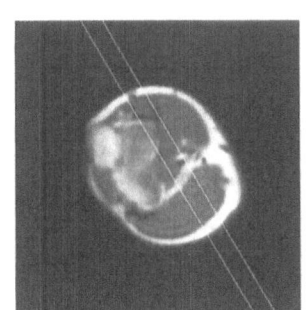

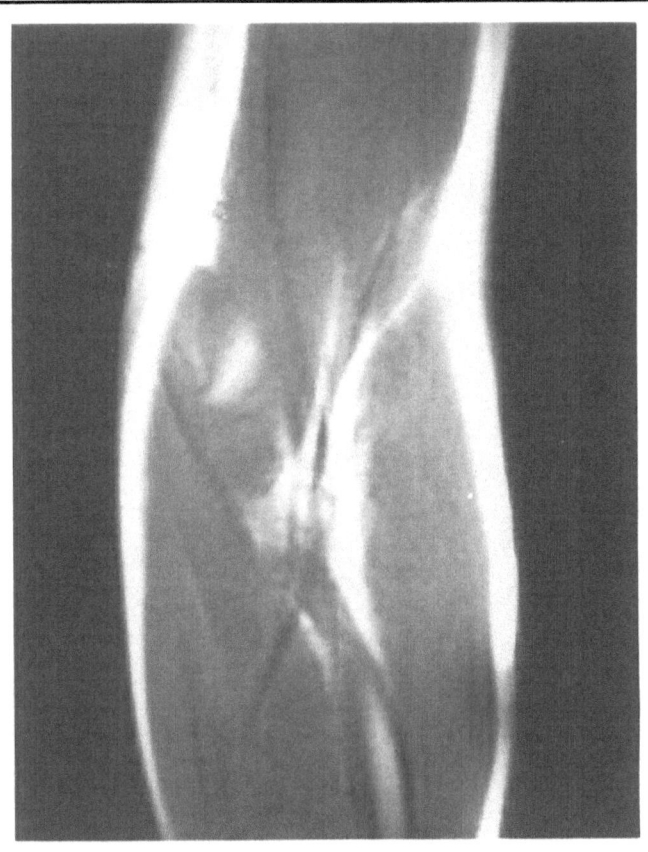

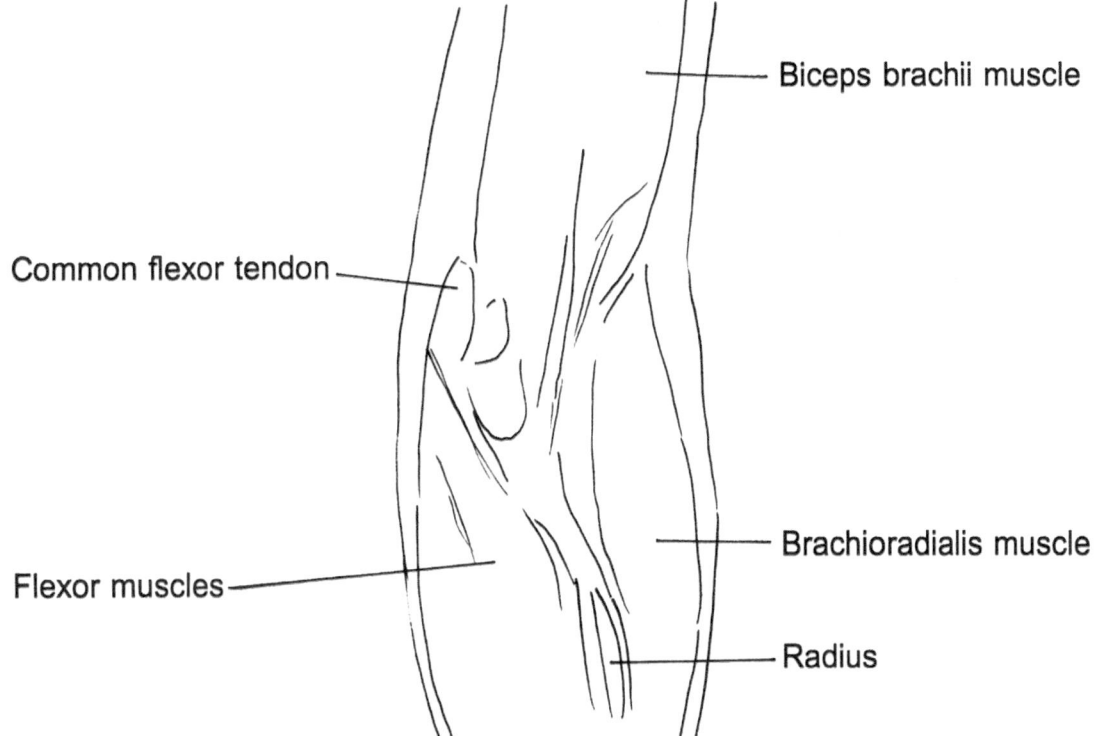

Biceps brachii muscle

Common flexor tendon

Brachioradialis muscle

Flexor muscles

Radius

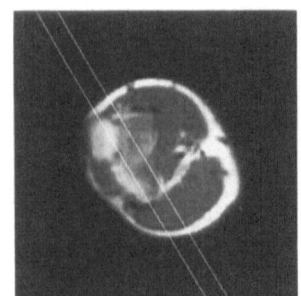

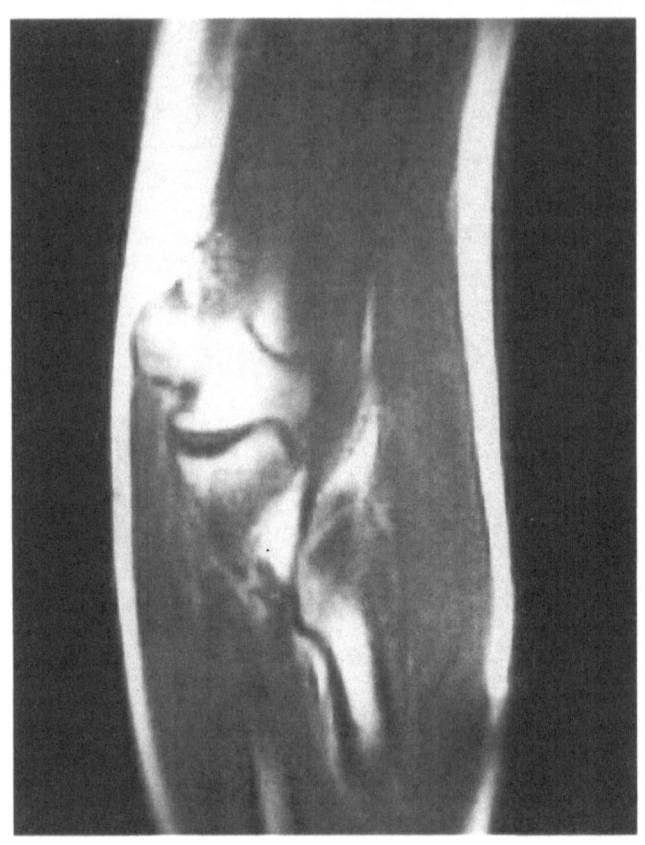

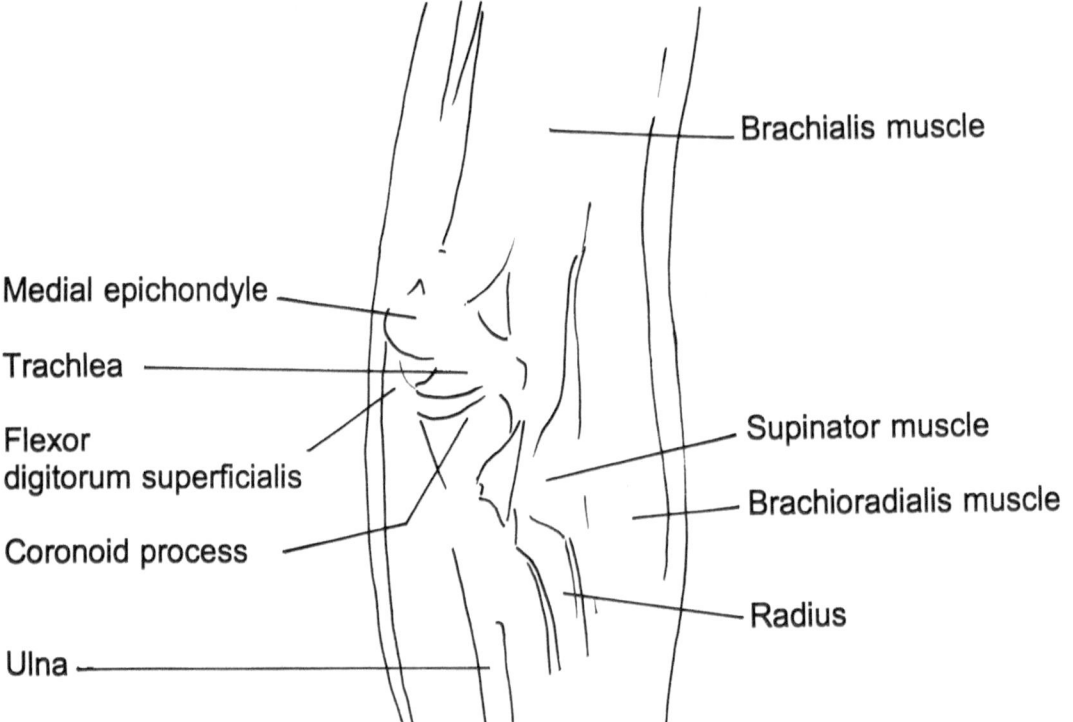

Brachialis muscle

Medial epichondyle

Trachlea

Flexor
digitorum superficialis

Coronoid process

Ulna

Supinator muscle

Brachioradialis muscle

Radius

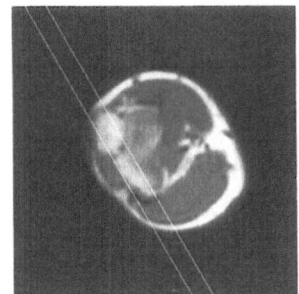

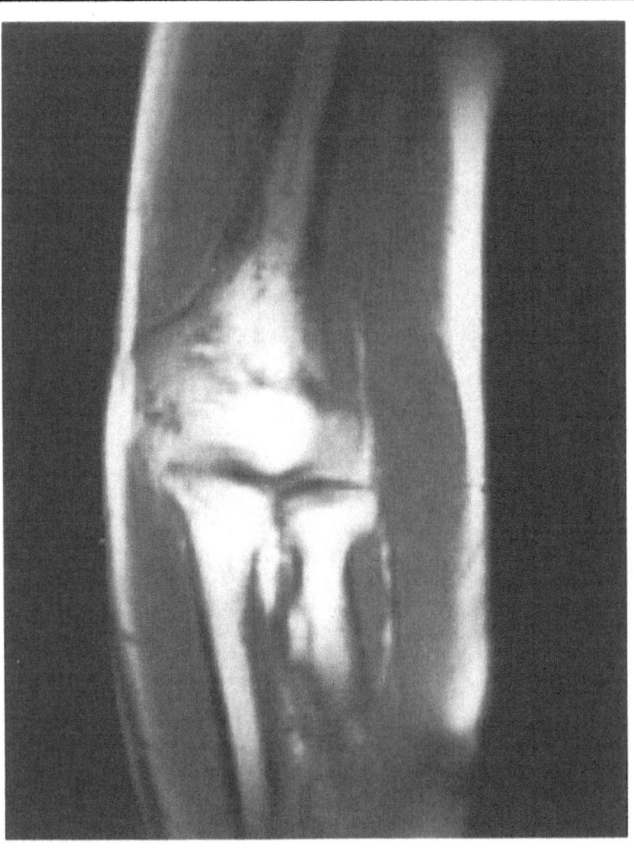

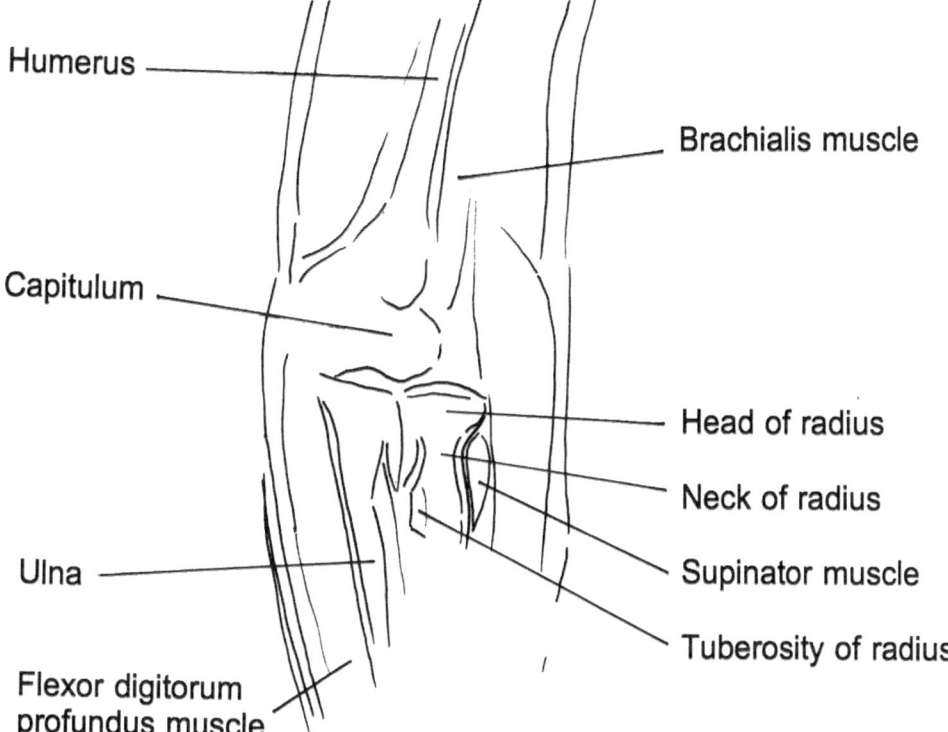

Humerus

Brachialis muscle

Capitulum

Head of radius

Neck of radius

Ulna

Supinator muscle

Tuberosity of radius

Flexor digitorum
profundus muscle

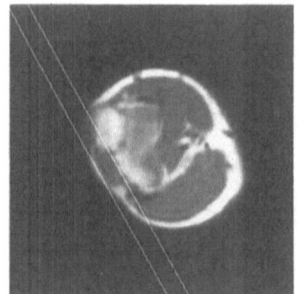

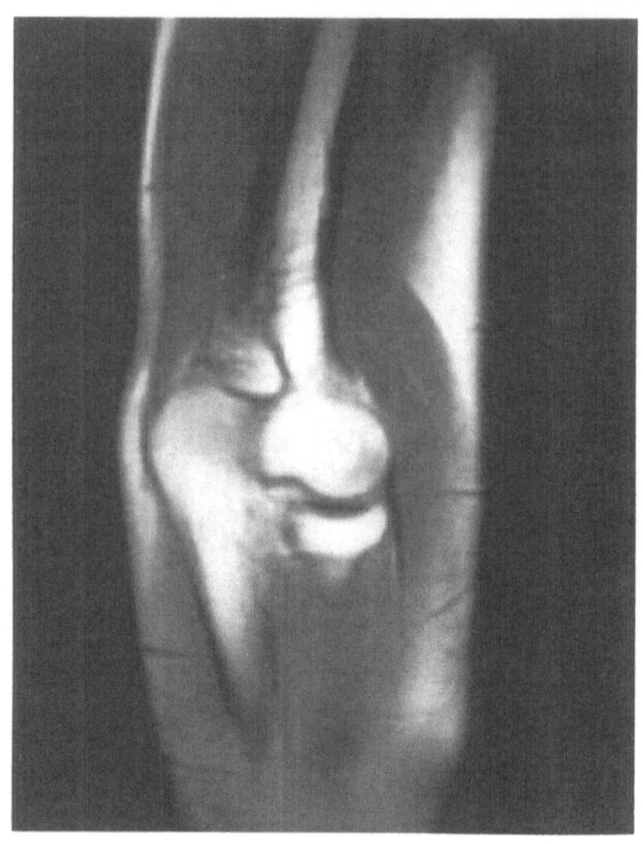

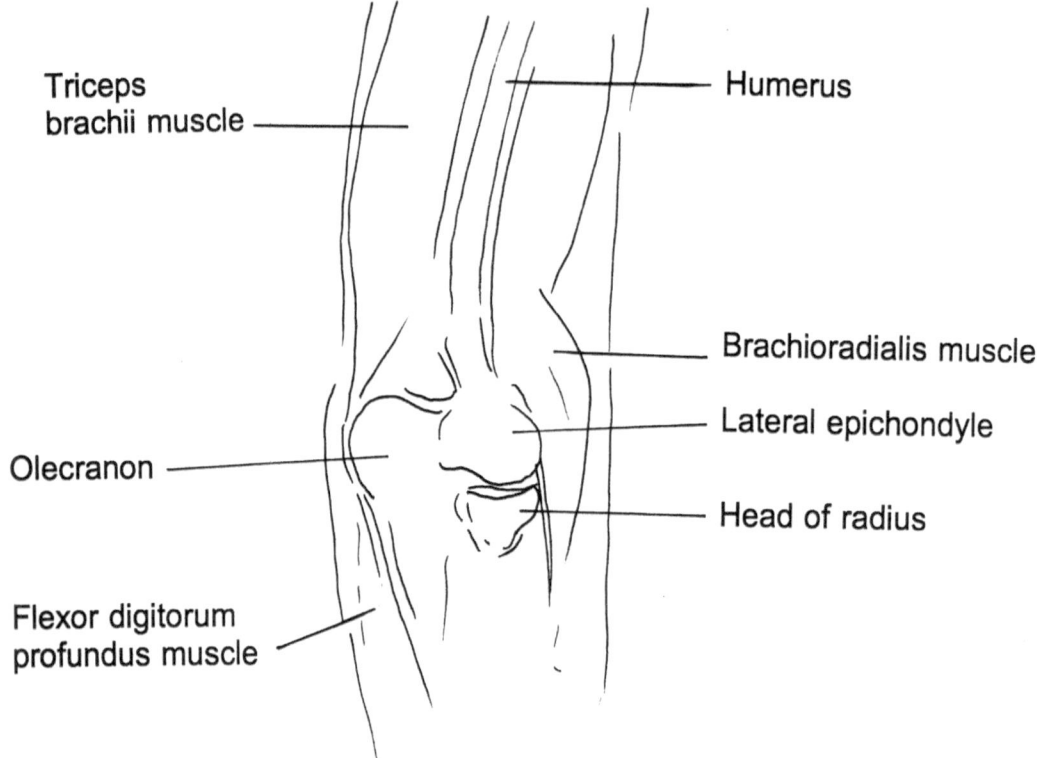

Triceps
brachii muscle ——

Humerus

Brachioradialis muscle

Lateral epichondyle

Olecranon ——

Head of radius

Flexor digitorum
profundus muscle

HAND

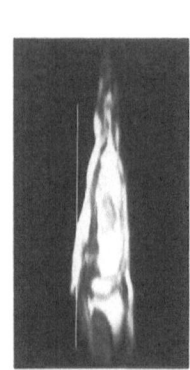

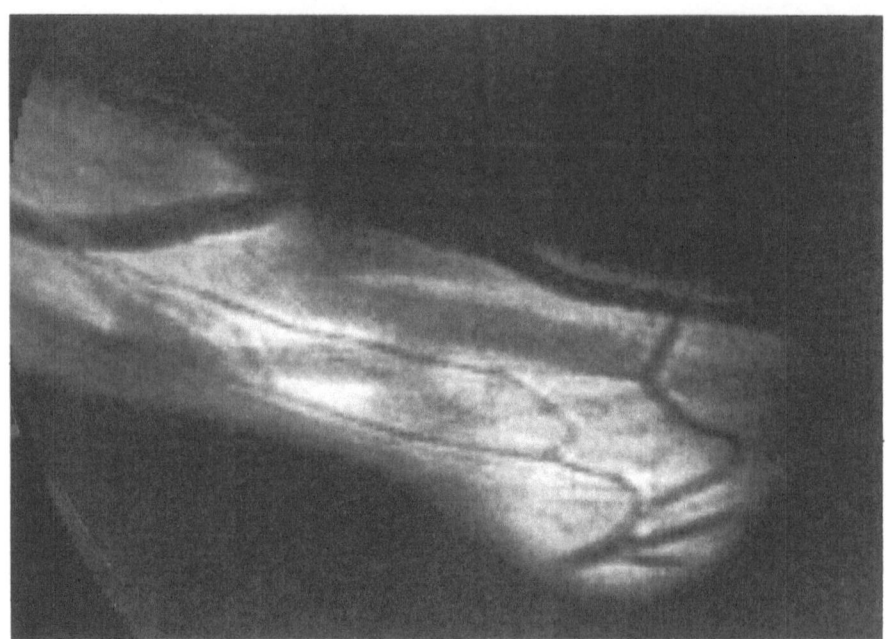

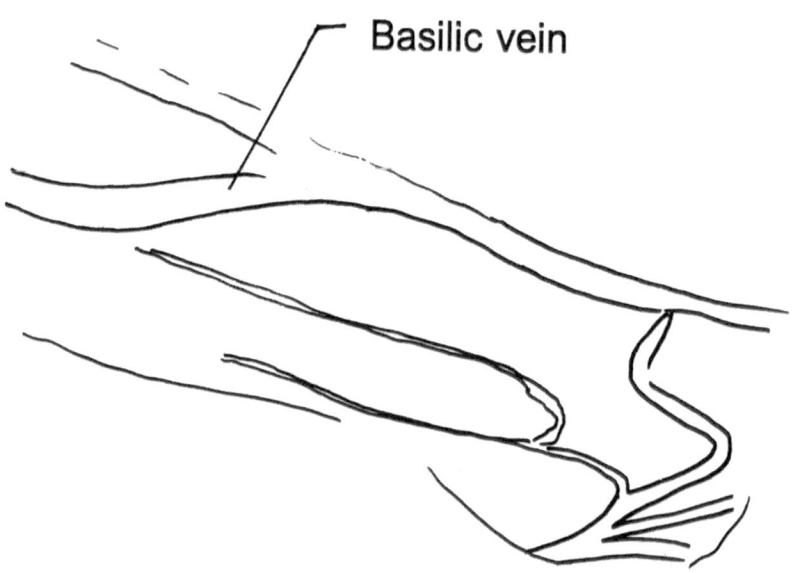

Basilic vein

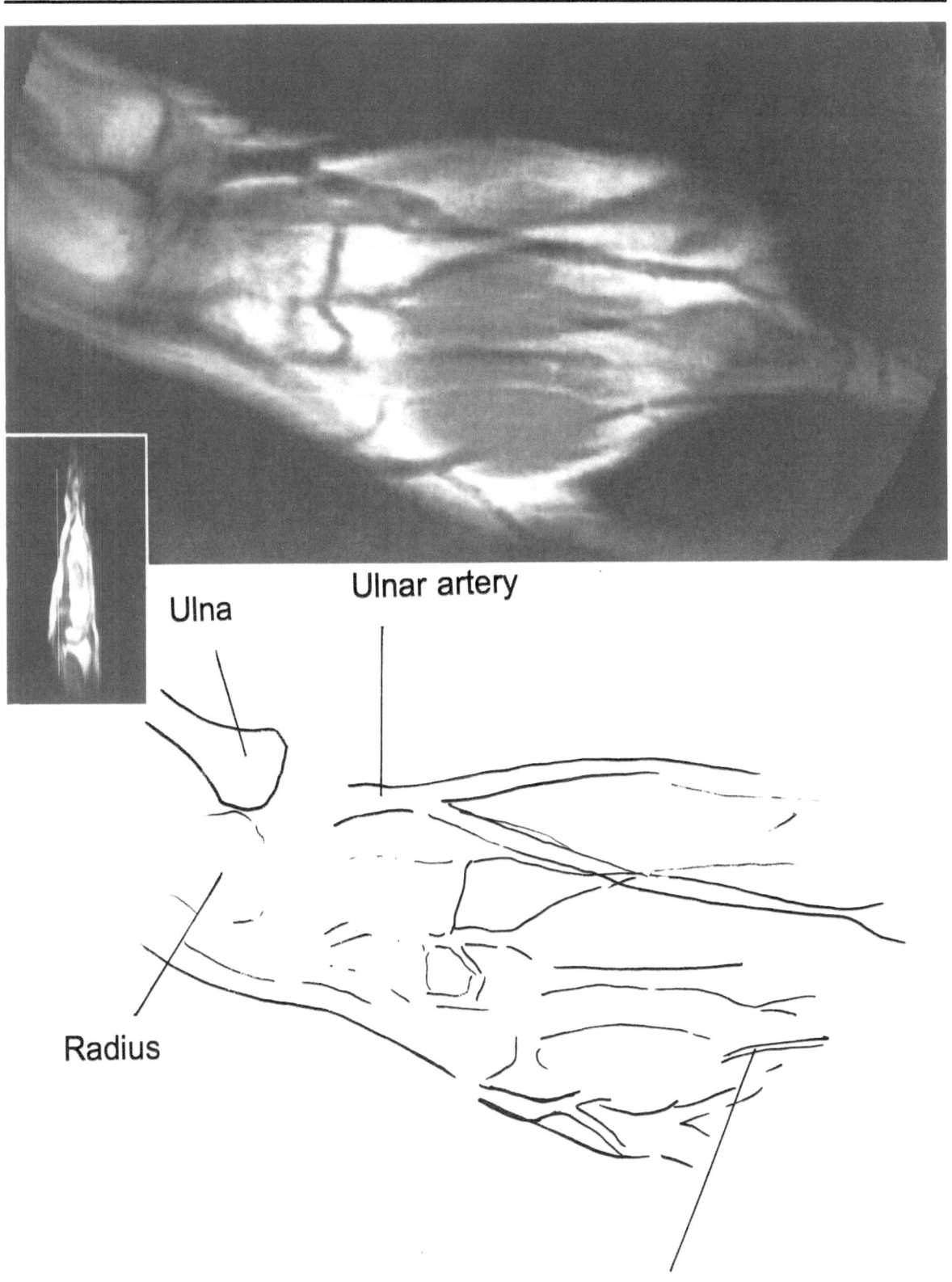

Ulna

Ulnar artery

Radius

Dorsalis indicis artery

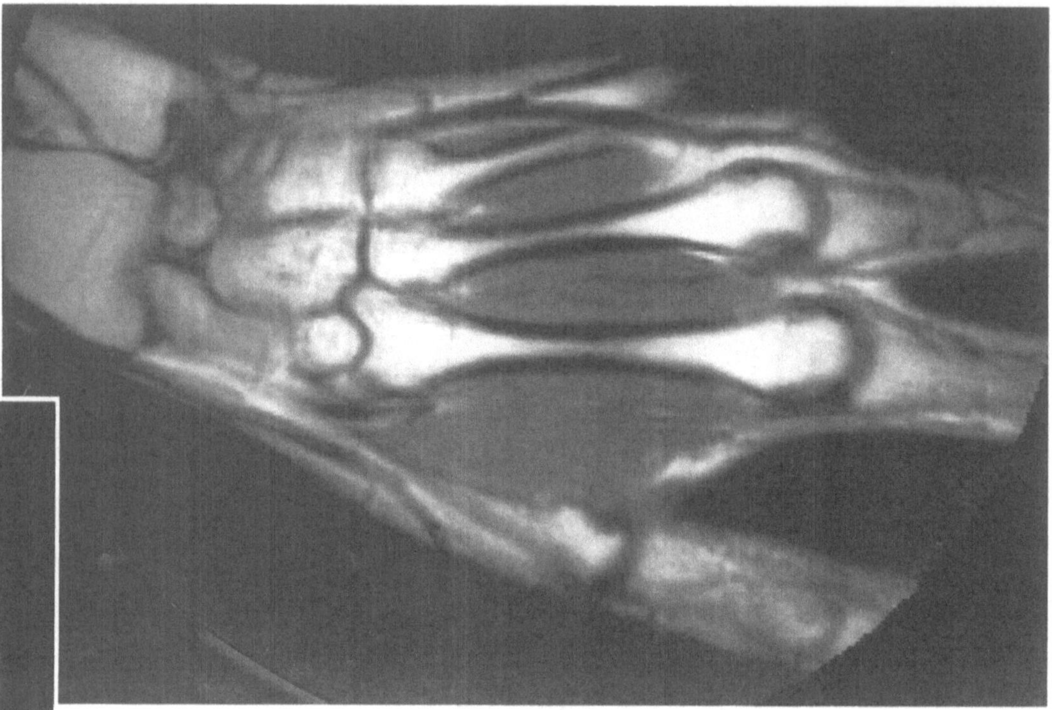

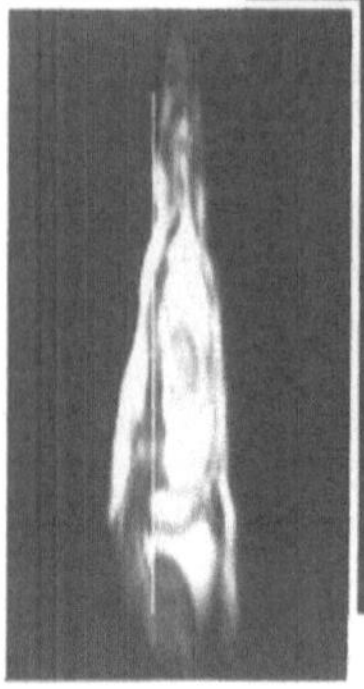

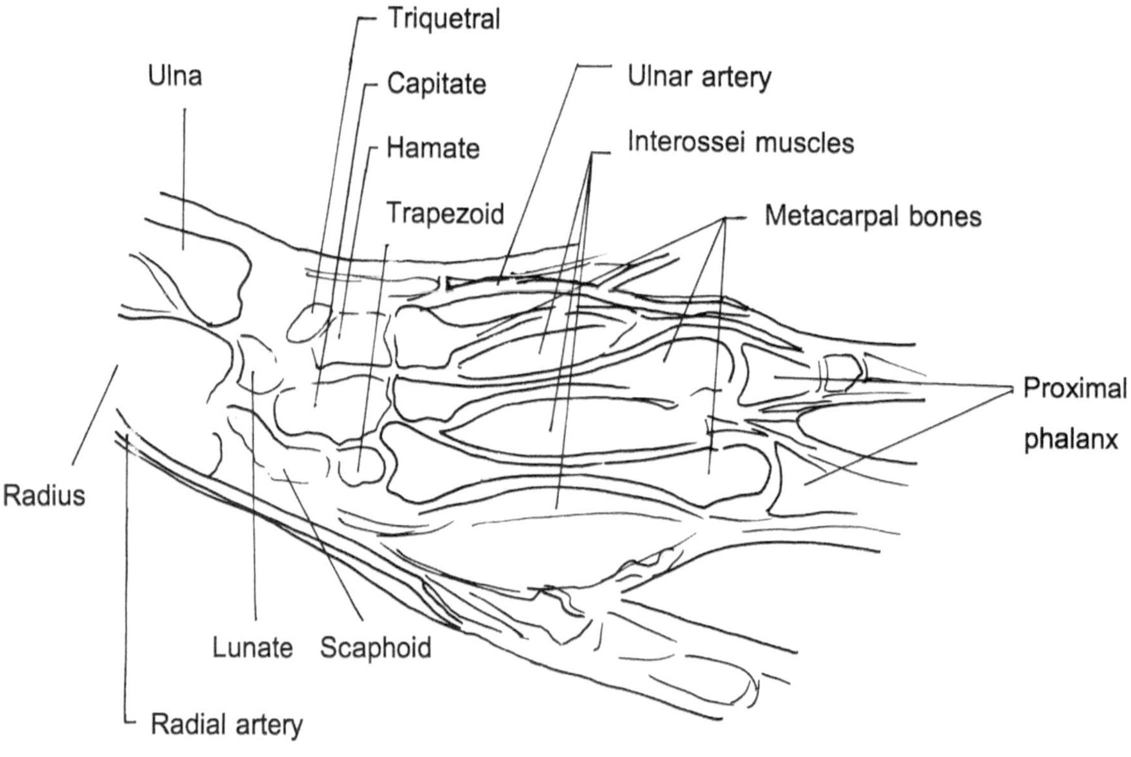

Ulna

Triquetral

Capitate

Ulnar artery

Hamate

Interossei muscles

Trapezoid

Metacarpal bones

Radius

Proximal
phalanx

Lunate Scaphoid

Radial artery

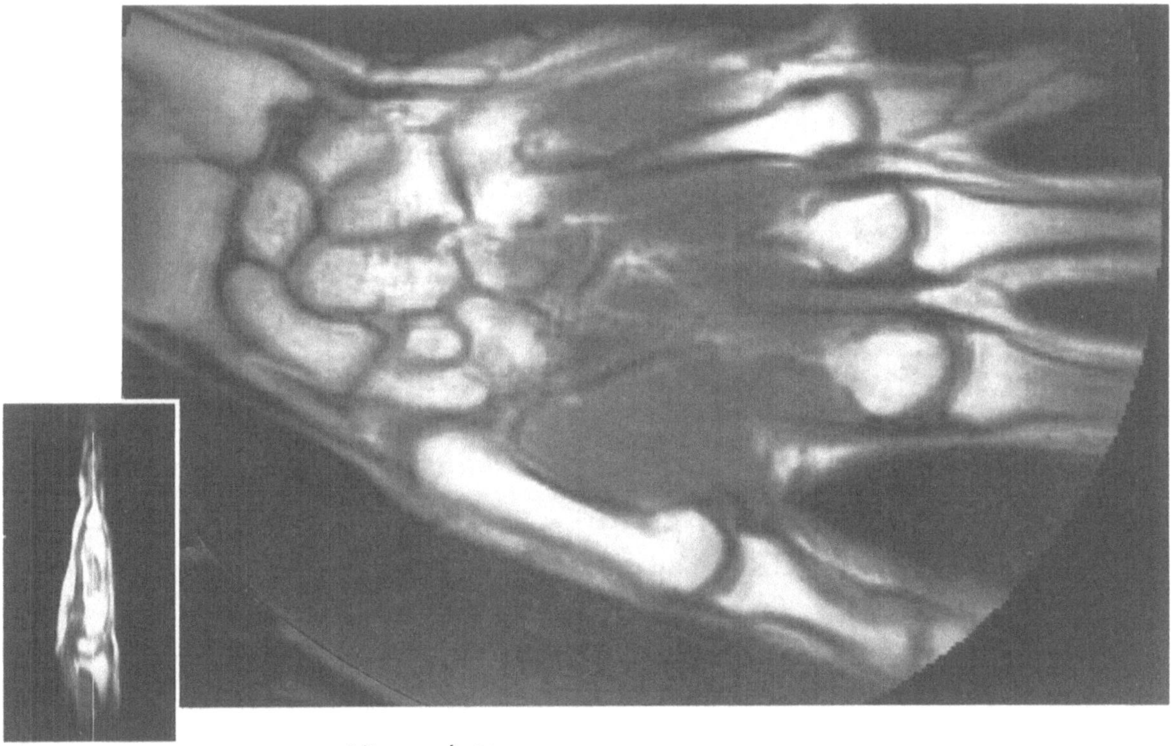

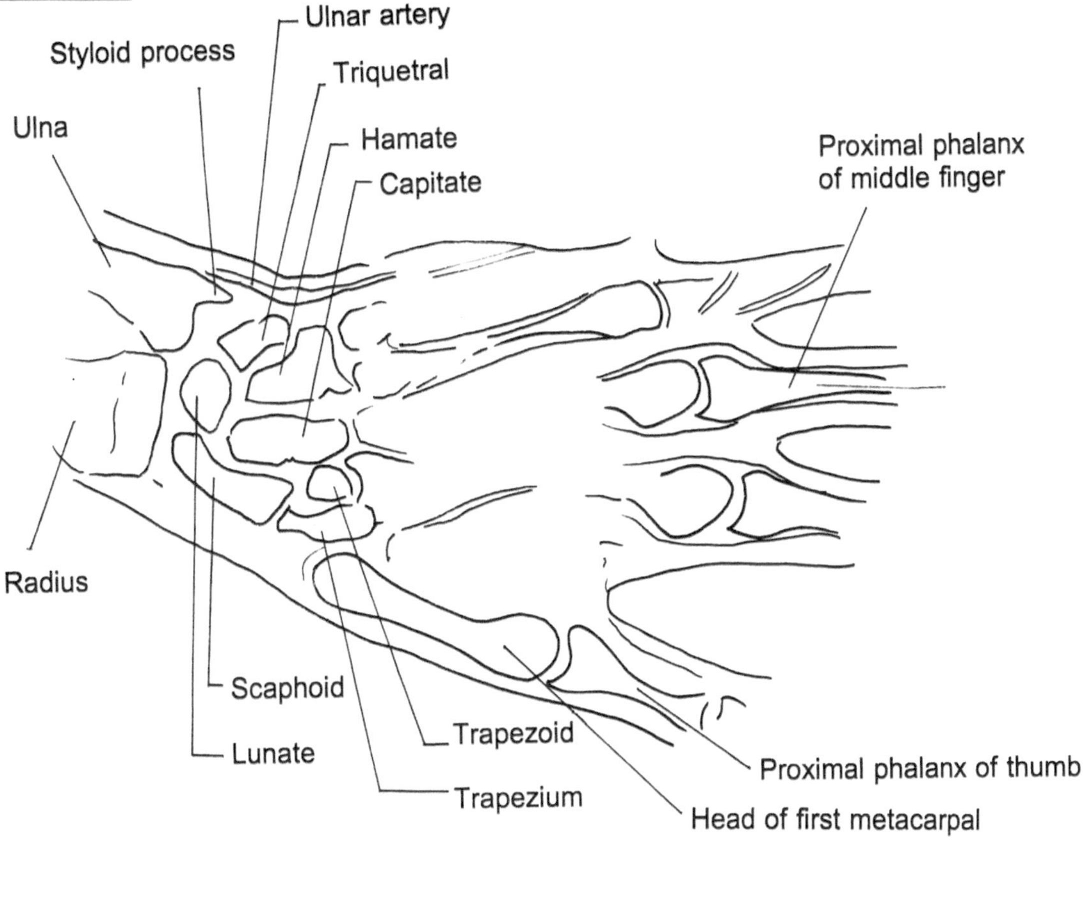

Ulnar artery

Styloid process

Triquetral

Ulna

Hamate

Capitate

Proximal phalanx
of middle finger

Radius

Scaphoid

Lunate

Trapezoid

Trapezium

Proximal phalanx of thumb

Head of first metacarpal

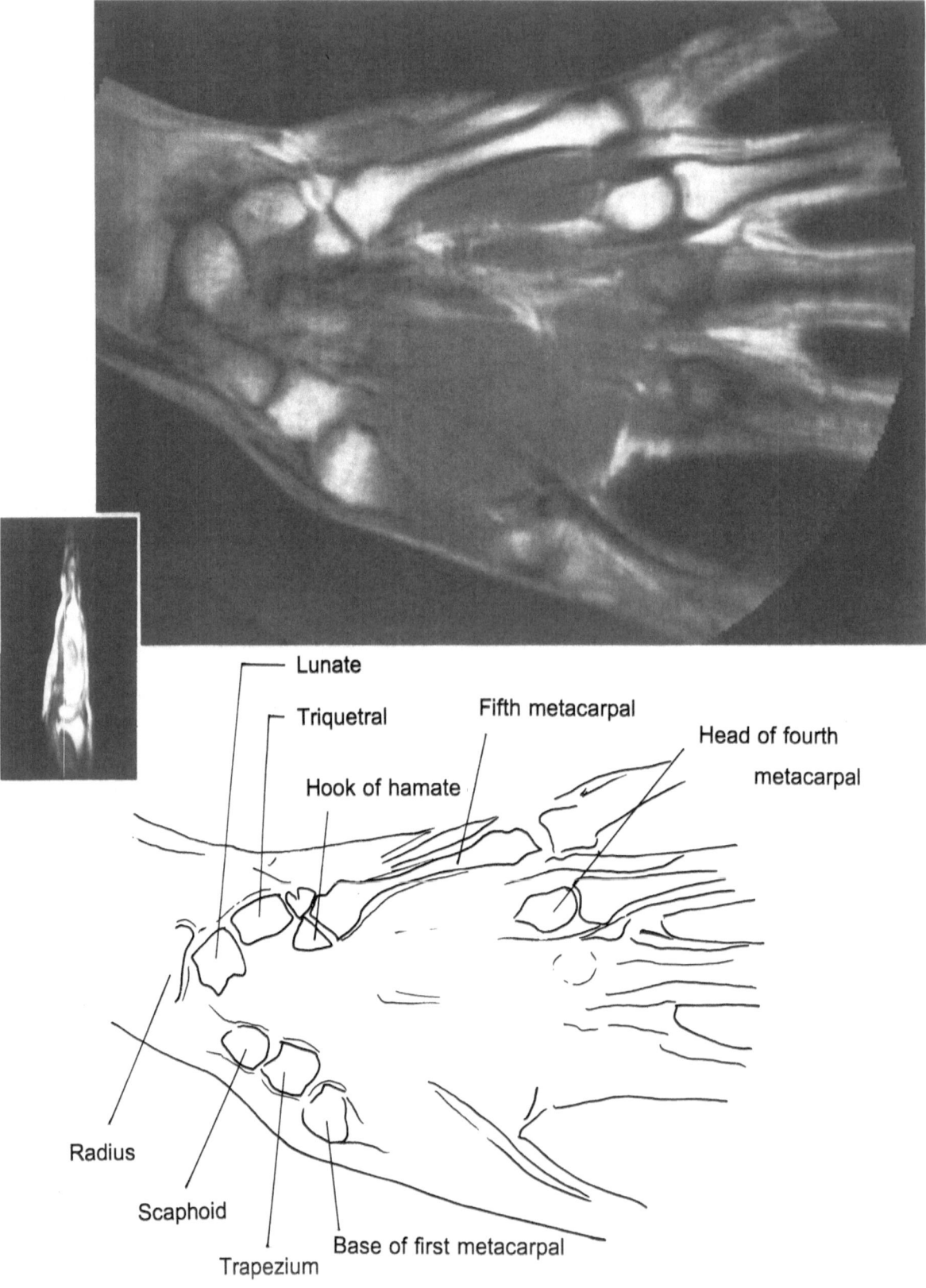

Lunate

Triquetral

Fifth metacarpal

Head of fourth
metacarpal

Hook of hamate

Radius

Scaphoid

Trapezium

Base of first metacarpal

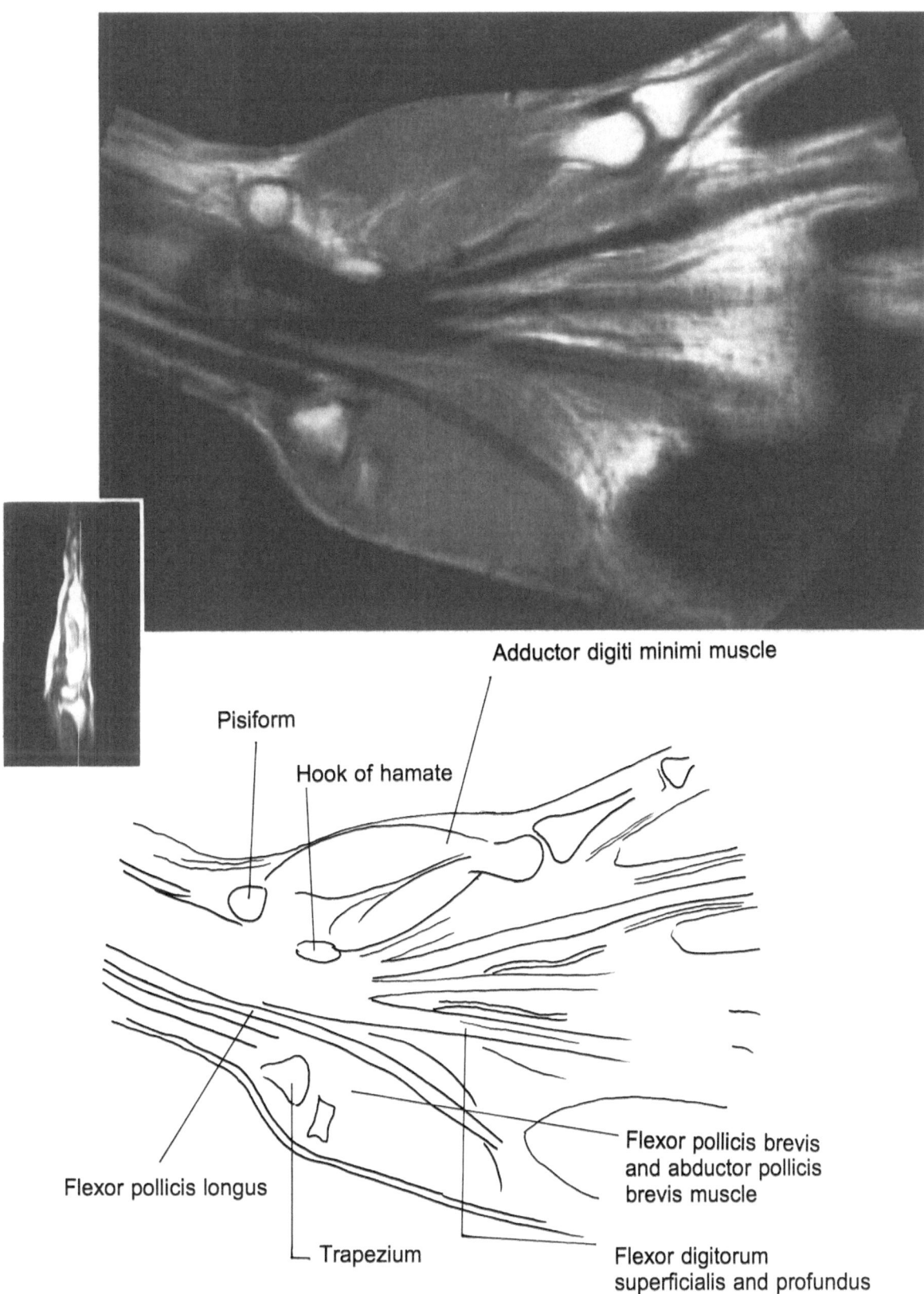

Adductor digiti minimi muscle

Pisiform

Hook of hamate

Flexor pollicis longus

Trapezium

Flexor pollicis brevis
and abductor pollicis
brevis muscle

Flexor digitorum
superficialis and profundus

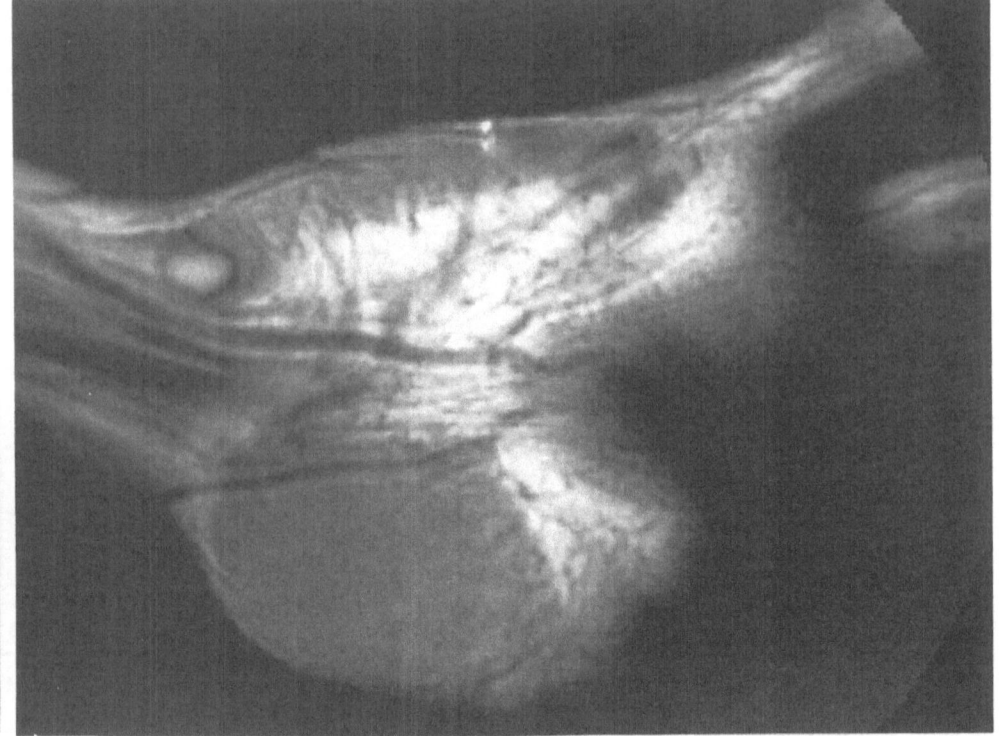

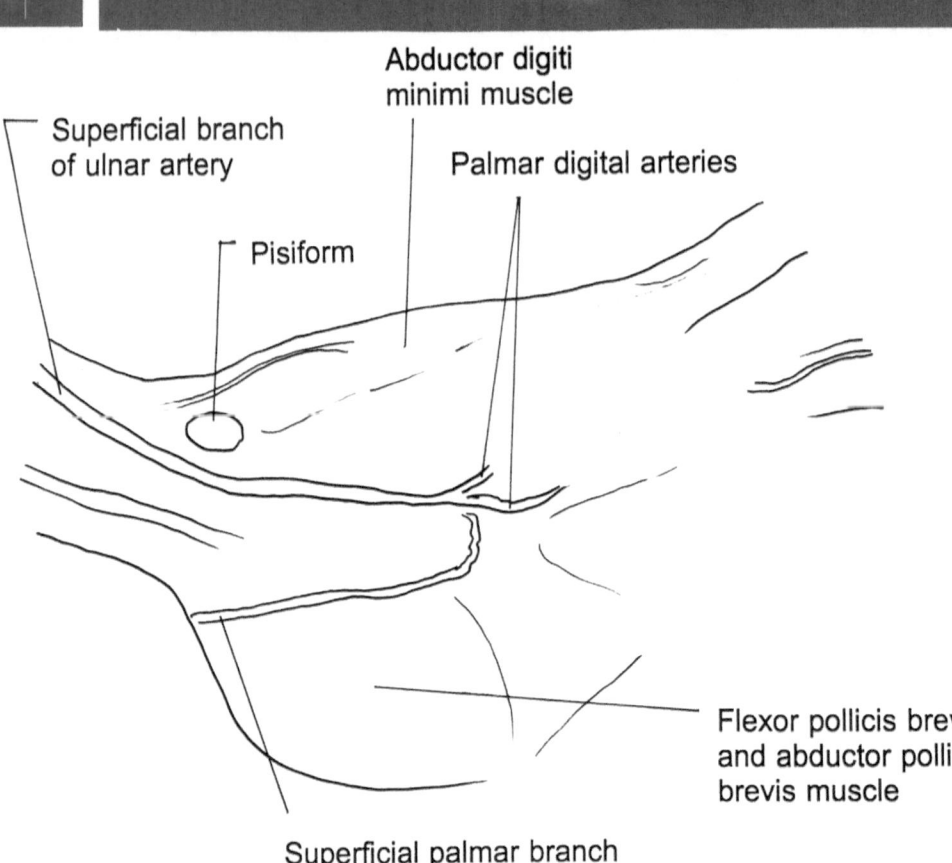

Abductor digiti
minimi muscle

Superficial branch
of ulnar artery

Palmar digital arteries

Pisiform

Flexor pollicis brevis
and abductor pollicis
brevis muscle

Superficial palmar branch
(of radial artery)

PELVIS

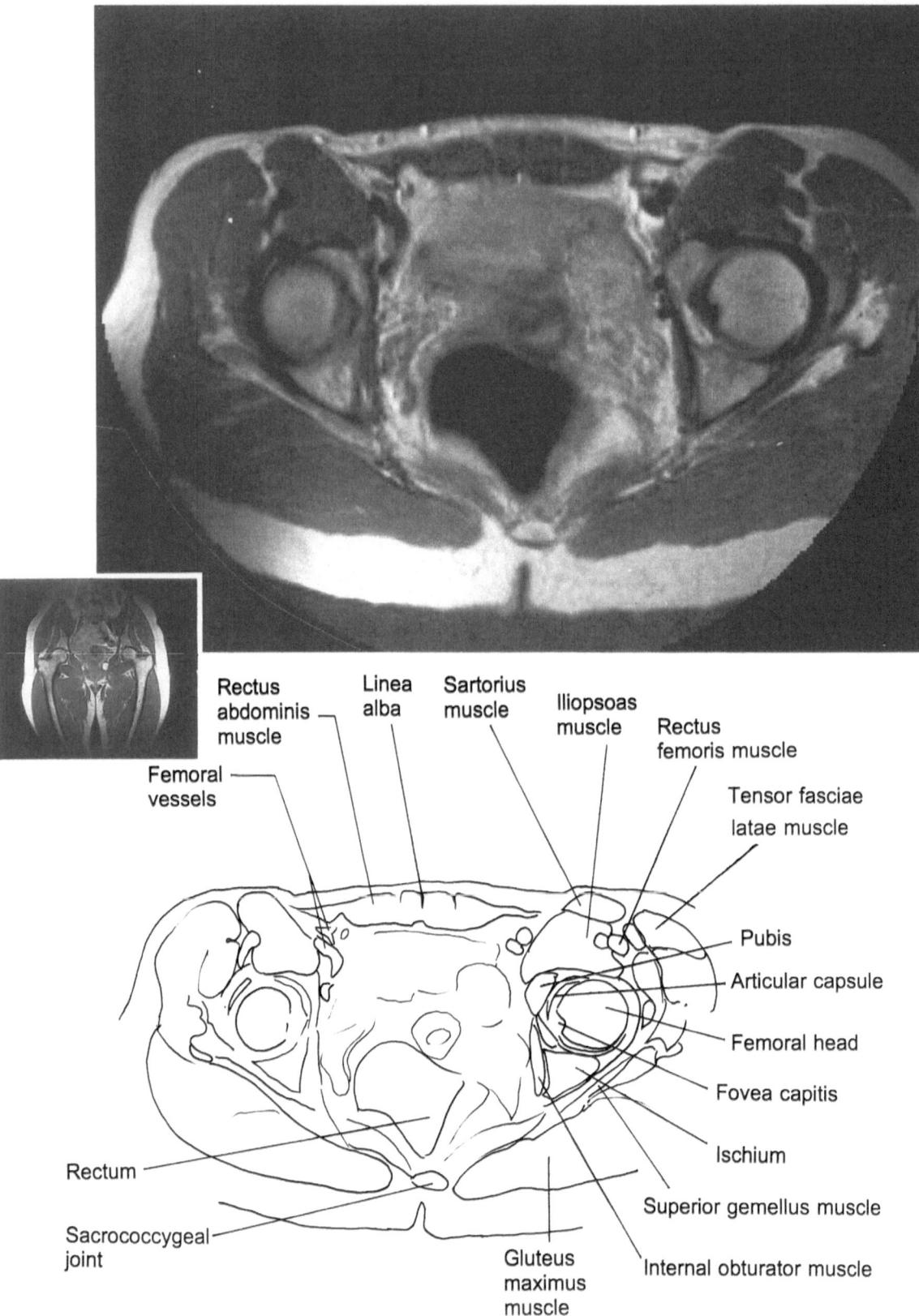

Rectus abdominis muscle

Linea alba

Sartorius muscle

Iliopsoas muscle

Rectus femoris muscle

Femoral vessels

Tensor fasciae latae muscle

Pubis

Articular capsule

Femoral head

Fovea capitis

Ischium

Rectum

Superior gemellus muscle

Sacrococcygeal joint

Gluteus maximus muscle

Internal obturator muscle

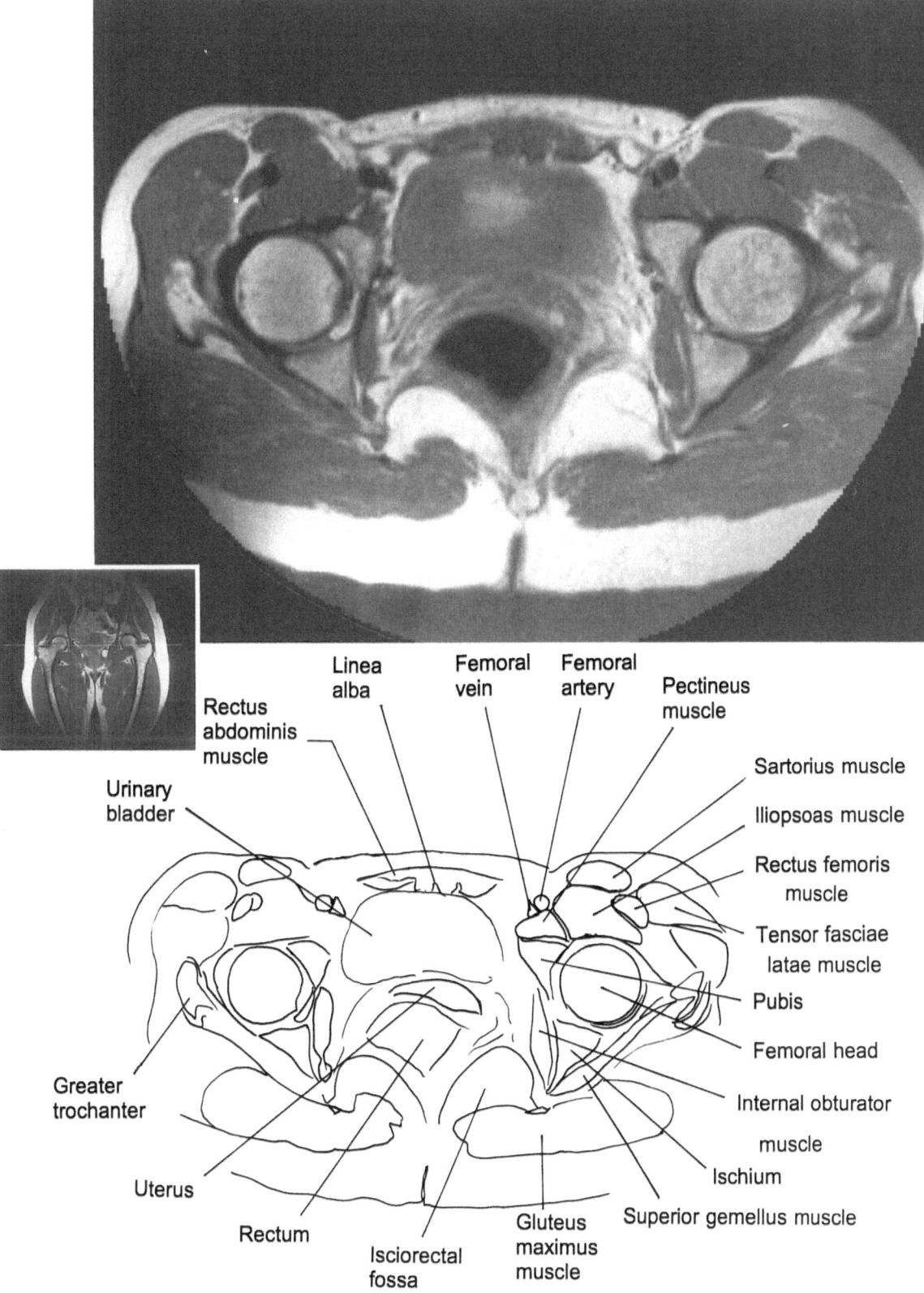

Linea
alba

Femoral
vein

Femoral
artery

Pectineus
muscle

Rectus
abdominis
muscle

Urinary
bladder

Sartorius muscle

Iliopsoas muscle

Rectus femoris
muscle

Tensor fasciae
latae muscle

Pubis

Femoral head

Internal obturator

muscle

Ischium

Superior gemellus muscle

Greater
trochanter

Uterus

Rectum

Isciorectal
fossa

Gluteus
maximus
muscle

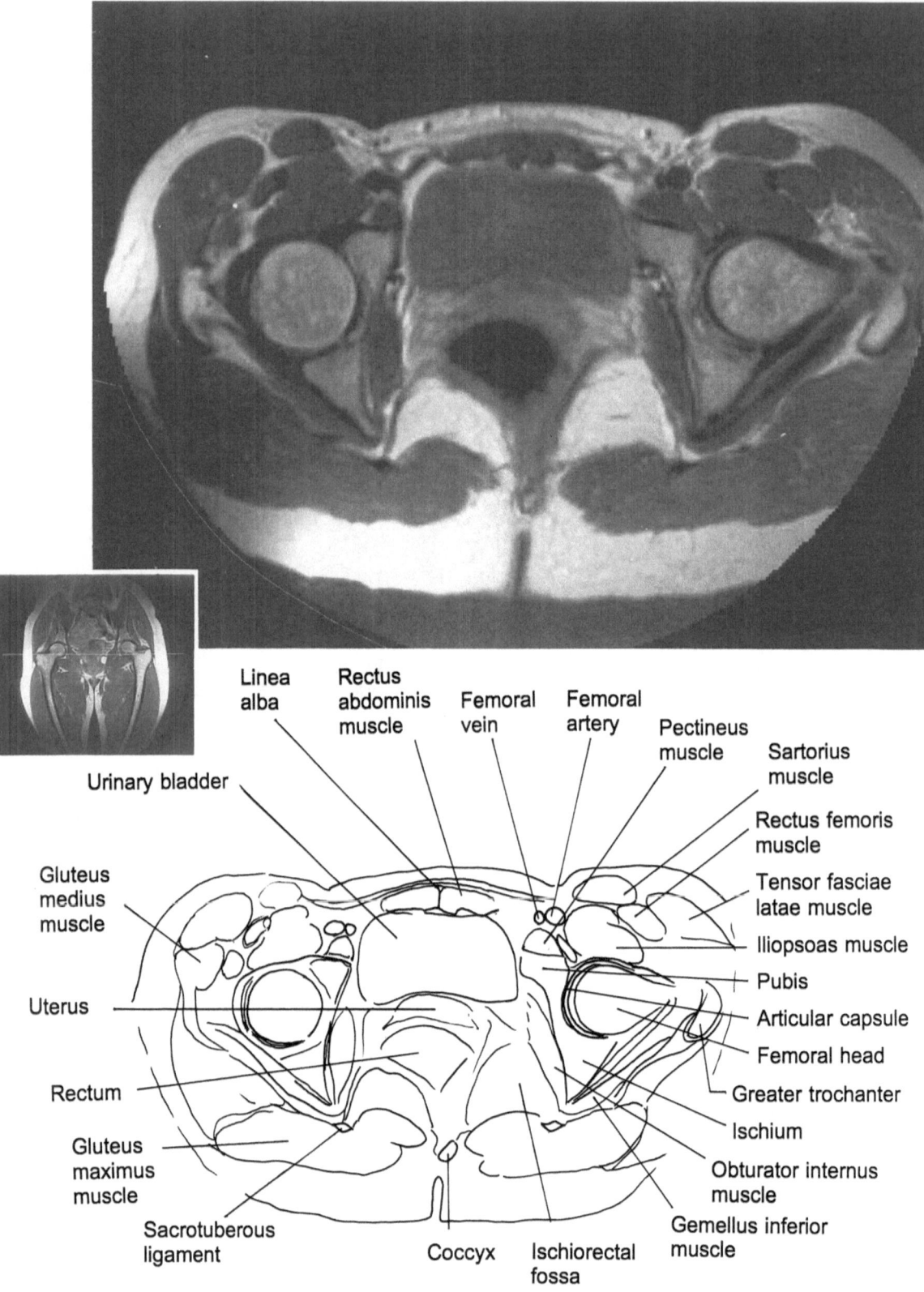

Linea alba

Rectus abdominis muscle

Femoral vein

Femoral artery

Pectineus muscle

Sartorius muscle

Urinary bladder

Rectus femoris muscle

Tensor fasciae latae muscle

Iliopsoas muscle

Gluteus medius muscle

Pubis

Articular capsule

Uterus

Femoral head

Greater trochanter

Rectum

Ischium

Gluteus maximus muscle

Obturator internus muscle

Sacrotuberous ligament

Gemellus inferior muscle

Coccyx

Ischiorectal fossa

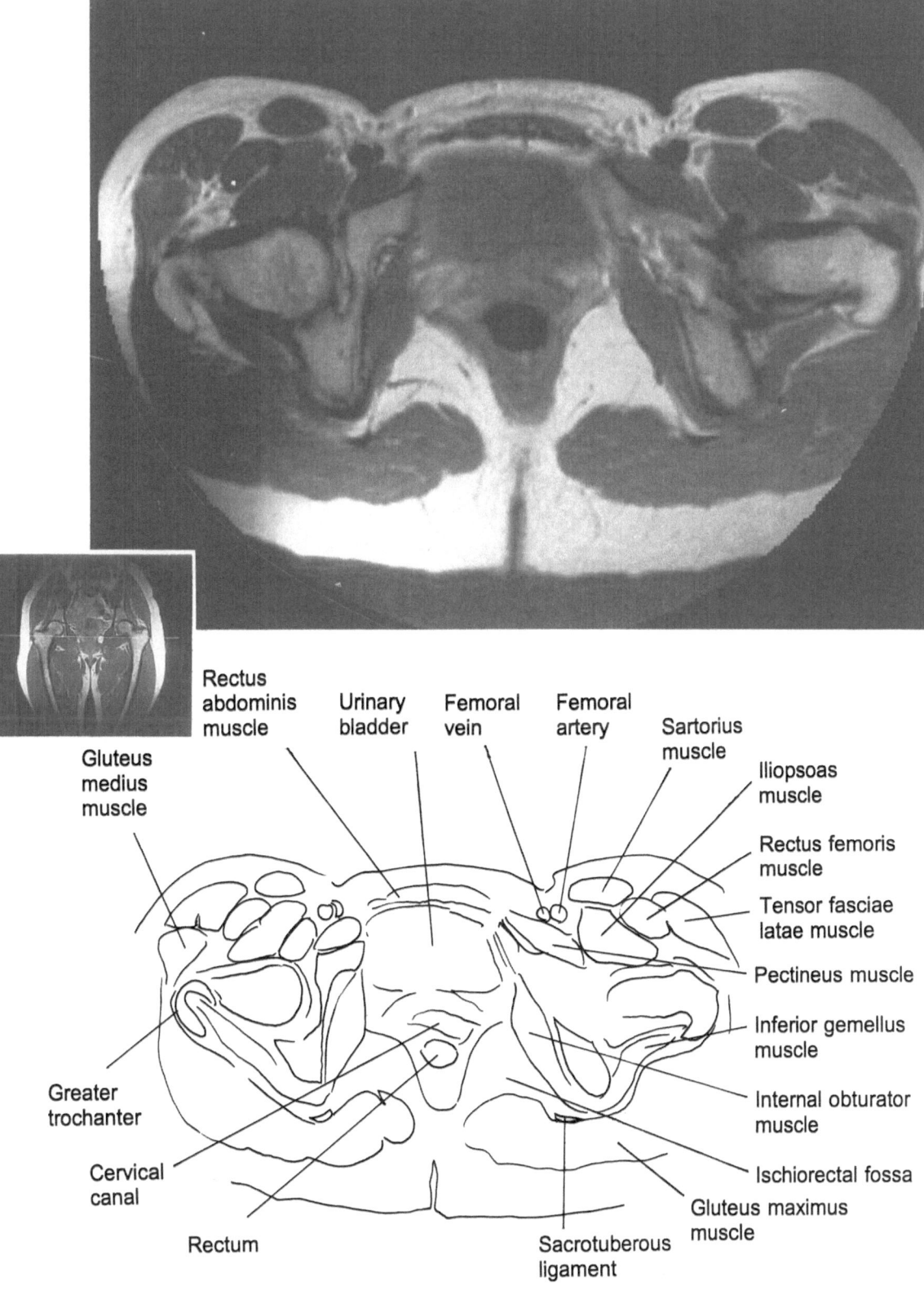

Rectus
abdominis
muscle

Urinary
bladder

Femoral
vein

Femoral
artery

Sartorius
muscle

Gluteus
medius
muscle

Iliopsoas
muscle

Rectus femoris
muscle

Tensor fasciae
latae muscle

Pectineus muscle

Inferior gemellus
muscle

Internal obturator
muscle

Greater
trochanter

Ischiorectal fossa

Cervical
canal

Gluteus maximus
muscle

Rectum

Sacrotuberous
ligament

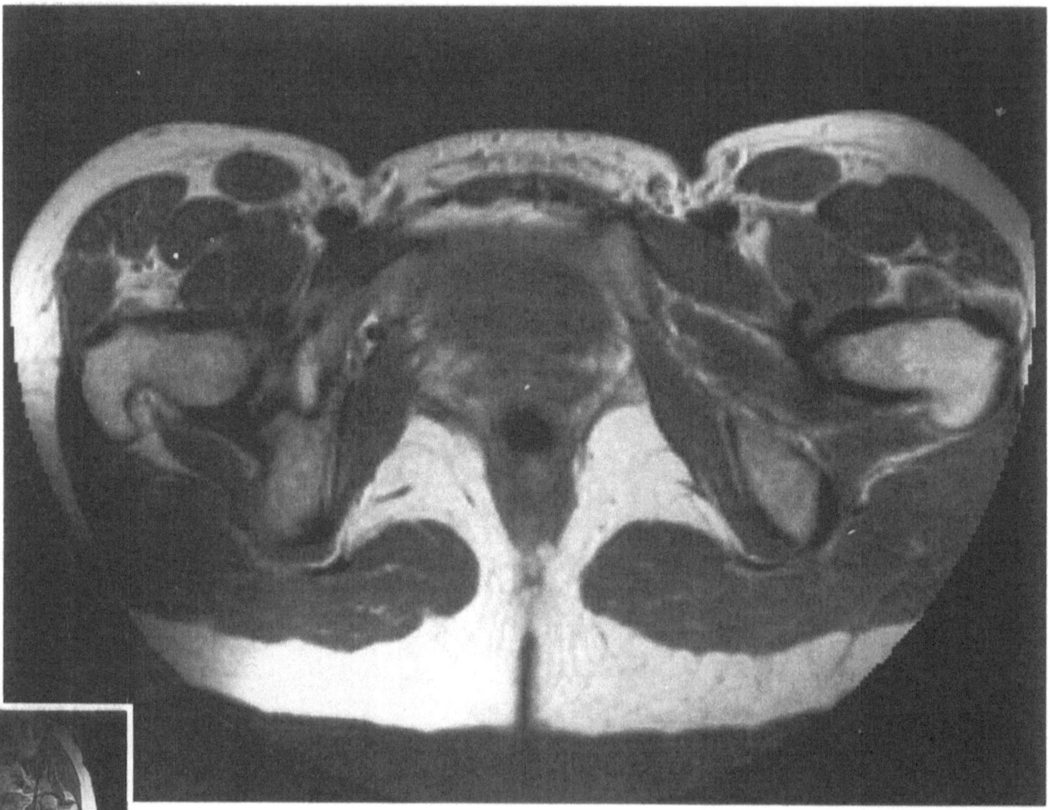

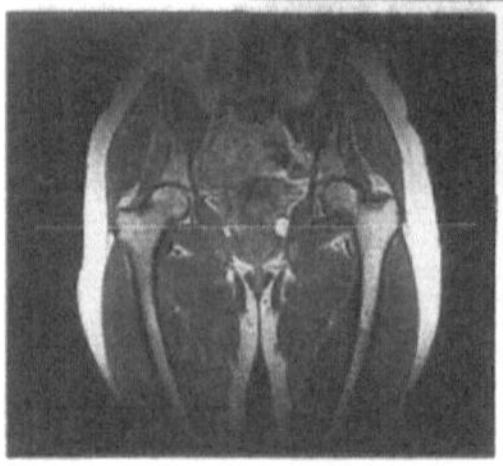

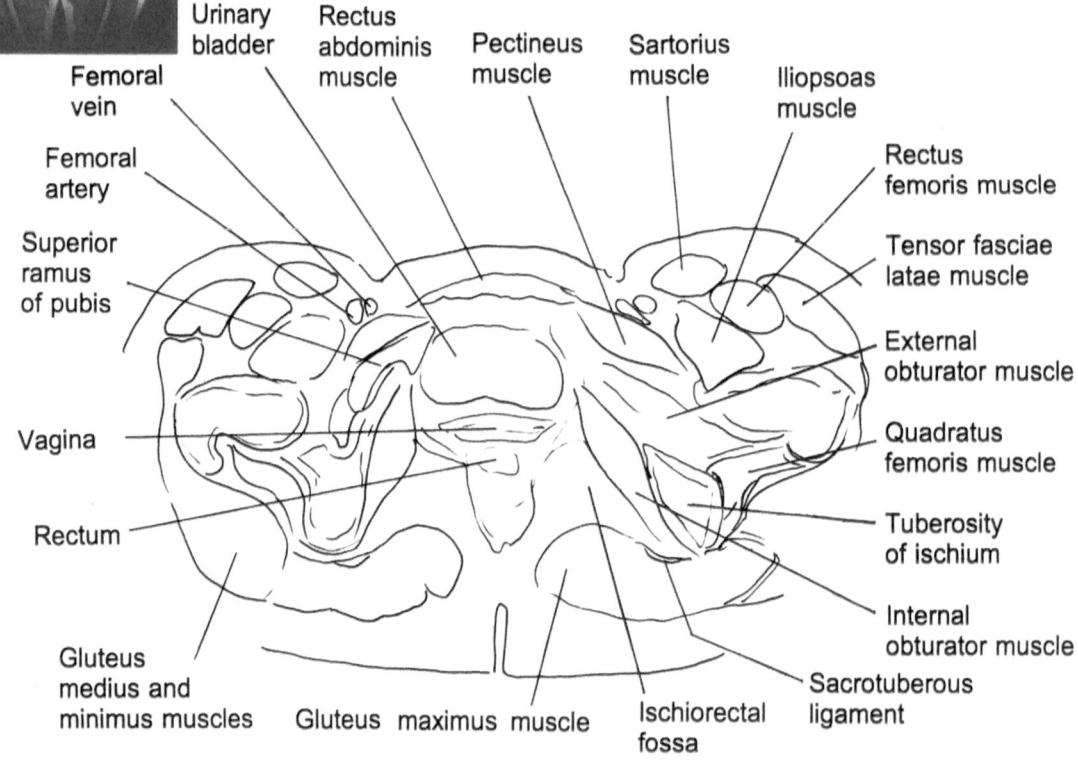

Urinary bladder

Rectus abdominis muscle

Pectineus muscle

Sartorius muscle

Iliopsoas muscle

Femoral vein

Femoral artery

Superior ramus of pubis

Vagina

Rectum

Gluteus medius and minimus muscles

Gluteus maximus muscle

Ischiorectal fossa

Sacrotuberous ligament

Internal obturator muscle

Tuberosity of ischium

Quadratus femoris muscle

External obturator muscle

Tensor fasciae latae muscle

Rectus femoris muscle

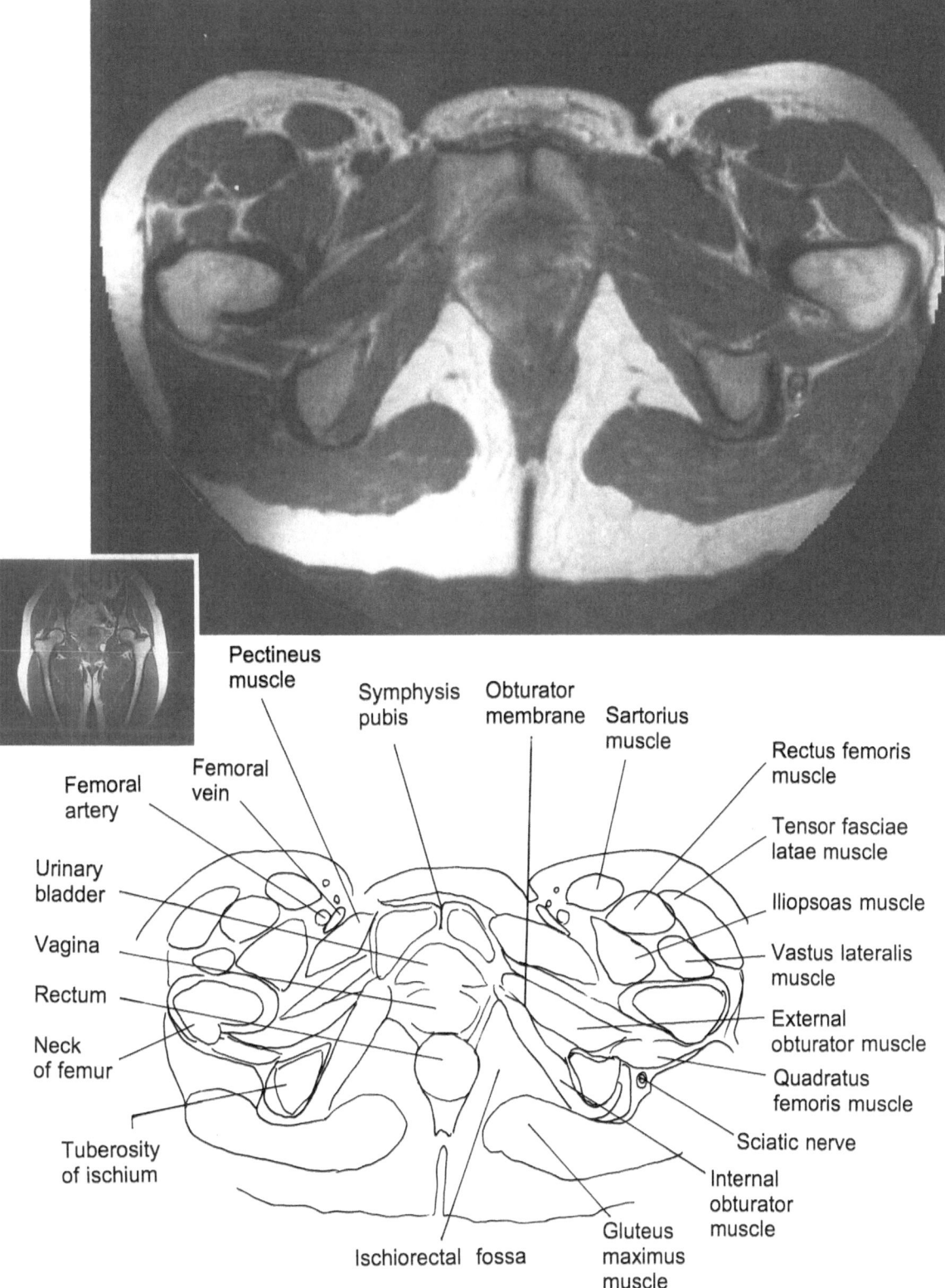

Pectineus
muscle

Symphysis
pubis

Obturator
membrane

Sartorius
muscle

Rectus femoris
muscle

Tensor fasciae
latae muscle

Iliopsoas muscle

Vastus lateralis
muscle

External
obturator muscle

Quadratus
femoris muscle

Sciatic nerve

Internal
obturator
muscle

Femoral
vein

Femoral
artery

Urinary
bladder

Vagina

Rectum

Neck
of femur

Tuberosity
of ischium

Ischiorectal fossa

Gluteus
maximus
muscle

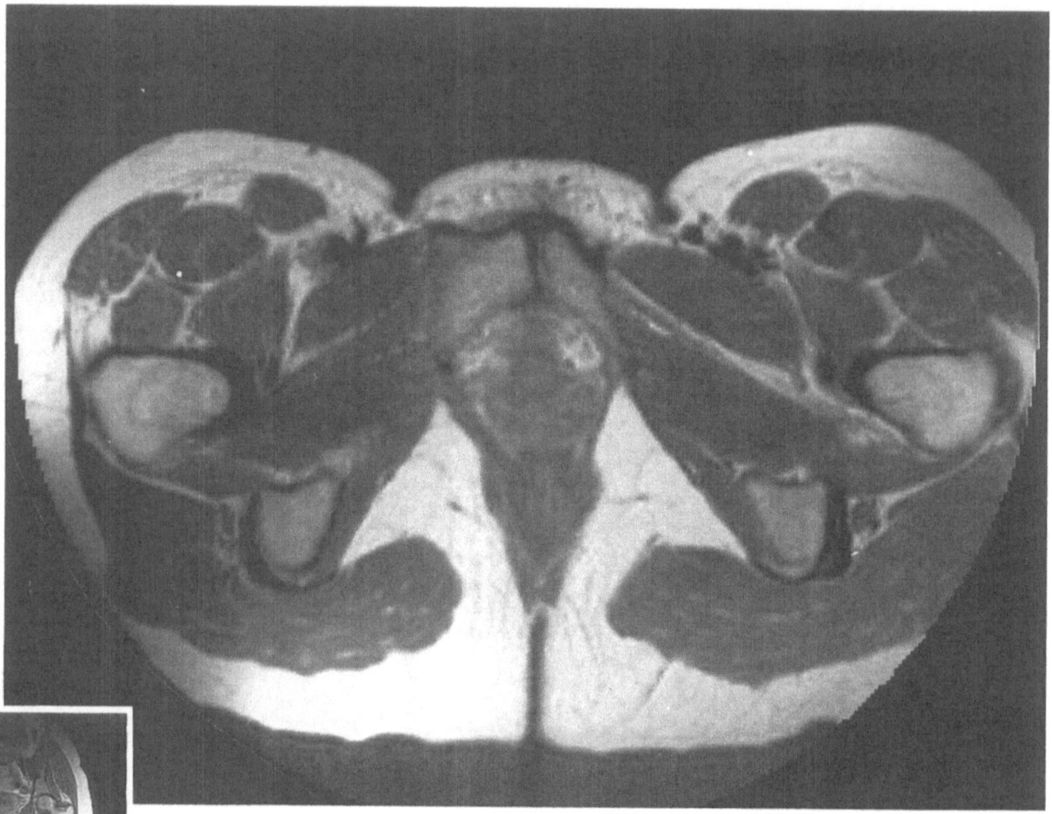

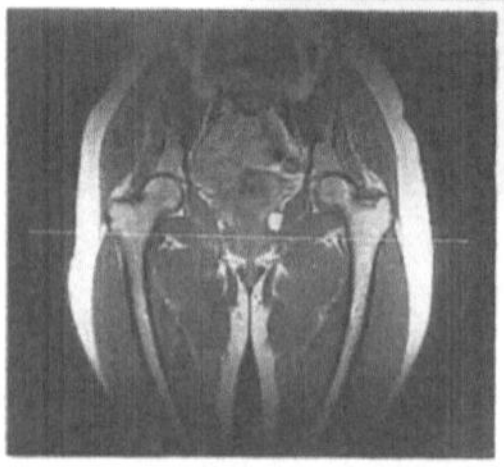

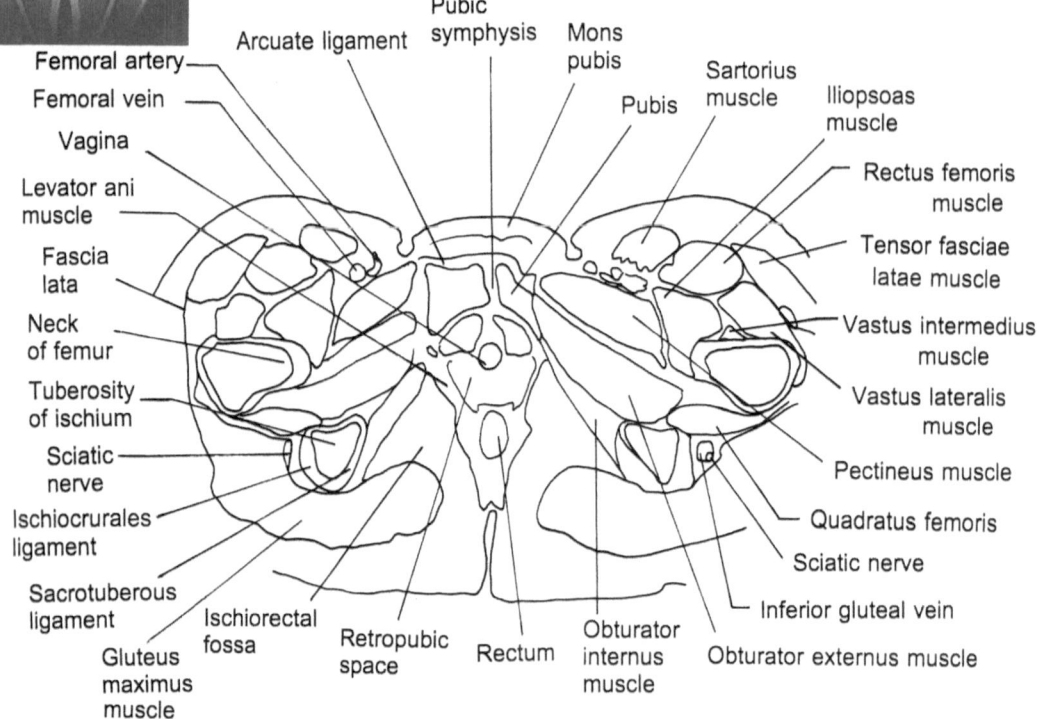

Femoral artery

Femoral vein

Vagina

Levator ani
muscle

Fascia
lata

Neck
of femur

Tuberosity
of ischium

Sciatic
nerve

Ischiocrurales
ligament

Sacrotuberous
ligament

Gluteus
maximus
muscle

Ischiorectal
fossa

Arcuate ligament

Pubic
symphysis

Mons
pubis

Pubis

Retropubic
space

Rectum

Obturator
internus
muscle

Sartorius
muscle

Iliopsoas
muscle

Rectus femoris
muscle

Tensor fasciae
latae muscle

Vastus intermedius
muscle

Vastus lateralis
muscle

Pectineus muscle

Quadratus femoris

Sciatic nerve

Inferior gluteal vein

Obturator externus muscle

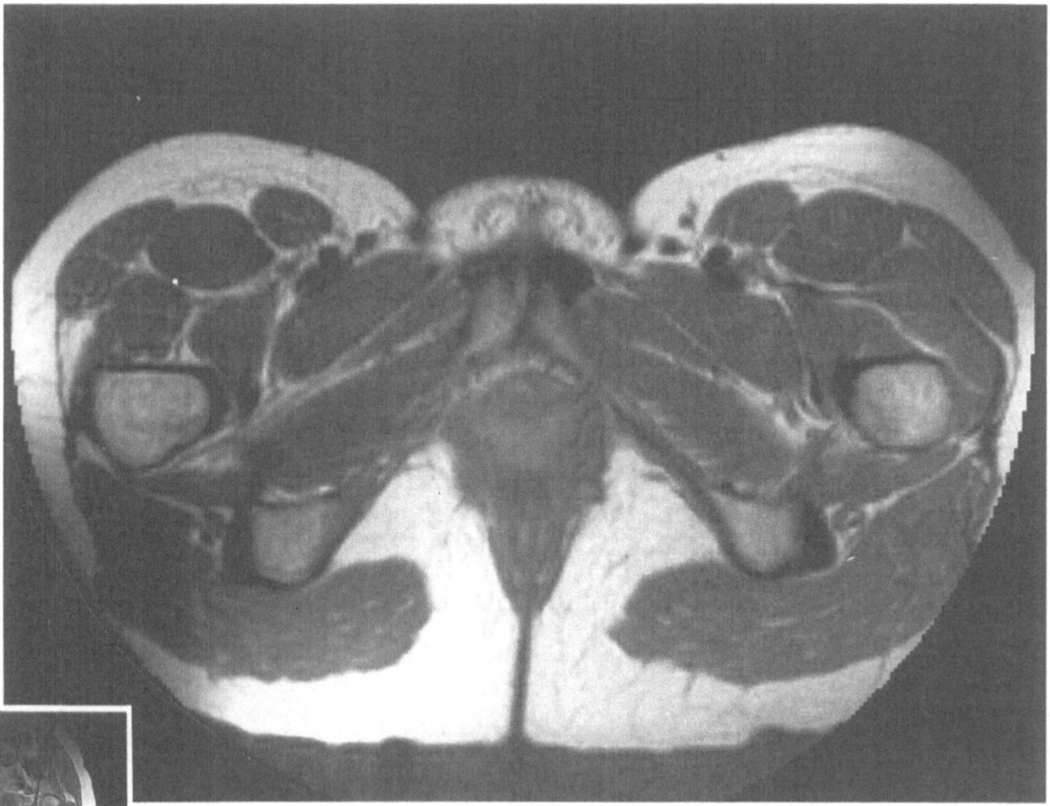

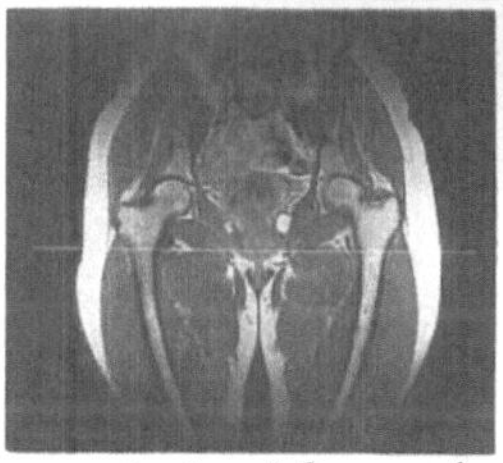

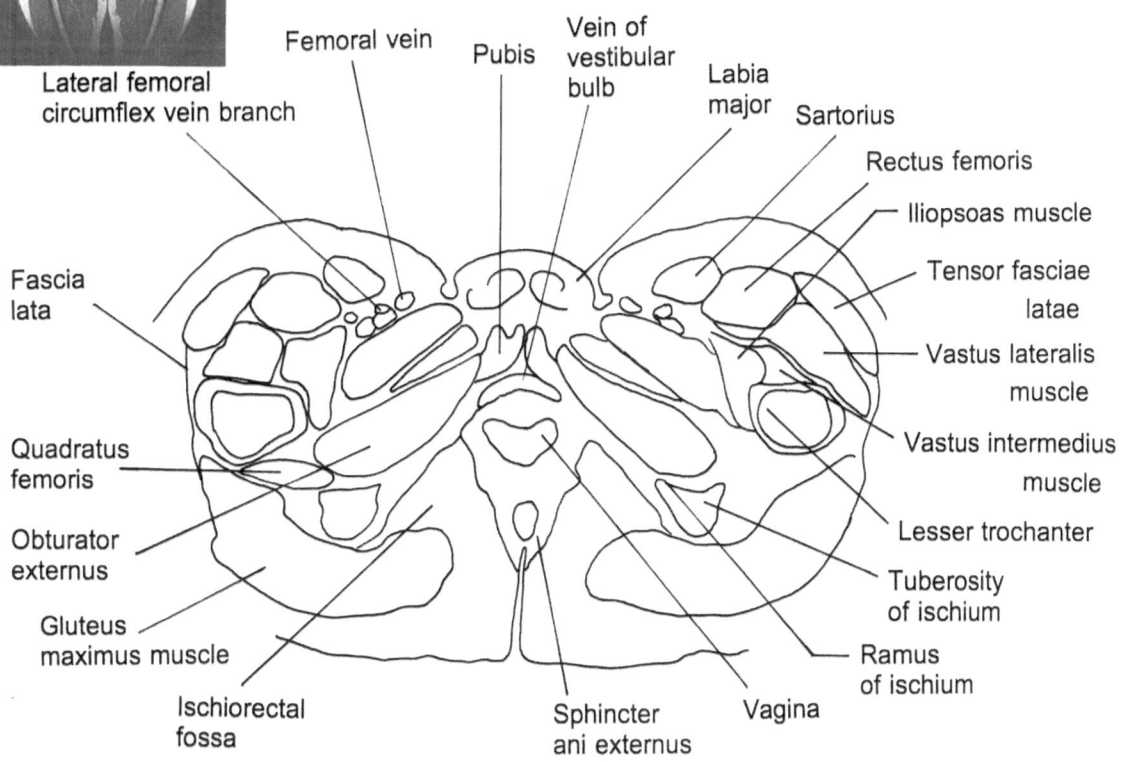

Lateral femoral circumflex vein branch
Femoral vein
Pubis
Vein of vestibular bulb
Labia major
Sartorius
Rectus femoris
Iliopsoas muscle
Tensor fasciae latae
Vastus lateralis muscle
Vastus intermedius muscle
Lesser trochanter
Tuberosity of ischium
Ramus of ischium
Fascia lata
Quadratus femoris
Obturator externus
Gluteus maximus muscle
Ischiorectal fossa
Sphincter ani externus
Vagina

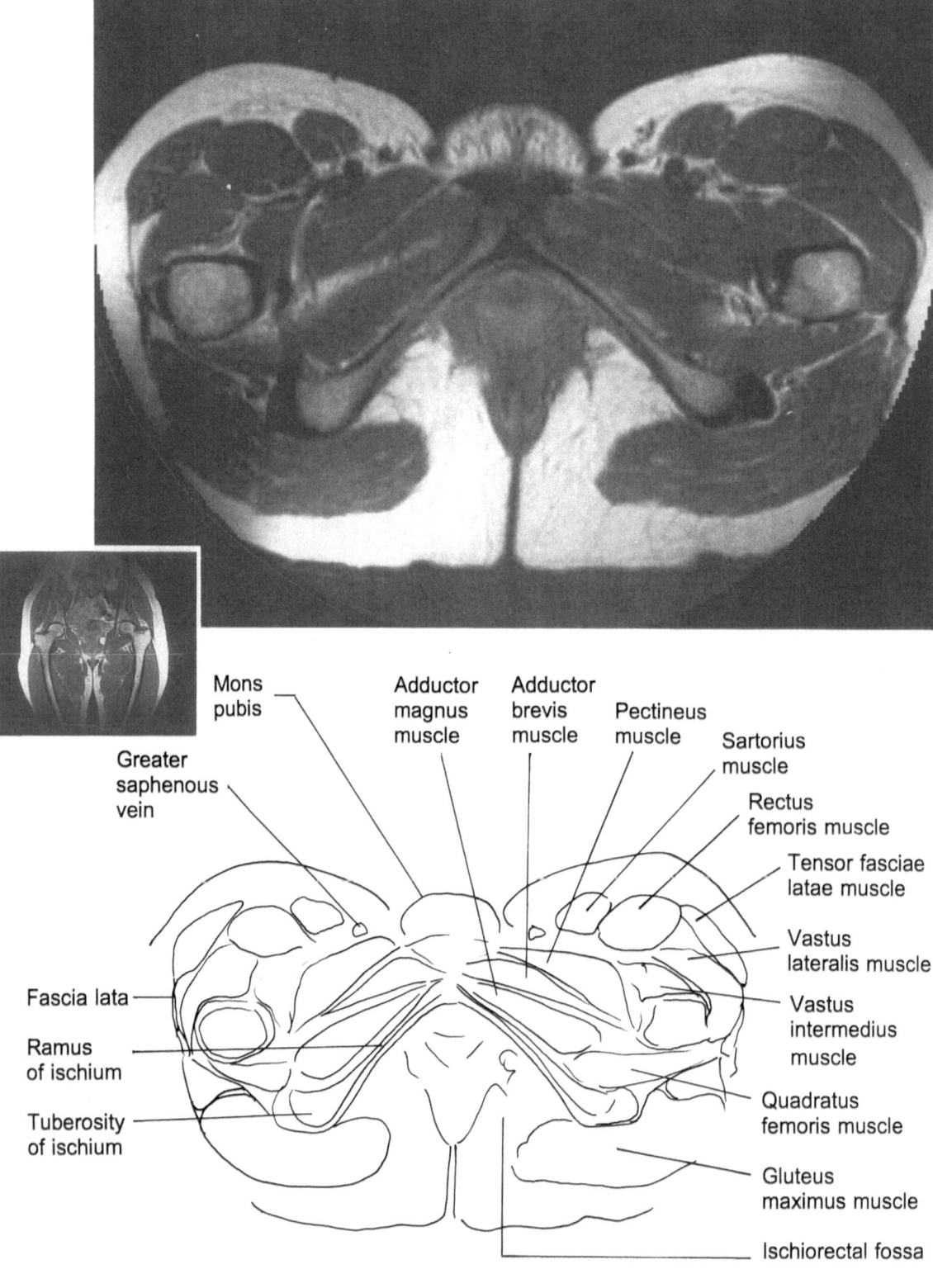

Mons pubis

Greater saphenous vein

Adductor magnus muscle

Adductor brevis muscle

Pectineus muscle

Sartorius muscle

Rectus femoris muscle

Tensor fasciae latae muscle

Vastus lateralis muscle

Vastus intermedius muscle

Fascia lata

Ramus of ischium

Tuberosity of ischium

Quadratus femoris muscle

Gluteus maximus muscle

Ischiorectal fossa

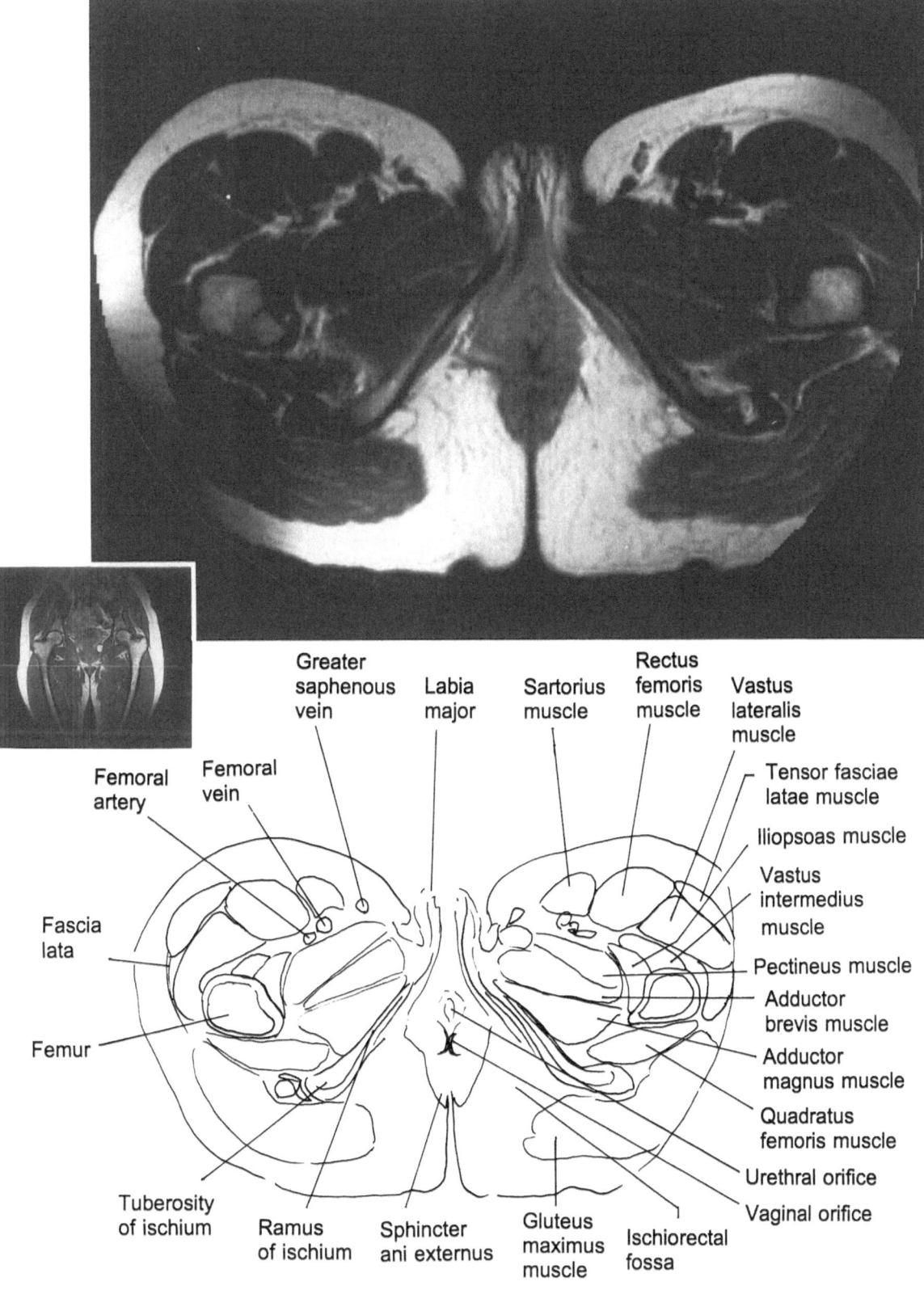

Greater saphenous vein • Labia major • Sartorius muscle • Rectus femoris muscle • Vastus lateralis muscle • Femoral artery • Femoral vein • Tensor fasciae latae muscle • Iliopsoas muscle • Vastus intermedius muscle • Fascia lata • Pectineus muscle • Adductor brevis muscle • Adductor magnus muscle • Femur • Quadratus femoris muscle • Urethral orifice • Vaginal orifice • Tuberosity of ischium • Ramus of ischium • Sphincter ani externus • Gluteus maximus muscle • Ischiorectal fossa

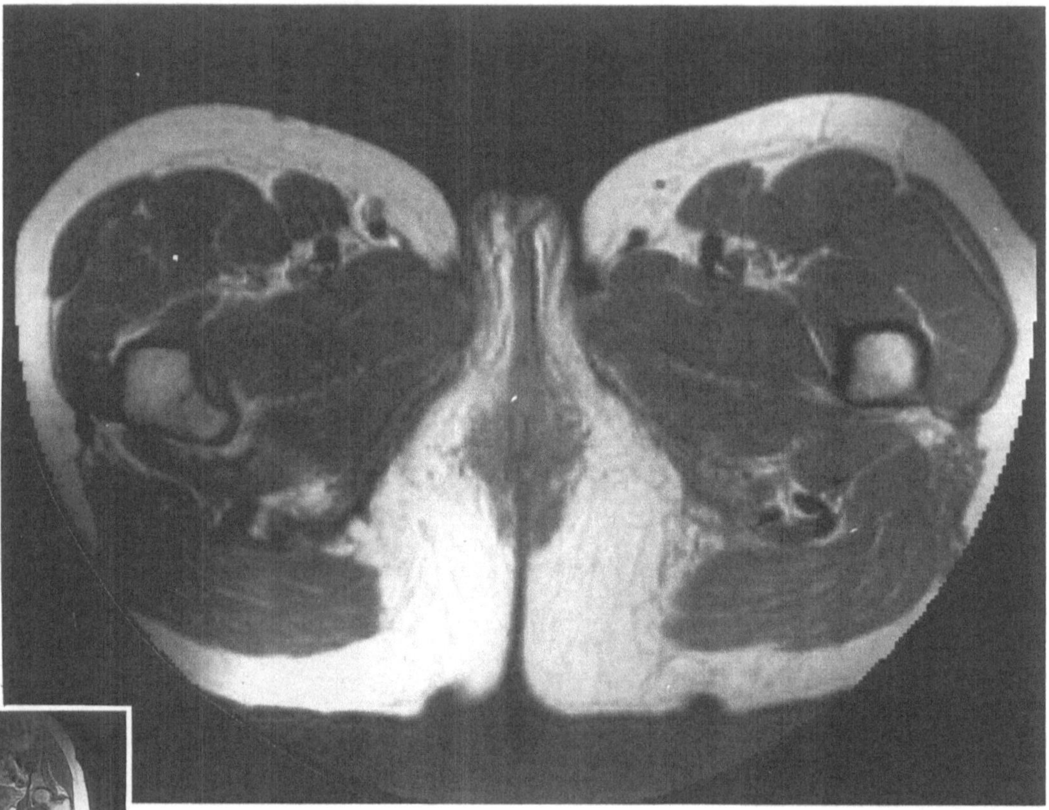

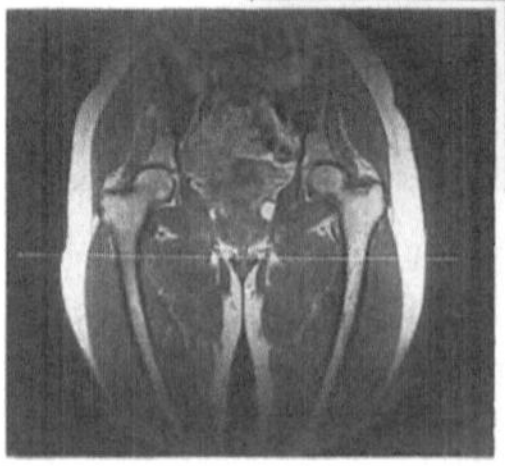

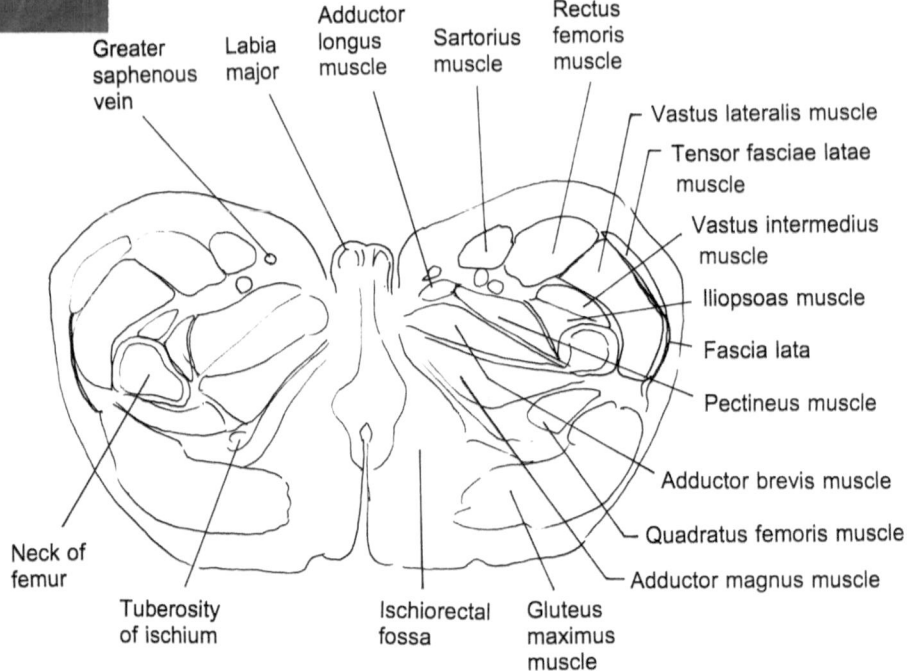

Greater saphenous vein

Labia major

Adductor longus muscle

Sartorius muscle

Rectus femoris muscle

Vastus lateralis muscle

Tensor fasciae latae muscle

Vastus intermedius muscle

Iliopsoas muscle

Fascia lata

Pectineus muscle

Adductor brevis muscle

Quadratus femoris muscle

Adductor magnus muscle

Neck of femur

Tuberosity of ischium

Ischiorectal fossa

Gluteus maximus muscle

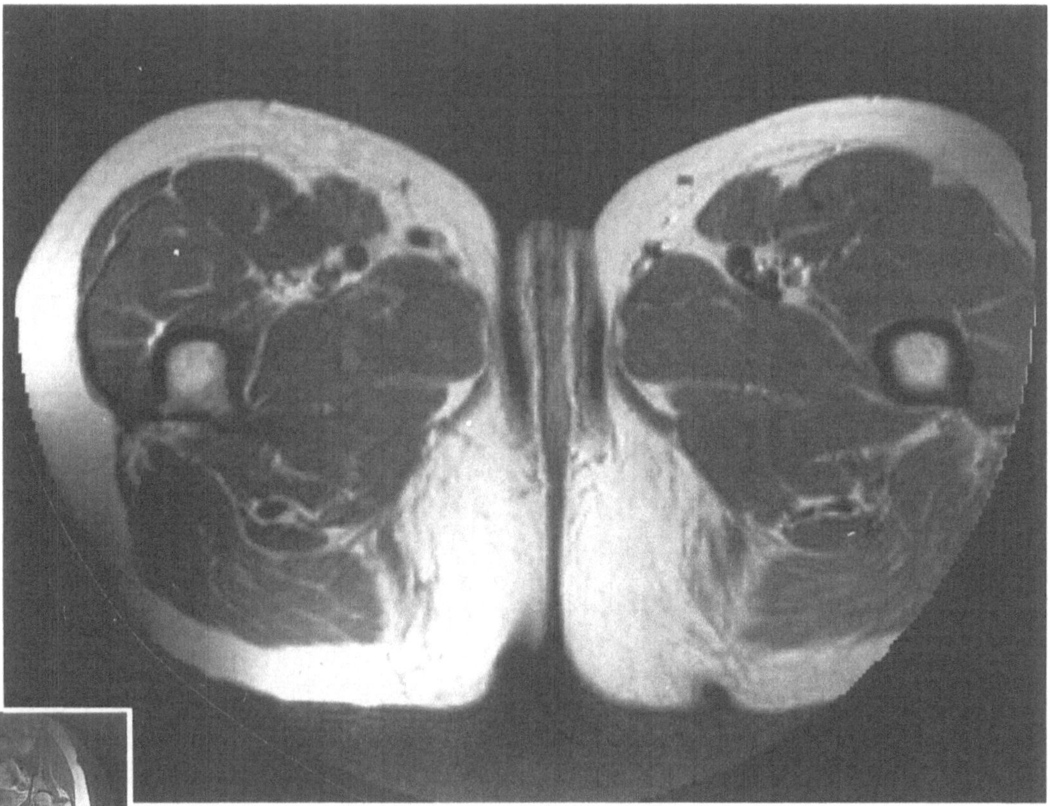

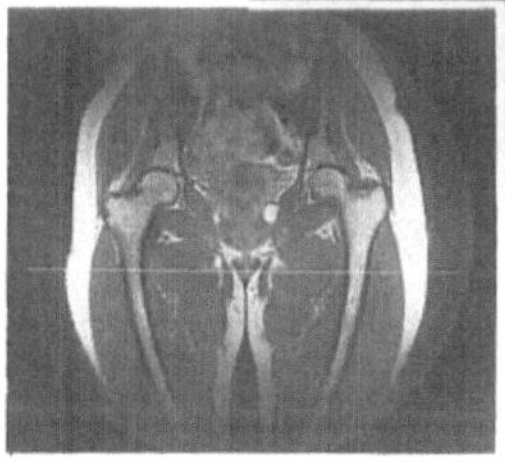

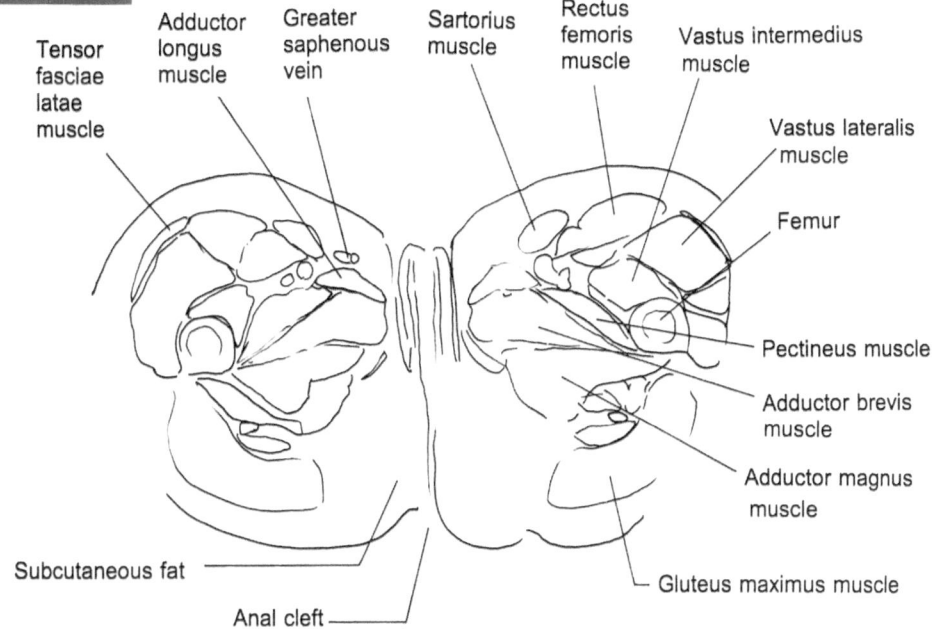

Tensor fasciae latae muscle

Adductor longus muscle

Greater saphenous vein

Sartorius muscle

Rectus femoris muscle

Vastus intermedius muscle

Vastus lateralis muscle

Femur

Pectineus muscle

Adductor brevis muscle

Adductor magnus muscle

Gluteus maximus muscle

Subcutaneous fat

Anal cleft

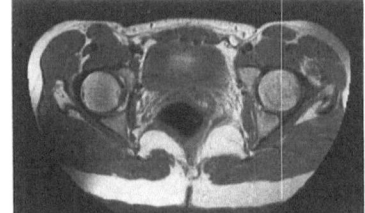

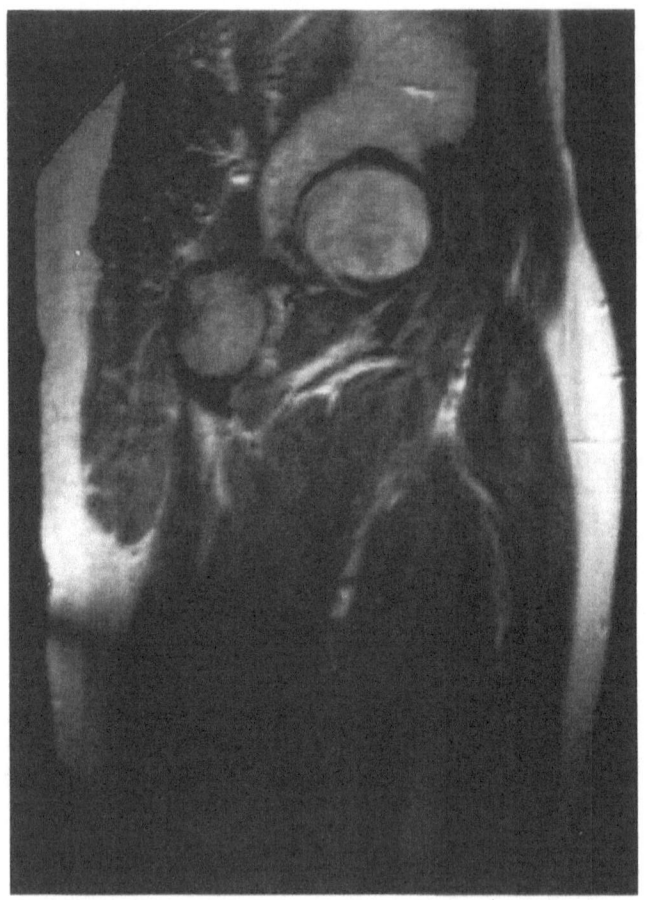

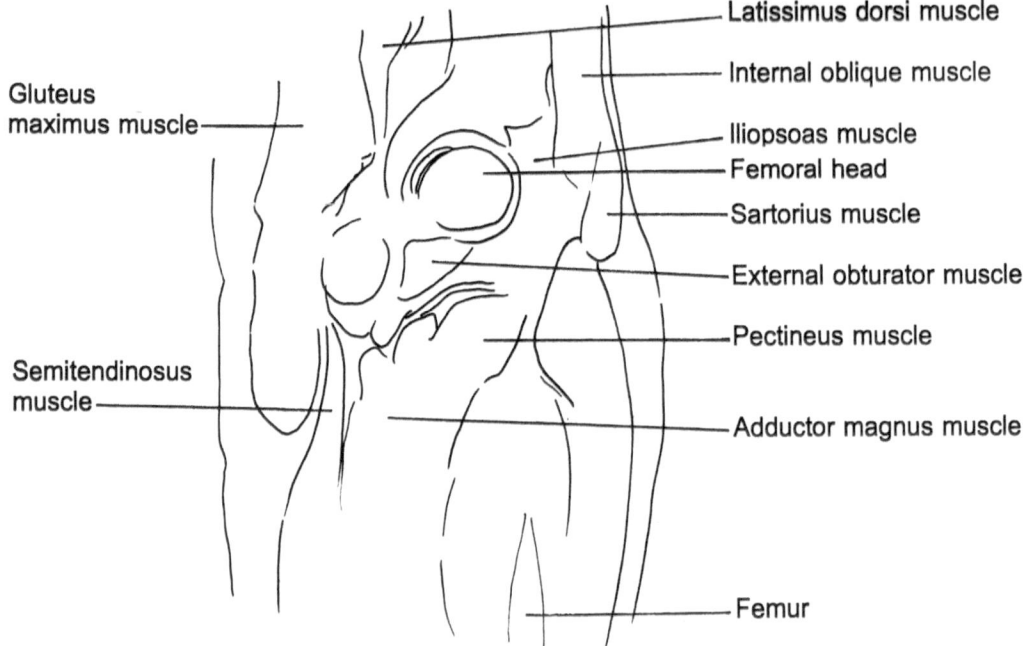

Gluteus
maximus muscle

Semitendinosus
muscle

Latissimus dorsi muscle

Internal oblique muscle

Iliopsoas muscle

Femoral head

Sartorius muscle

External obturator muscle

Pectineus muscle

Adductor magnus muscle

Femur

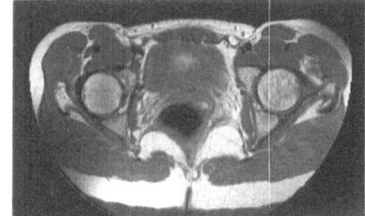

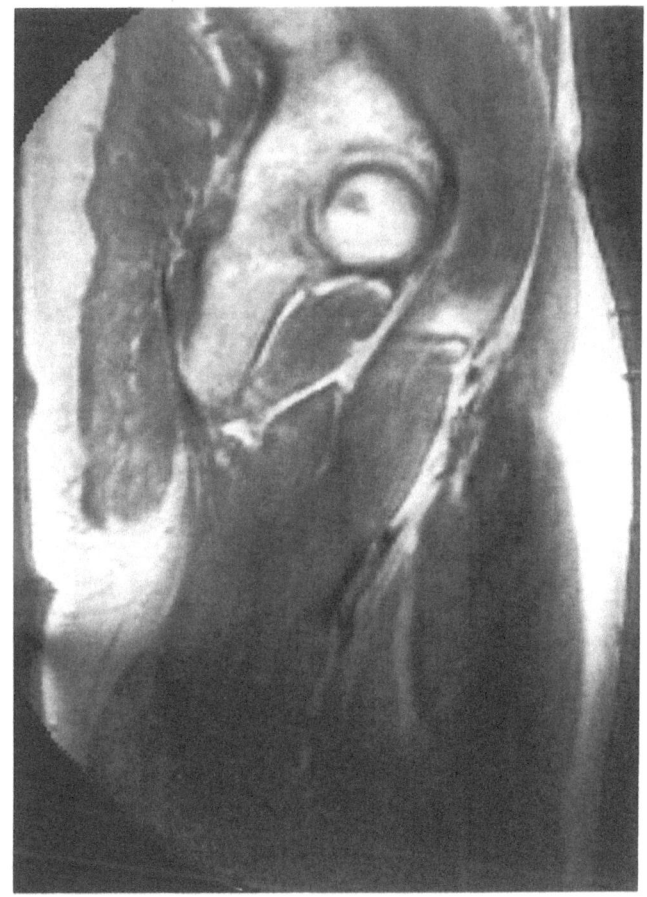

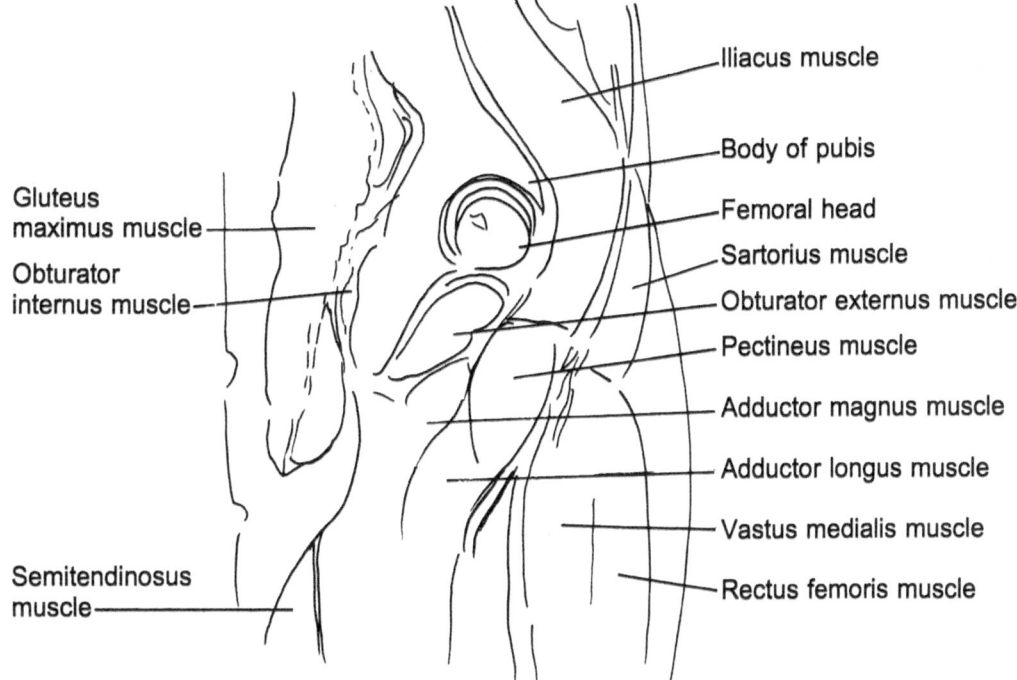

Iliacus muscle

Body of pubis

Femoral head

Sartorius muscle

Obturator externus muscle

Pectineus muscle

Adductor magnus muscle

Adductor longus muscle

Vastus medialis muscle

Rectus femoris muscle

Gluteus
maximus muscle

Obturator
internus muscle

Semitendinosus
muscle

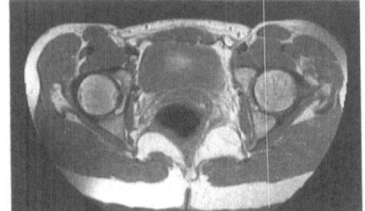

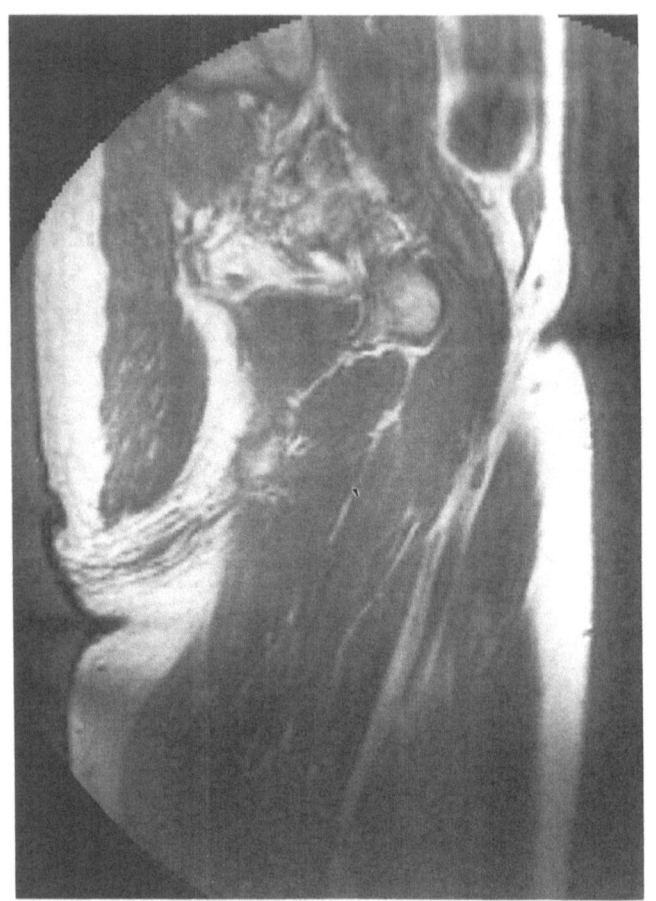

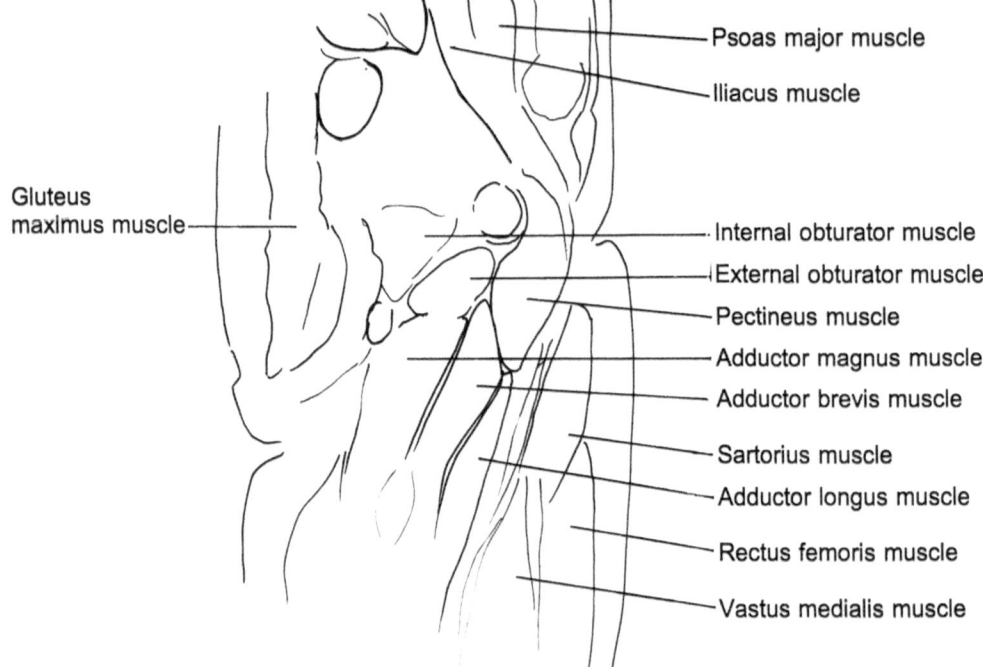

Psoas major muscle

Iliacus muscle

Gluteus maximus muscle

Internal obturator muscle

External obturator muscle

Pectineus muscle

Adductor magnus muscle

Adductor brevis muscle

Sartorius muscle

Adductor longus muscle

Rectus femoris muscle

Vastus medialis muscle

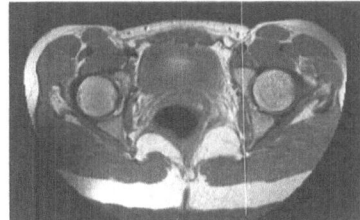

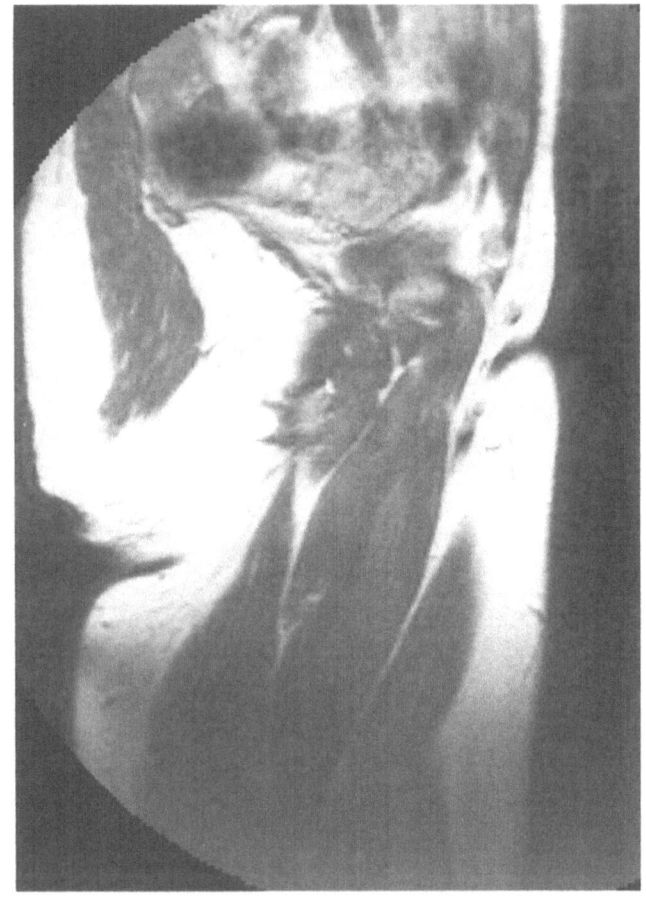

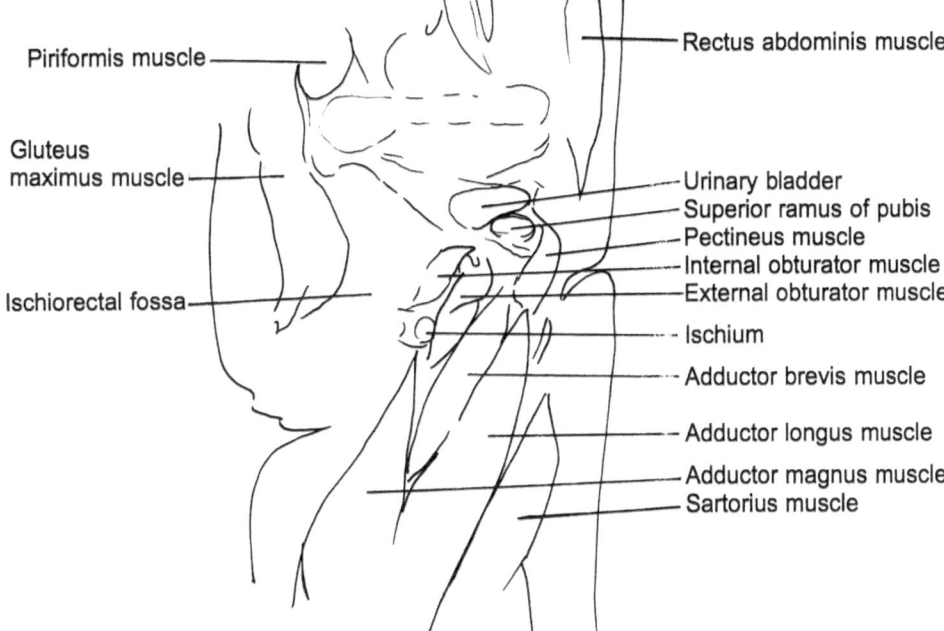

Piriformis muscle — Rectus abdominis muscle

Gluteus
maximus muscle —

Ischiorectal fossa —

Urinary bladder
Superior ramus of pubis
Pectineus muscle
Internal obturator muscle
External obturator muscle
Ischium
Adductor brevis muscle
Adductor longus muscle
Adductor magnus muscle
Sartorius muscle

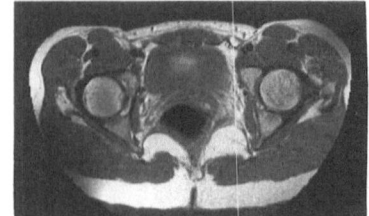

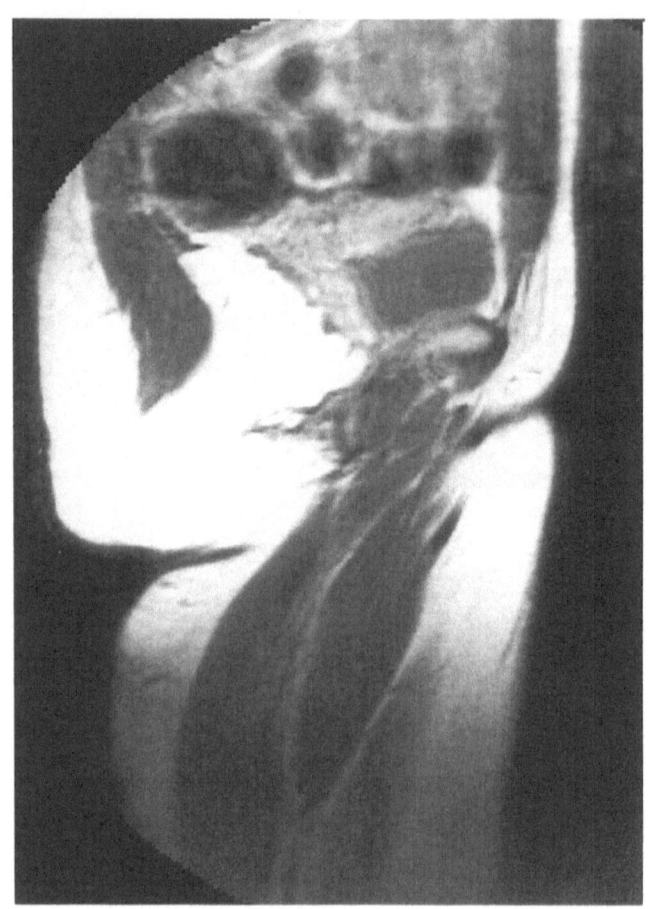

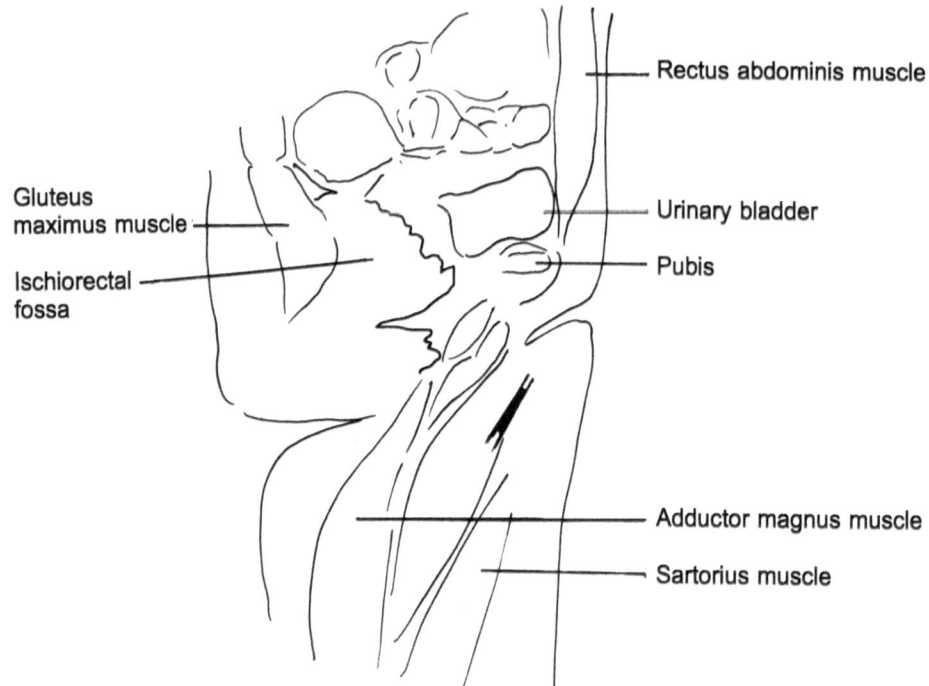

Rectus abdominis muscle

Gluteus maximus muscle

Urinary bladder

Ischiorectal fossa

Pubis

Adductor magnus muscle

Sartorius muscle

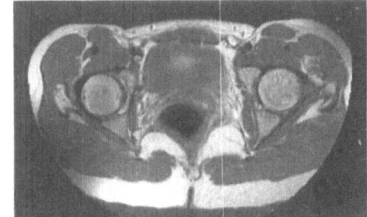

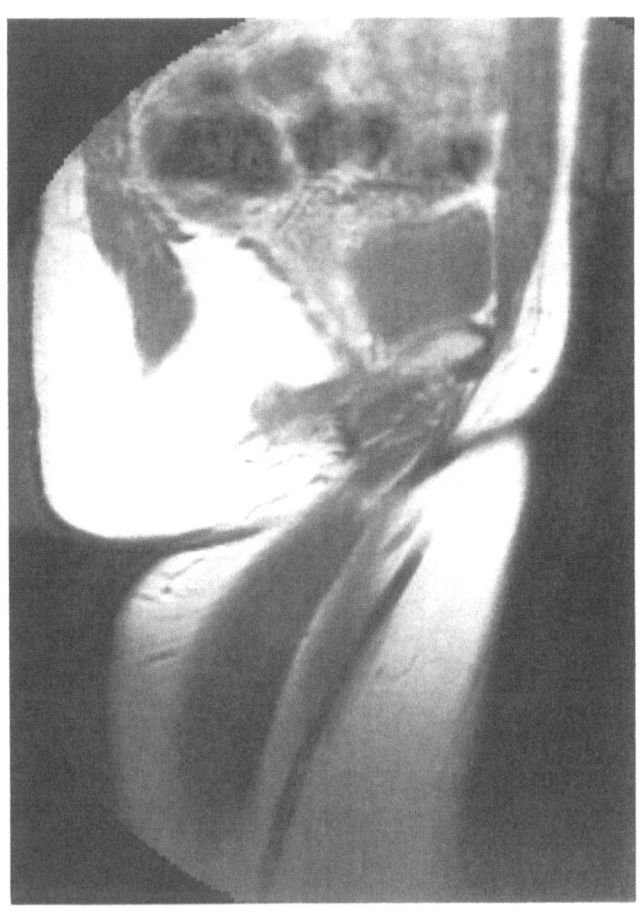

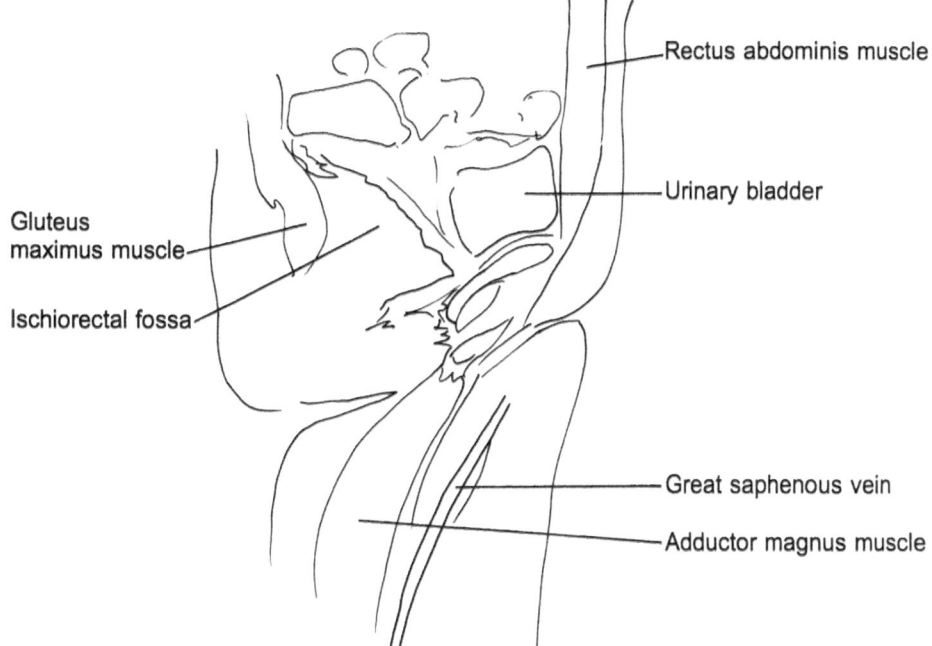

Rectus abdominis muscle

Urinary bladder

Gluteus
maximus muscle

Ischiorectal fossa

Great saphenous vein

Adductor magnus muscle

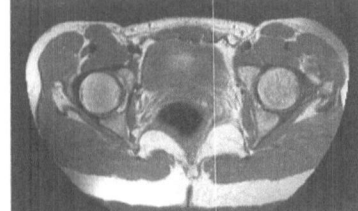

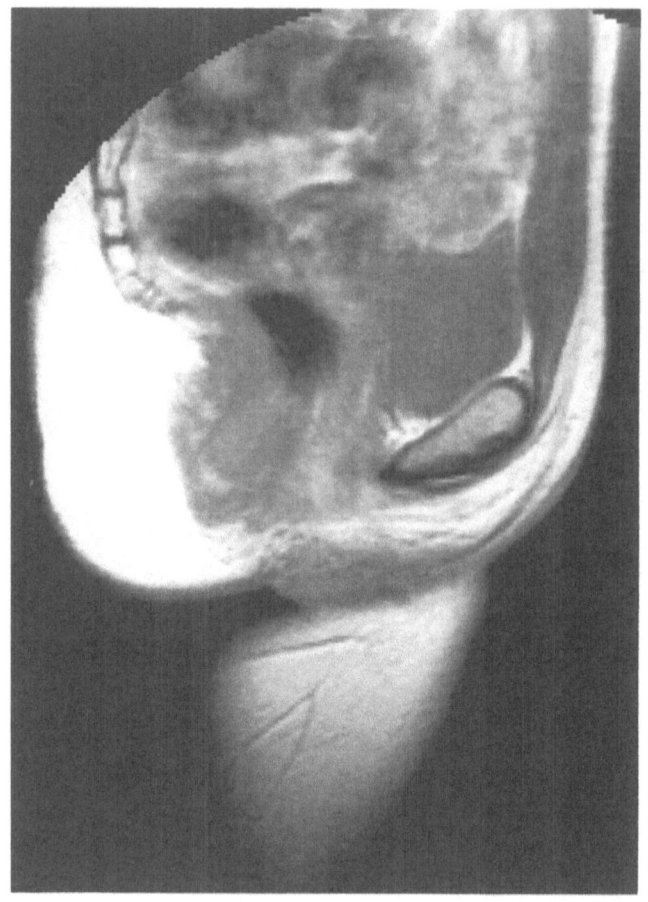

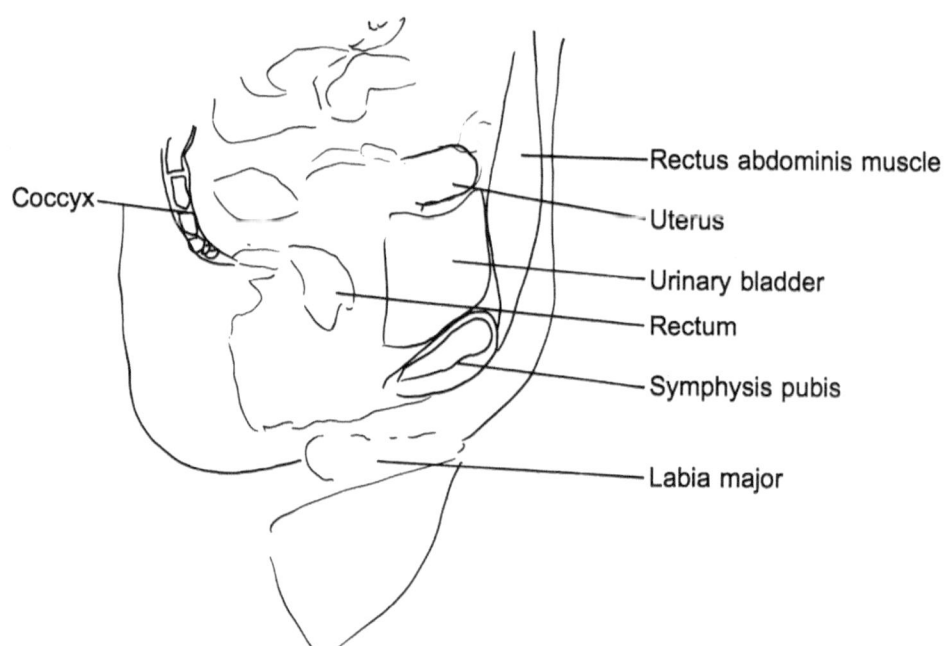

Coccyx

Rectus abdominis muscle

Uterus

Urinary bladder

Rectum

Symphysis pubis

Labia major

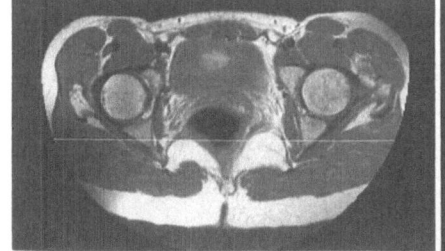

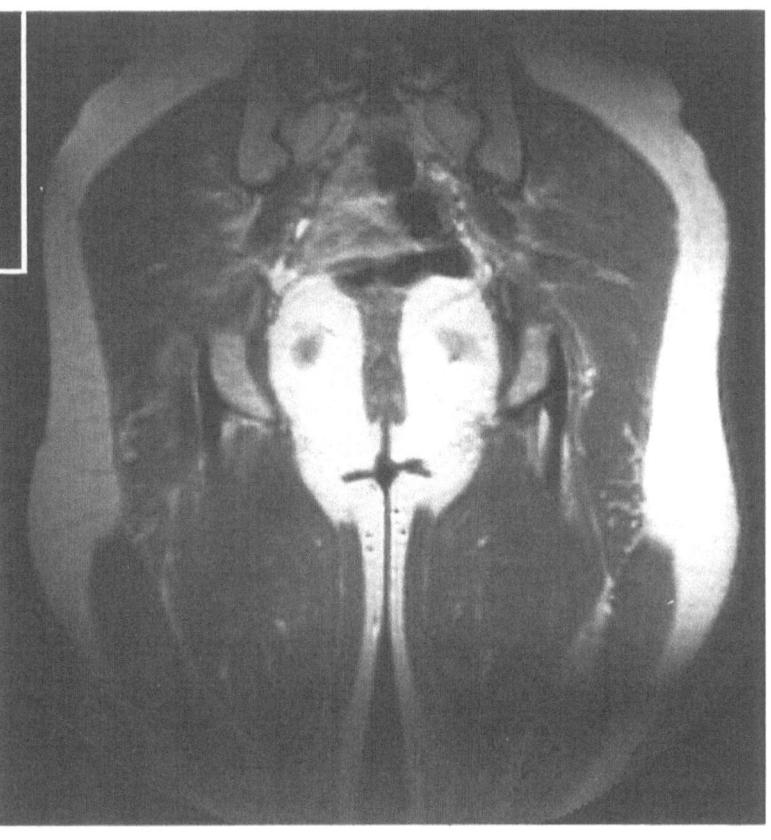

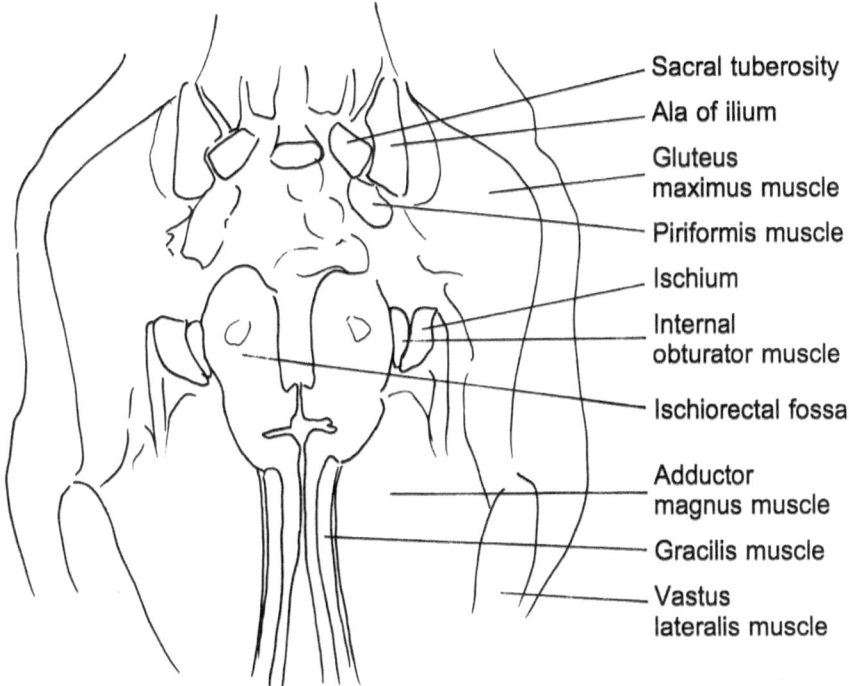

Sacral tuberosity

Ala of ilium

Gluteus
maximus muscle

Piriformis muscle

Ischium

Internal
obturator muscle

Ischiorectal fossa

Adductor
magnus muscle

Gracilis muscle

Vastus
lateralis muscle

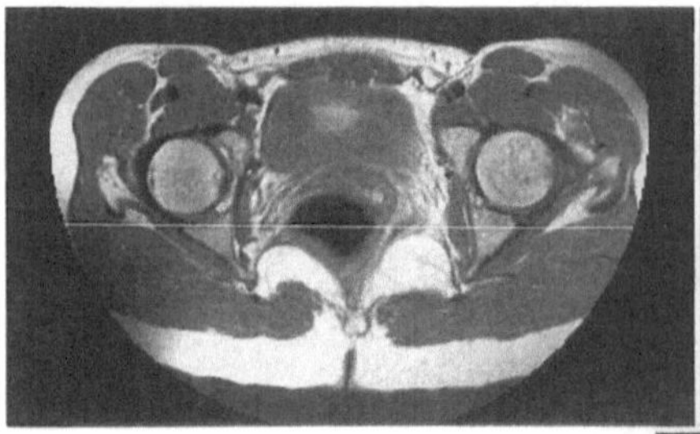

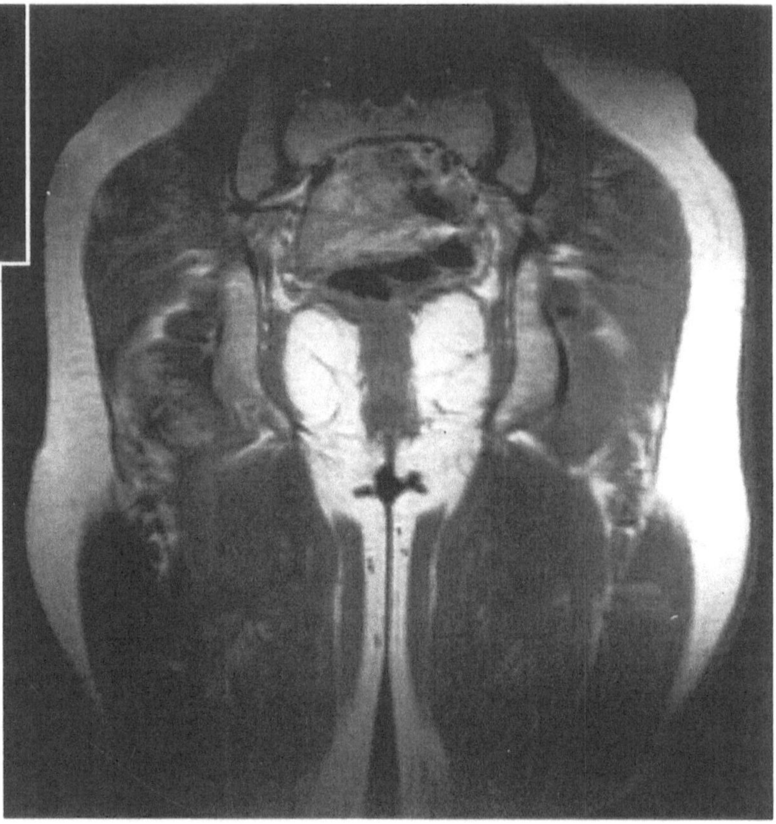

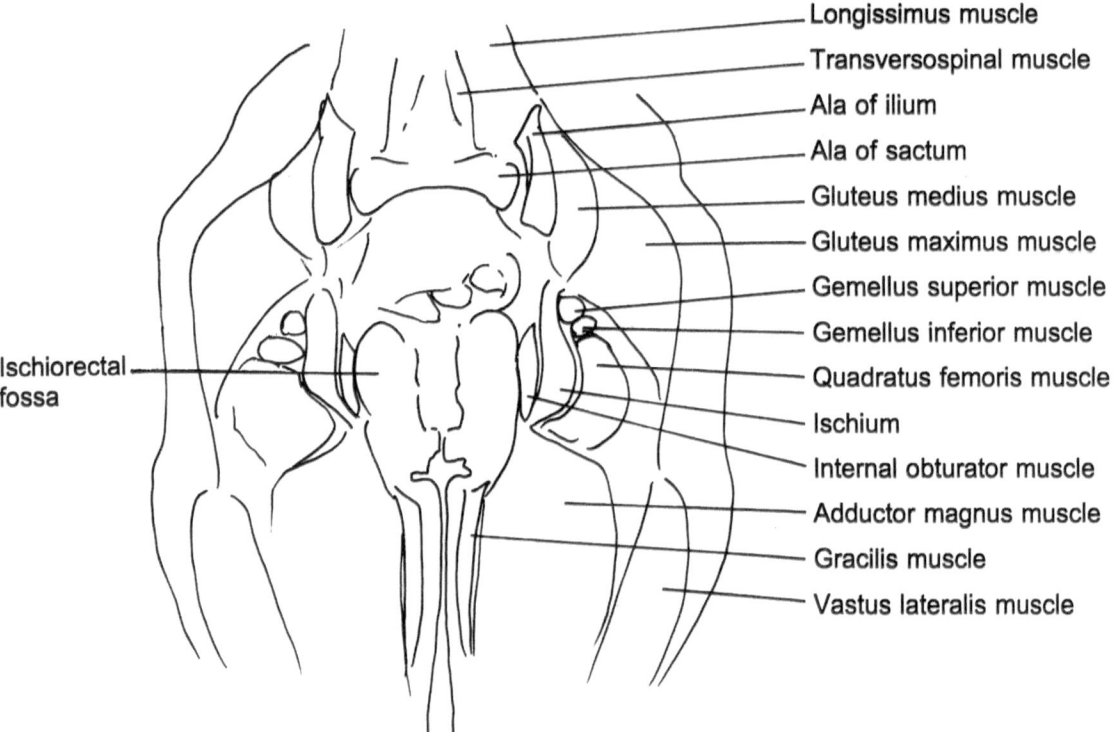

Longissimus muscle

Transversospinal muscle

Ala of ilium

Ala of sactum

Gluteus medius muscle

Gluteus maximus muscle

Gemellus superior muscle

Gemellus inferior muscle

Quadratus femoris muscle

Ischium

Internal obturator muscle

Adductor magnus muscle

Gracilis muscle

Vastus lateralis muscle

Ischiorectal fossa

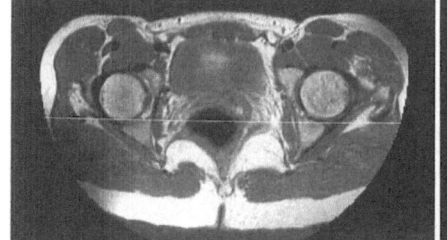

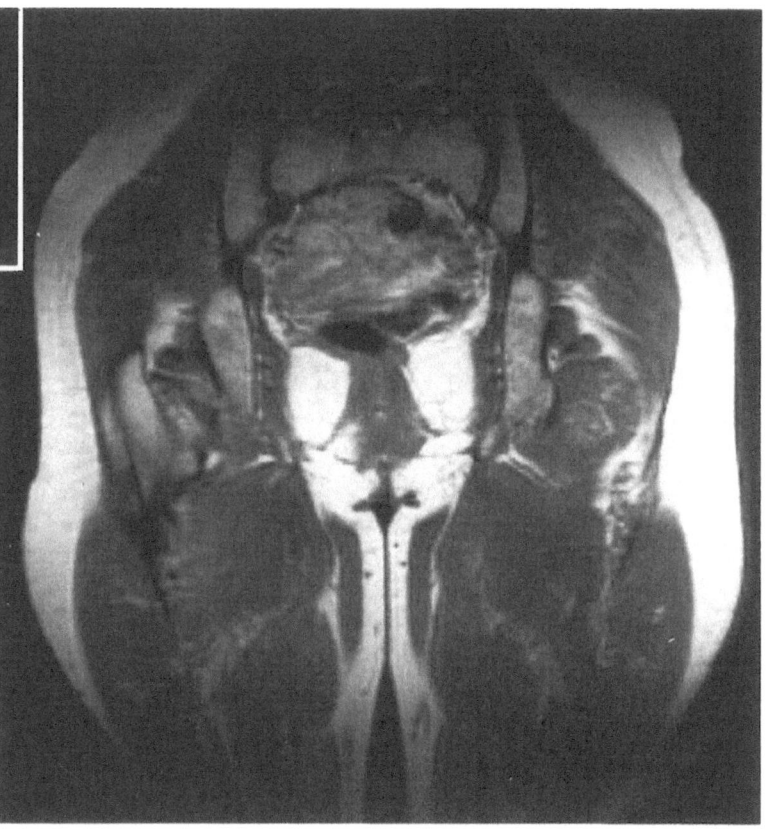

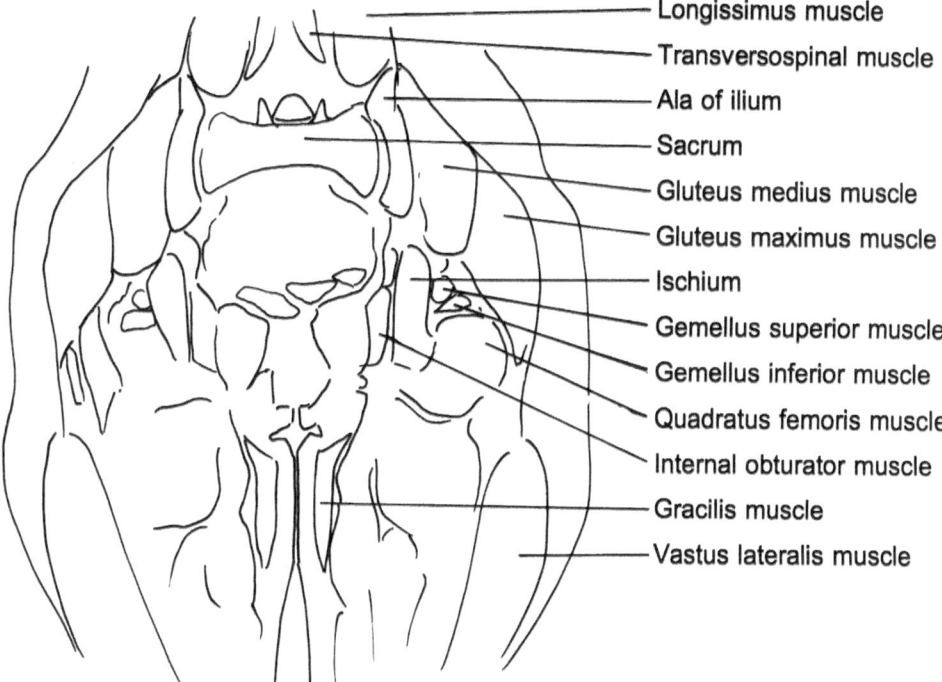

- Longissimus muscle
- Transversospinal muscle
- Ala of ilium
- Sacrum
- Gluteus medius muscle
- Gluteus maximus muscle
- Ischium
- Gemellus superior muscle
- Gemellus inferior muscle
- Quadratus femoris muscle
- Internal obturator muscle
- Gracilis muscle
- Vastus lateralis muscle

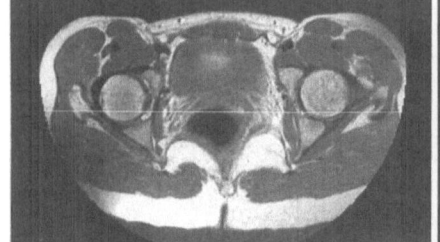

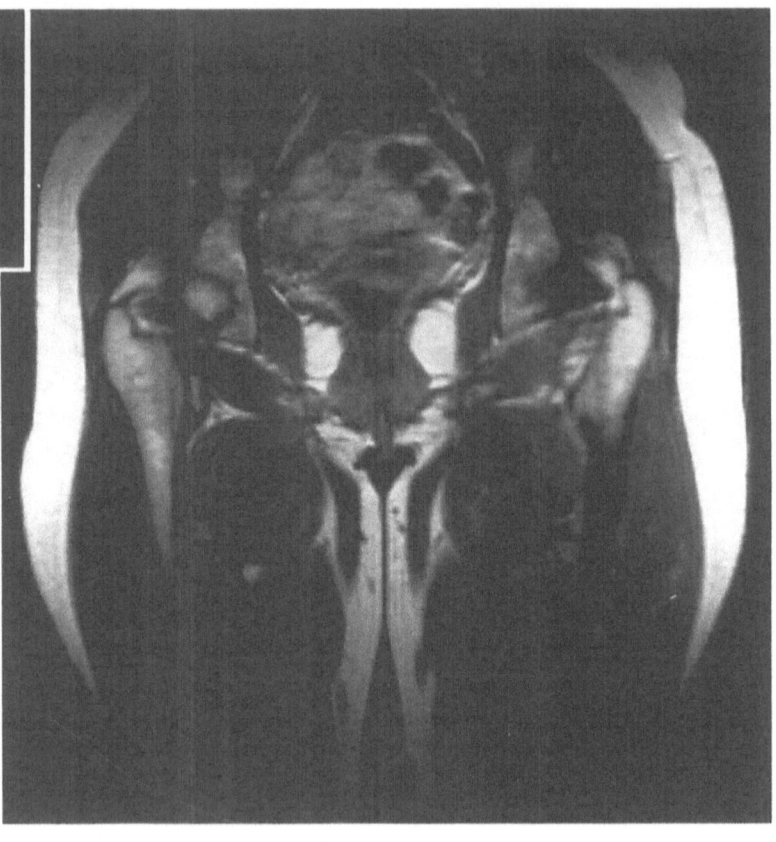

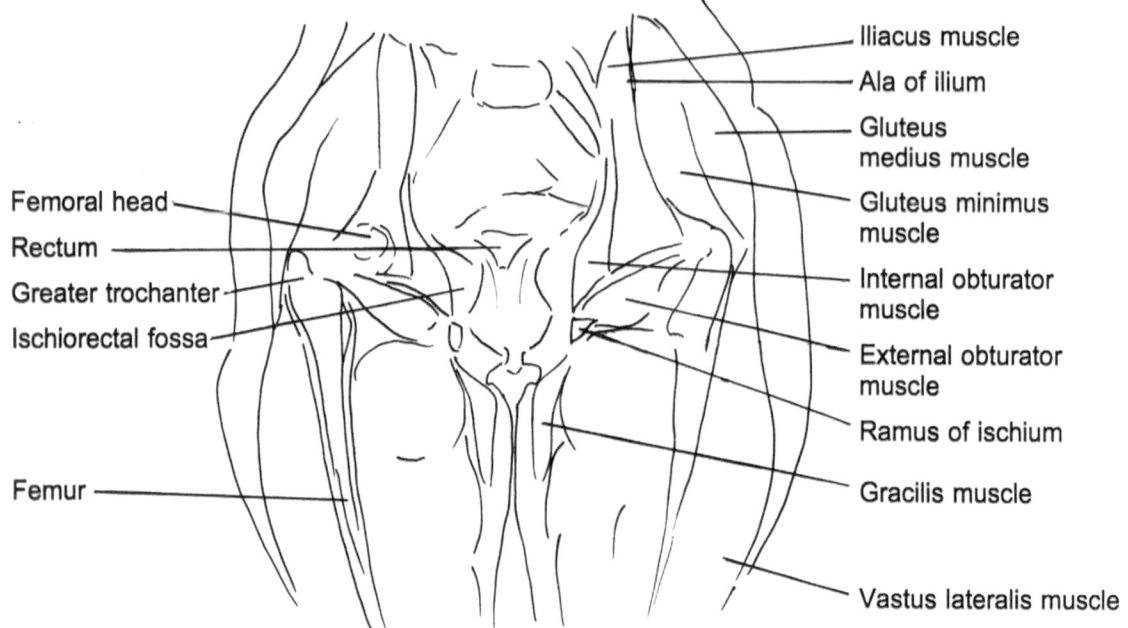

Iliacus muscle

Ala of ilium

Gluteus medius muscle

Gluteus minimus muscle

Femoral head

Rectum

Greater trochanter

Ischiorectal fossa

Internal obturator muscle

External obturator muscle

Ramus of ischium

Gracilis muscle

Femur

Vastus lateralis muscle

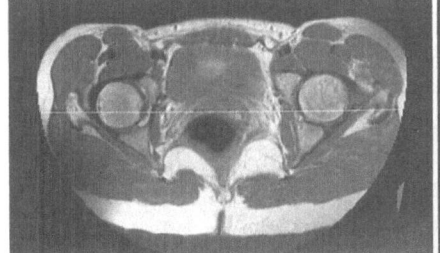

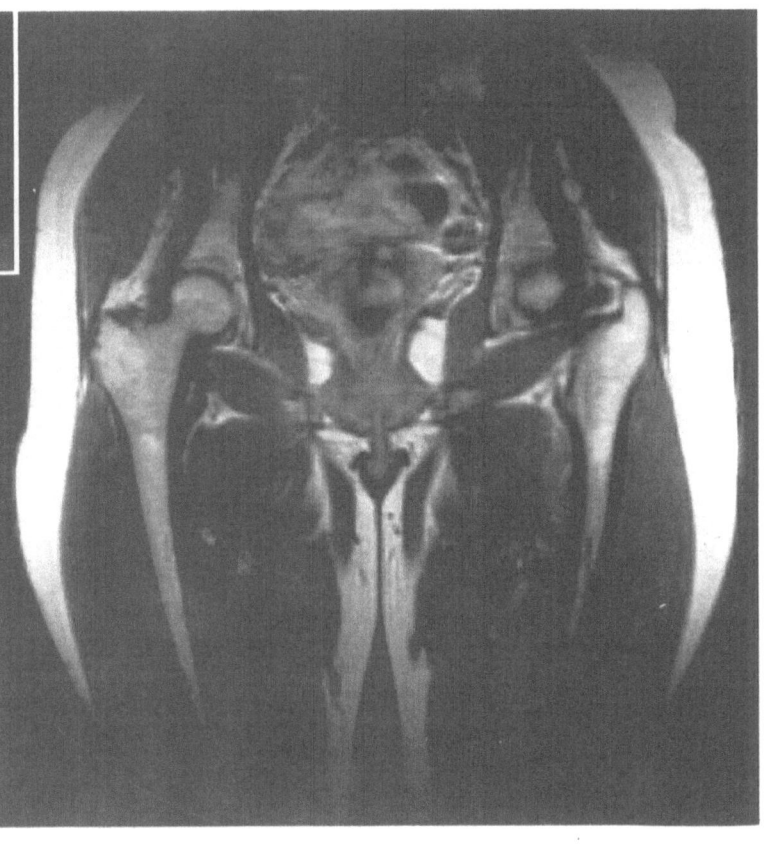

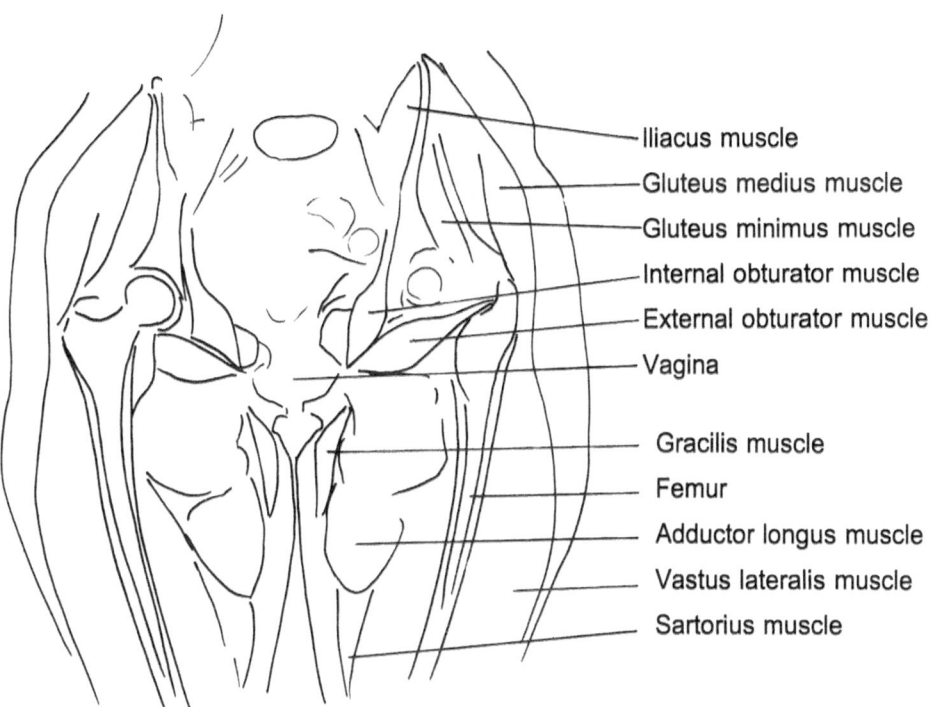

Iliacus muscle

Gluteus medius muscle

Gluteus minimus muscle

Internal obturator muscle

External obturator muscle

Vagina

Gracilis muscle

Femur

Adductor longus muscle

Vastus lateralis muscle

Sartorius muscle

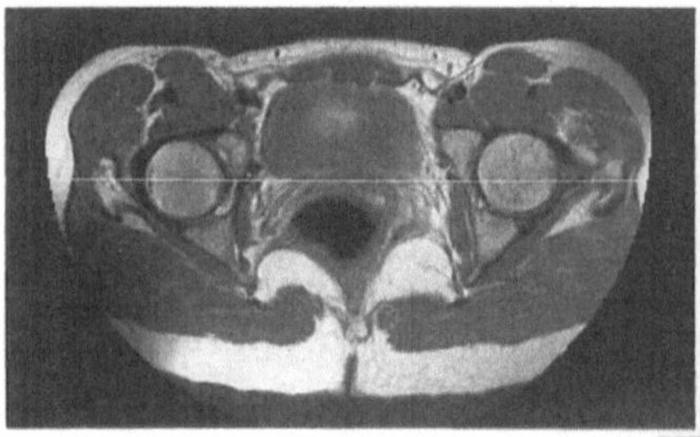

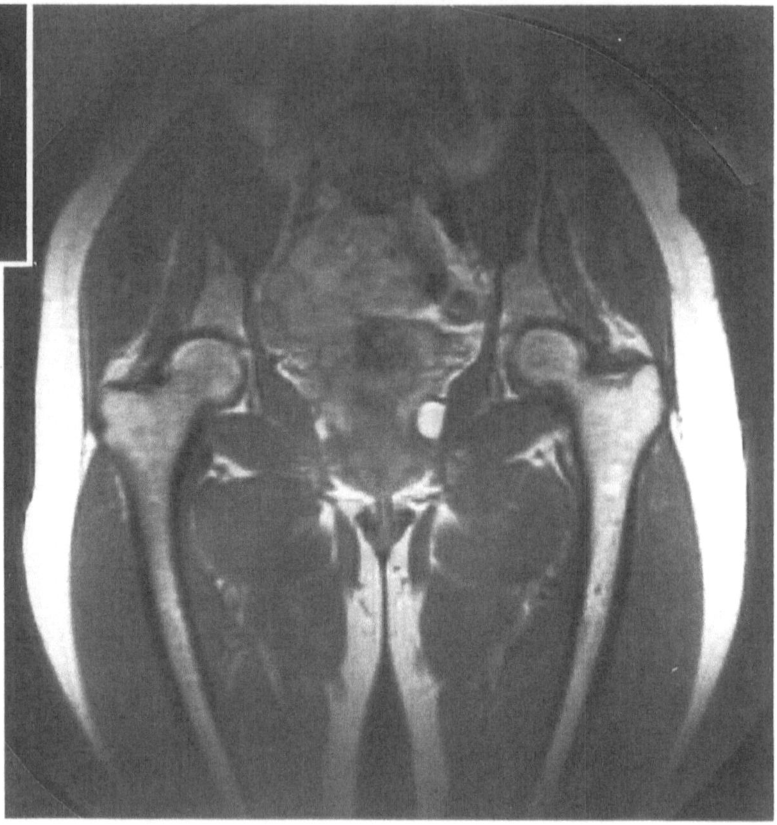

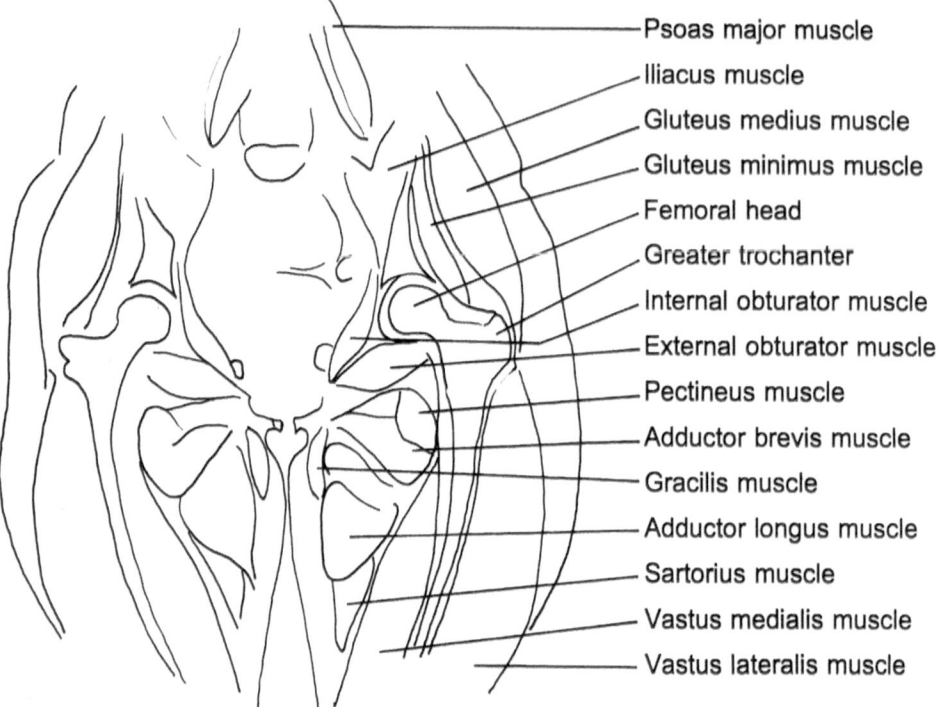

Psoas major muscle

Iliacus muscle

Gluteus medius muscle

Gluteus minimus muscle

Femoral head

Greater trochanter

Internal obturator muscle

External obturator muscle

Pectineus muscle

Adductor brevis muscle

Gracilis muscle

Adductor longus muscle

Sartorius muscle

Vastus medialis muscle

Vastus lateralis muscle

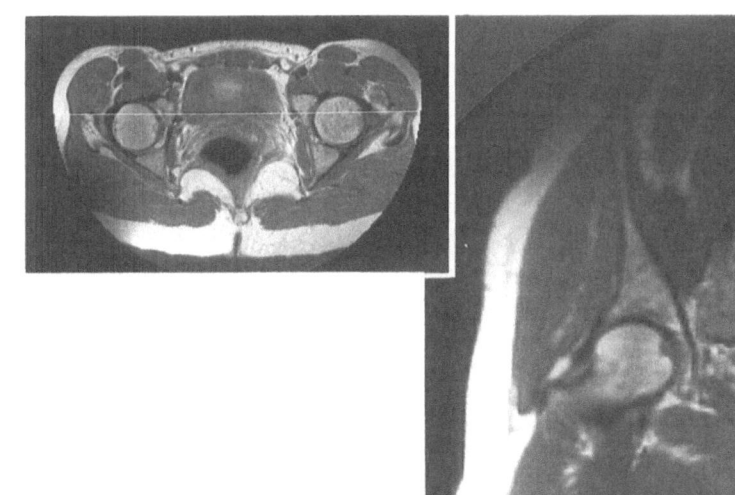

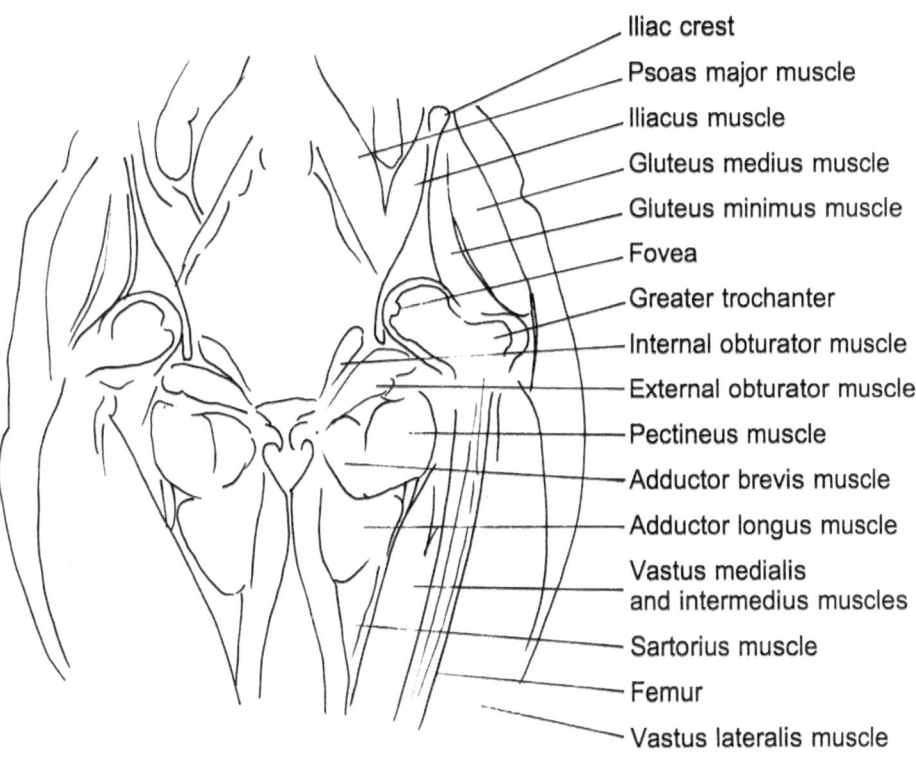

Iliac crest

Psoas major muscle

Iliacus muscle

Gluteus medius muscle

Gluteus minimus muscle

Fovea

Greater trochanter

Internal obturator muscle

External obturator muscle

Pectineus muscle

Adductor brevis muscle

Adductor longus muscle

Vastus medialis
and intermedius muscles

Sartorius muscle

Femur

Vastus lateralis muscle

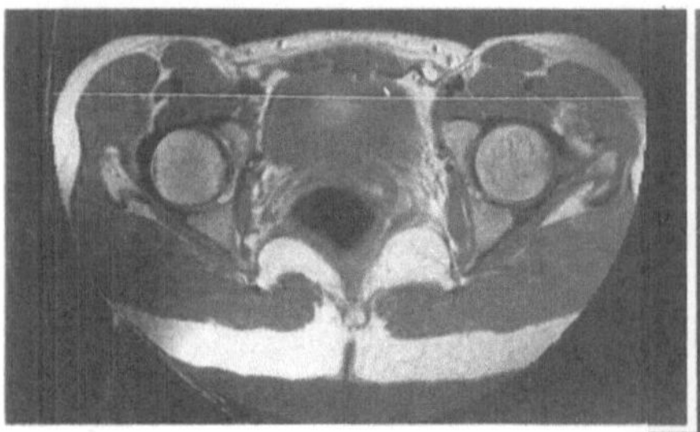

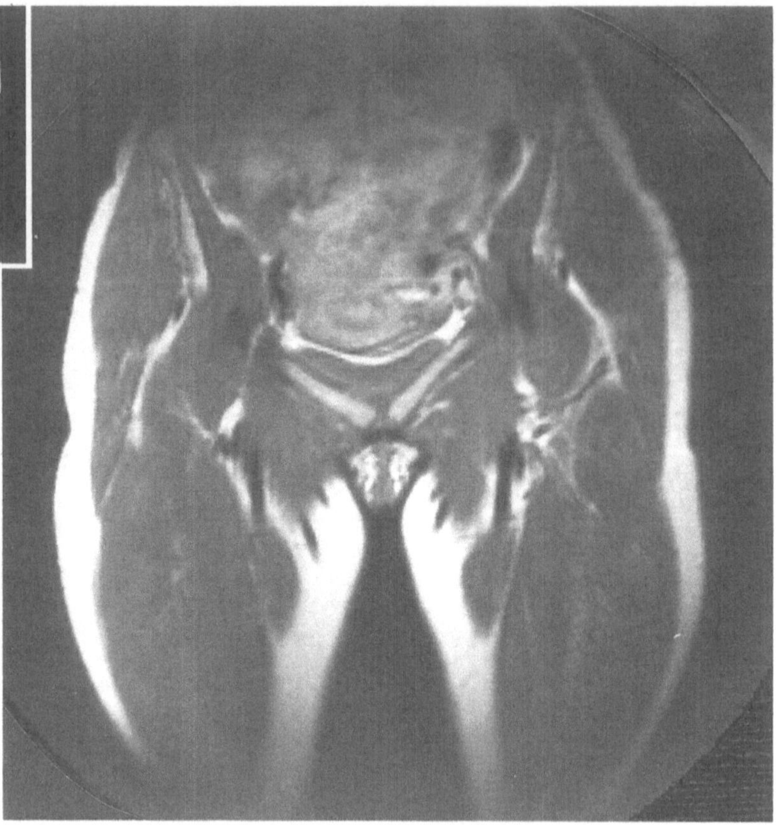

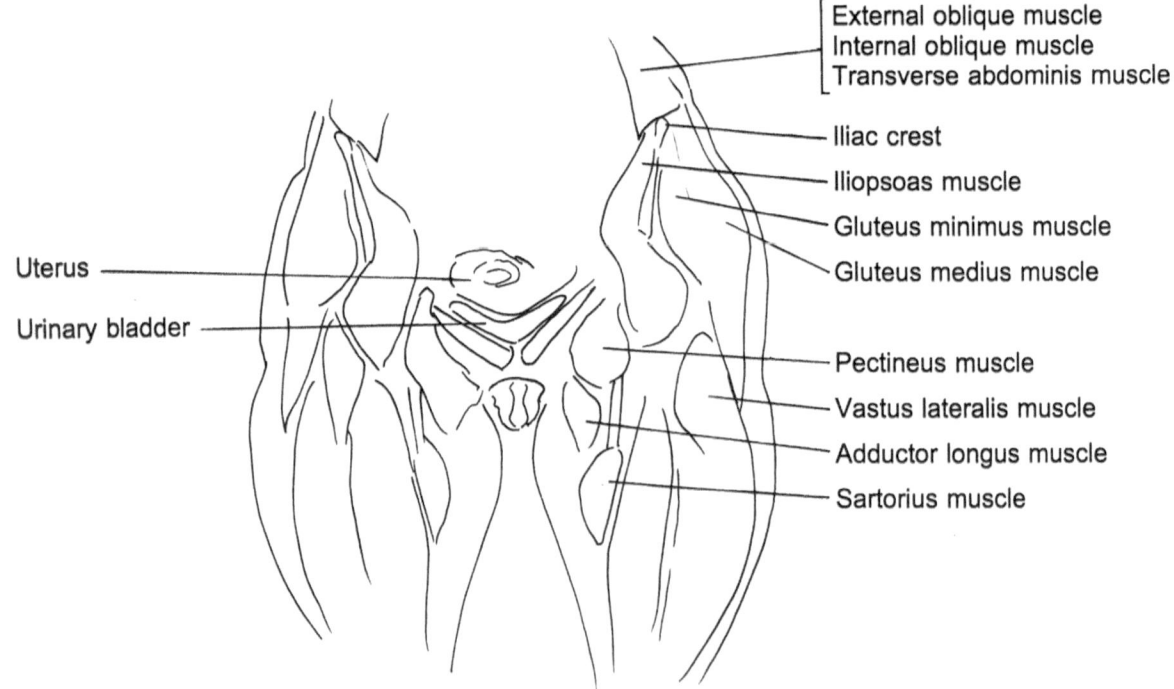

External oblique muscle
Internal oblique muscle
Transverse abdominis muscle

Iliac crest

Iliopsoas muscle

Gluteus minimus muscle

Gluteus medius muscle

Uterus

Urinary bladder

Pectineus muscle

Vastus lateralis muscle

Adductor longus muscle

Sartorius muscle

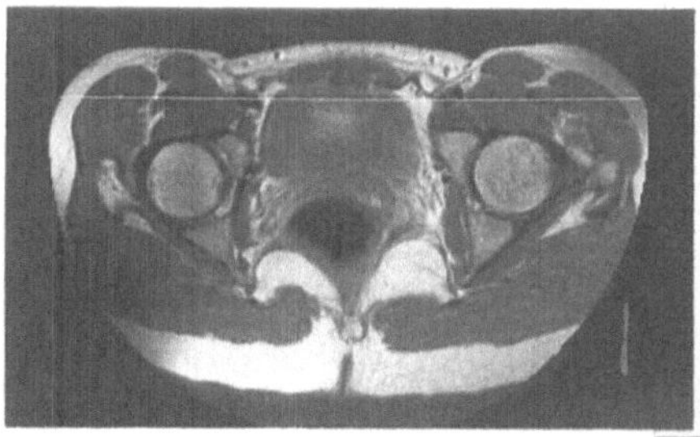

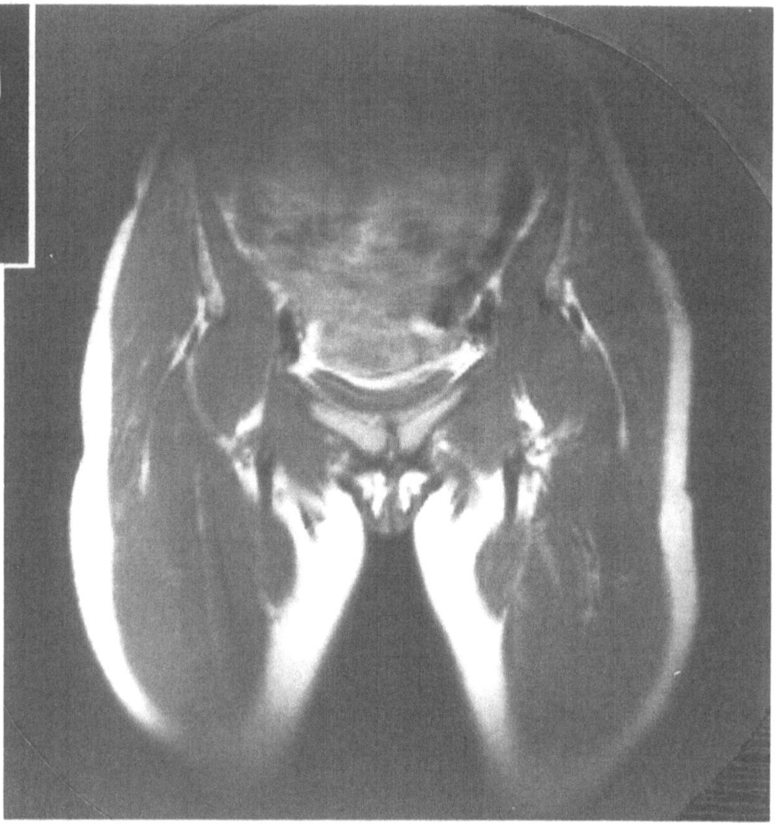

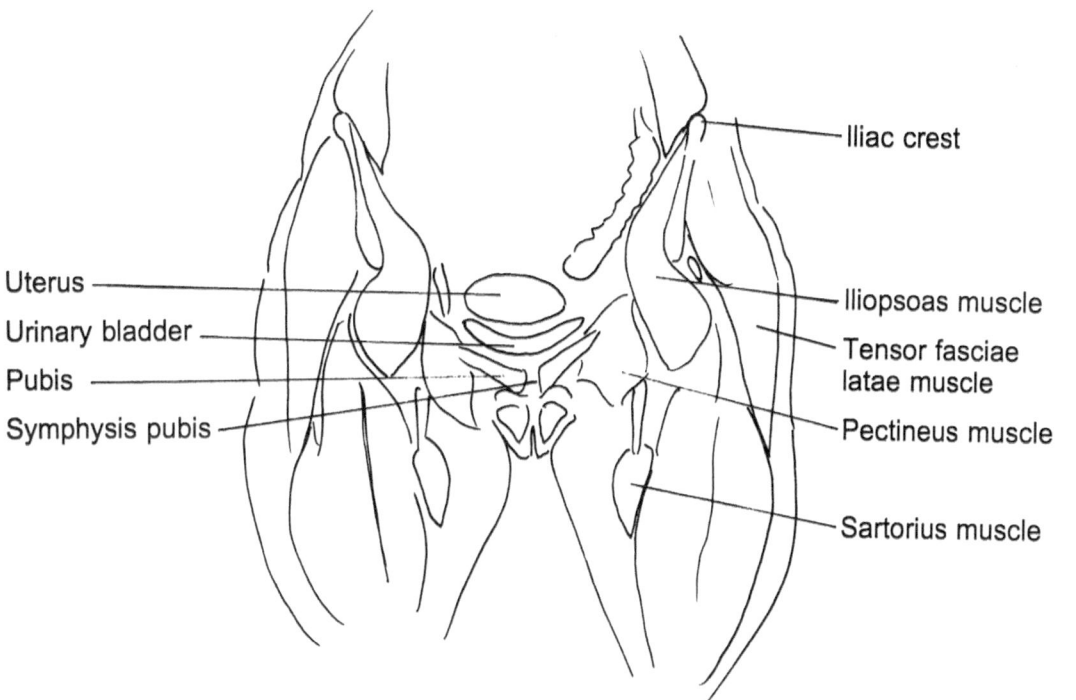

Iliac crest

Uterus

Iliopsoas muscle

Urinary bladder

Tensor fasciae latae muscle

Pubis

Symphysis pubis

Pectineus muscle

Sartorius muscle

KNEE

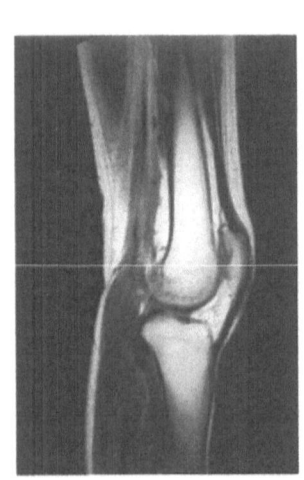

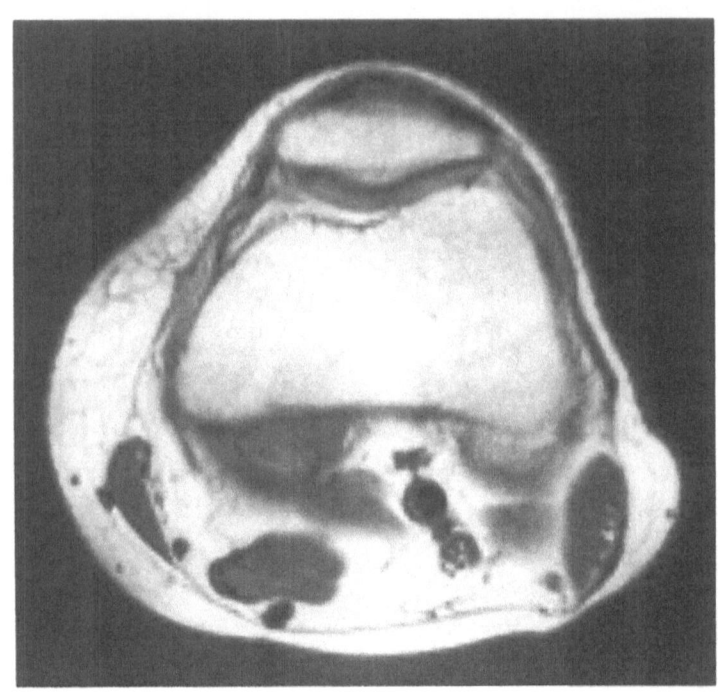

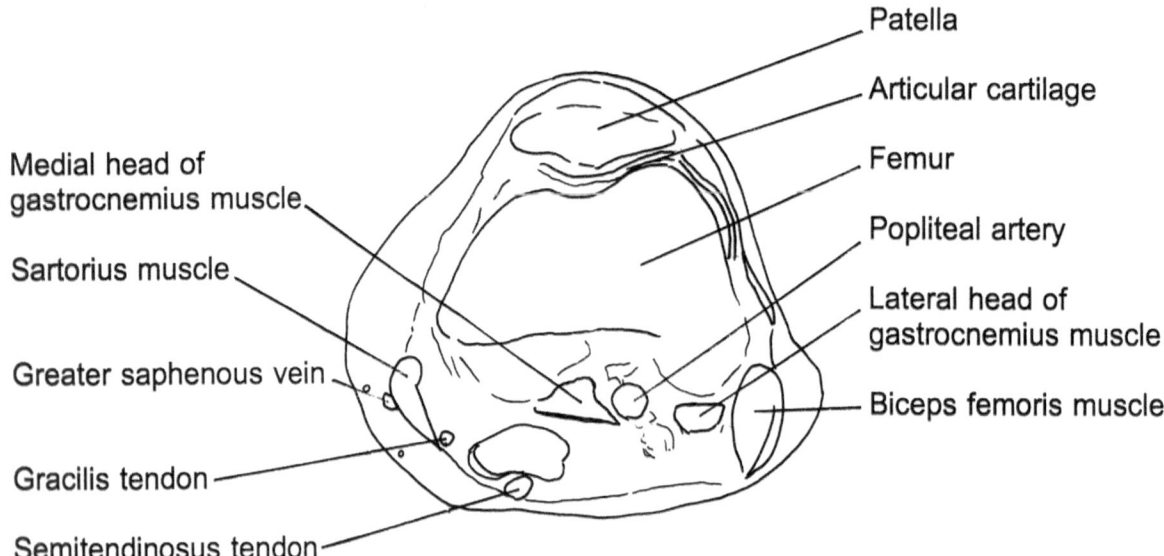

Patella

Articular cartilage

Medial head of
gastrocnemius muscle

Femur

Sartorius muscle

Popliteal artery

Lateral head of
gastrocnemius muscle

Greater saphenous vein

Biceps femoris muscle

Gracilis tendon

Semitendinosus tendon

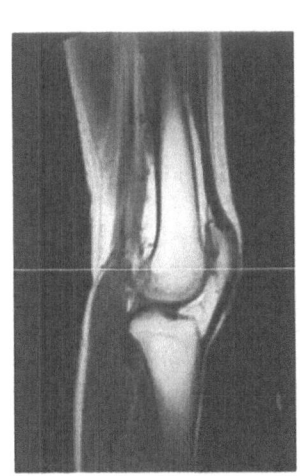

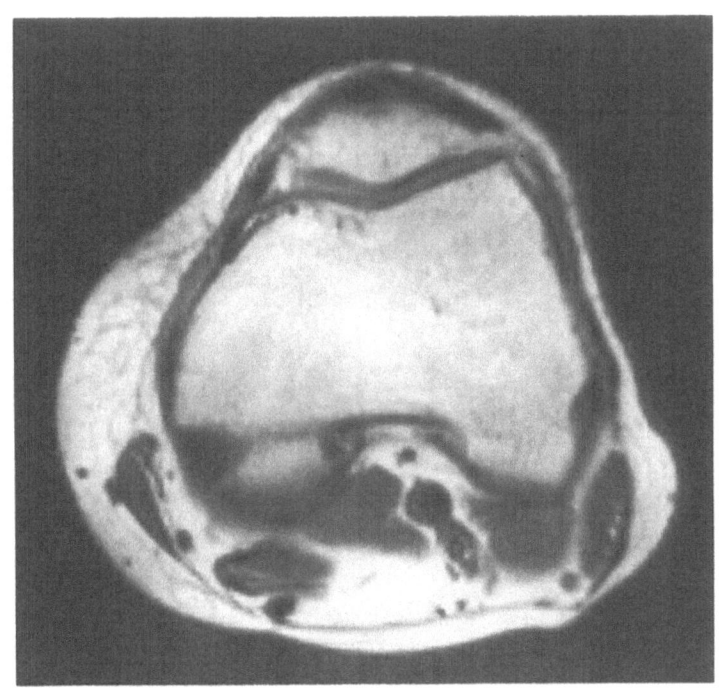

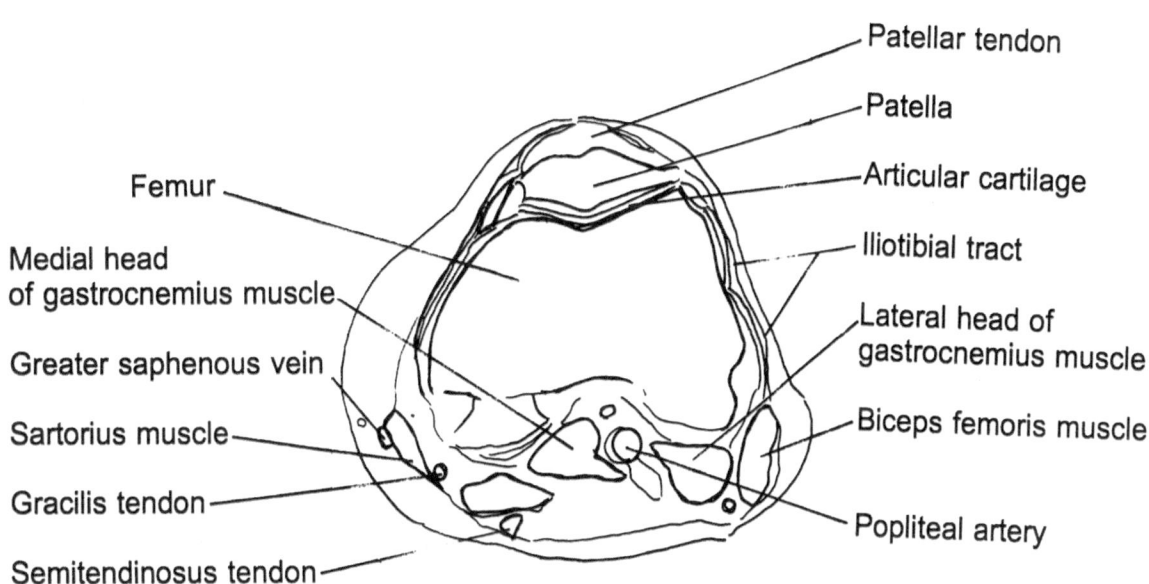

Patellar tendon

Patella

Articular cartilage

Iliotibial tract

Lateral head of gastrocnemius muscle

Biceps femoris muscle

Popliteal artery

Femur

Medial head of gastrocnemius muscle

Greater saphenous vein

Sartorius muscle

Gracilis tendon

Semitendinosus tendon

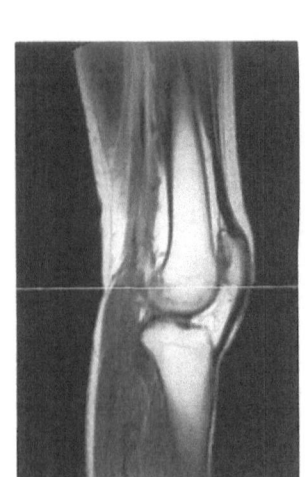

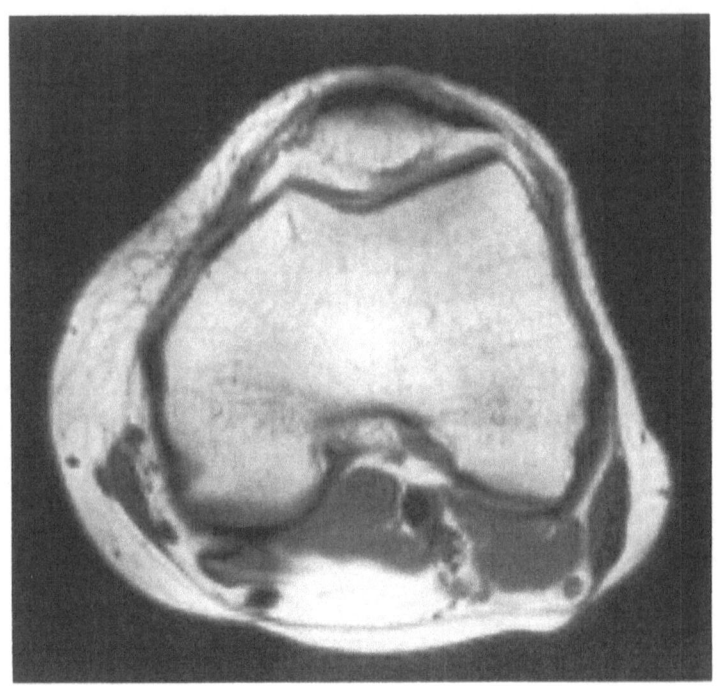

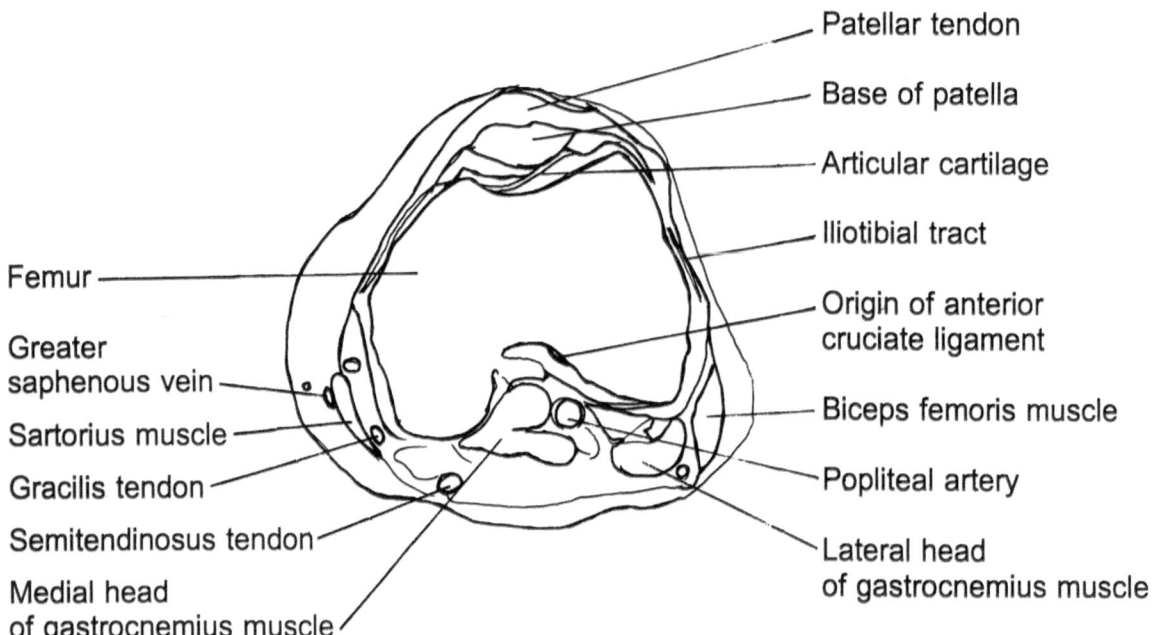

Patellar tendon

Base of patella

Articular cartilage

Iliotibial tract

Origin of anterior
cruciate ligament

Biceps femoris muscle

Popliteal artery

Lateral head
of gastrocnemius muscle

Femur

Greater
saphenous vein

Sartorius muscle

Gracilis tendon

Semitendinosus tendon

Medial head
of gastrocnemius muscle

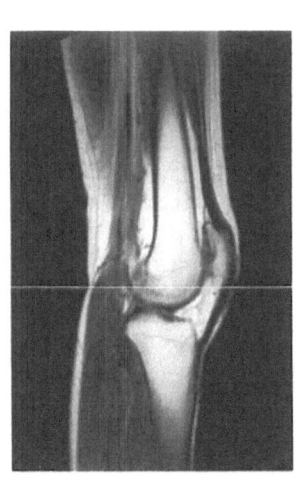

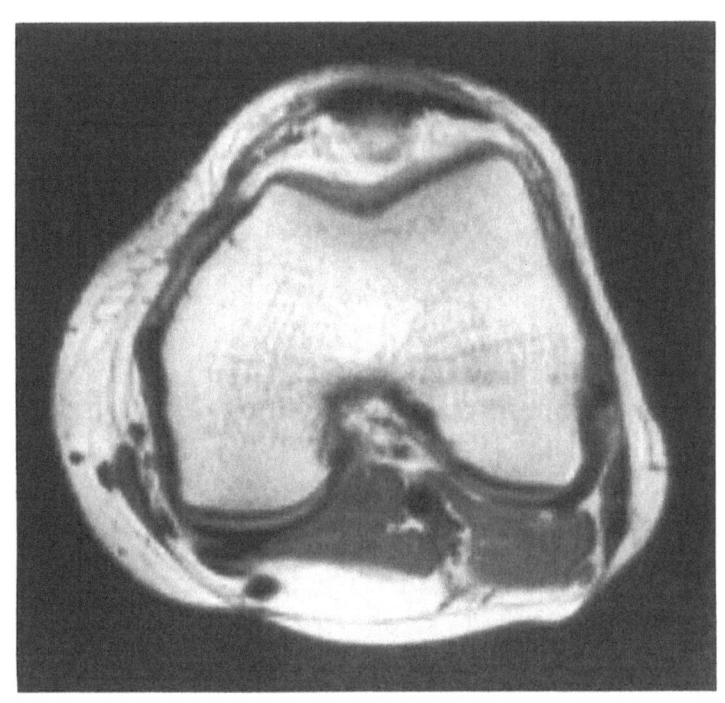

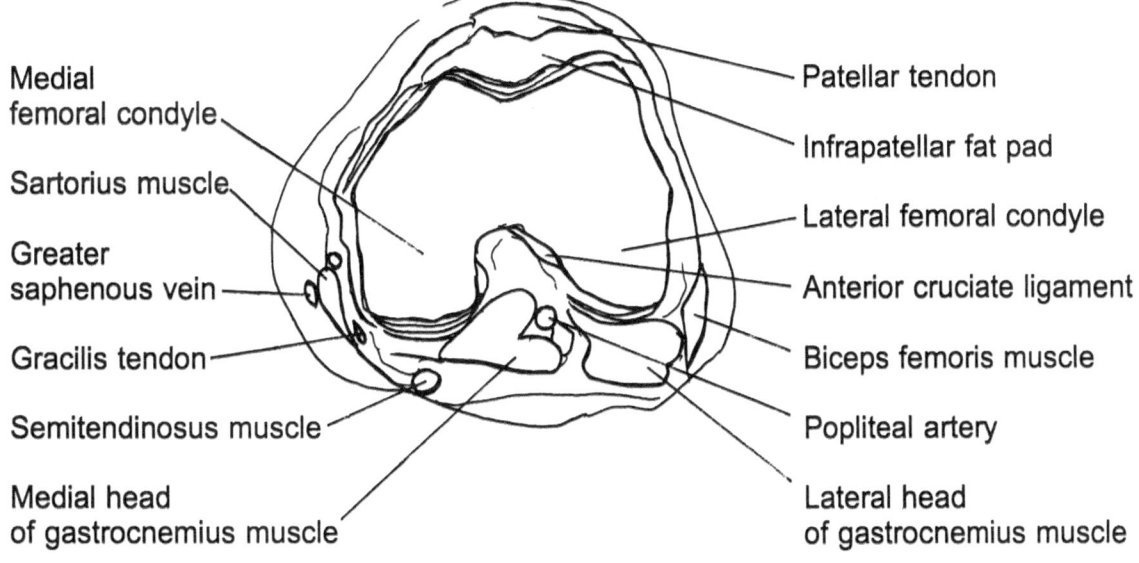

Medial
femoral condyle

Sartorius muscle

Greater
saphenous vein

Gracilis tendon

Semitendinosus muscle

Medial head
of gastrocnemius muscle

Patellar tendon

Infrapatellar fat pad

Lateral femoral condyle

Anterior cruciate ligament

Biceps femoris muscle

Popliteal artery

Lateral head
of gastrocnemius muscle

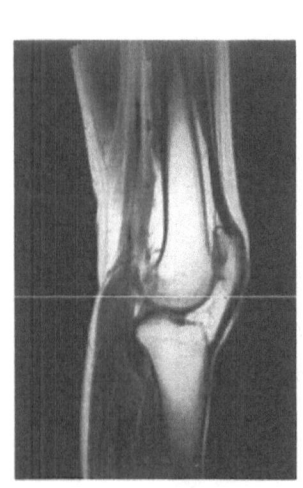

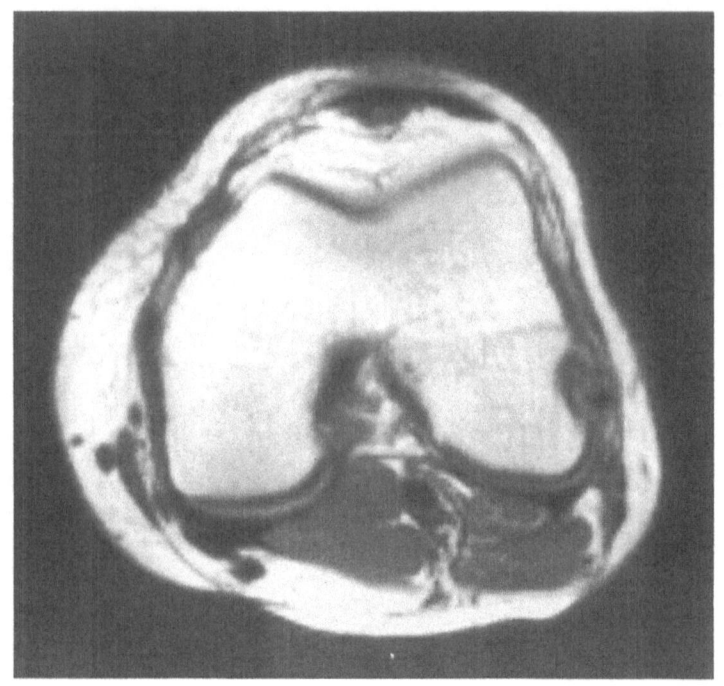

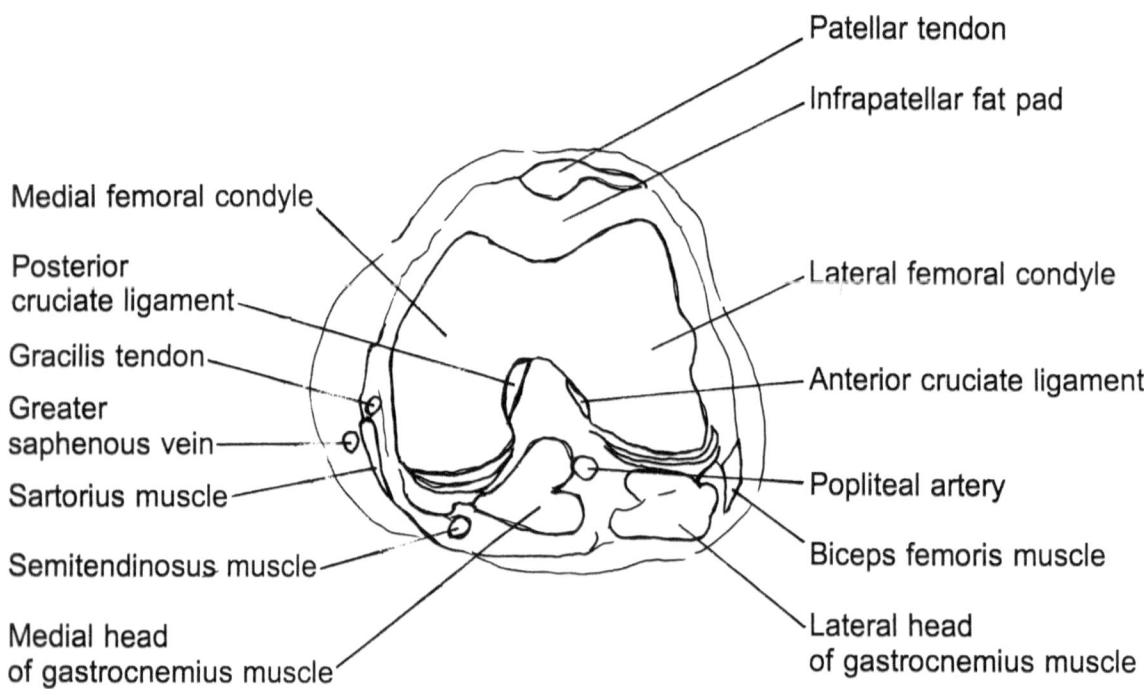

Patellar tendon

Infrapatellar fat pad

Medial femoral condyle

Posterior cruciate ligament

Gracilis tendon

Greater saphenous vein

Sartorius muscle

Semitendinosus muscle

Medial head of gastrocnemius muscle

Lateral femoral condyle

Anterior cruciate ligament

Popliteal artery

Biceps femoris muscle

Lateral head of gastrocnemius muscle

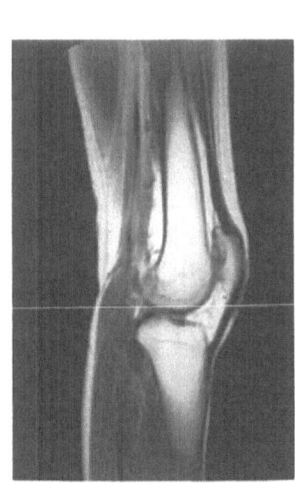
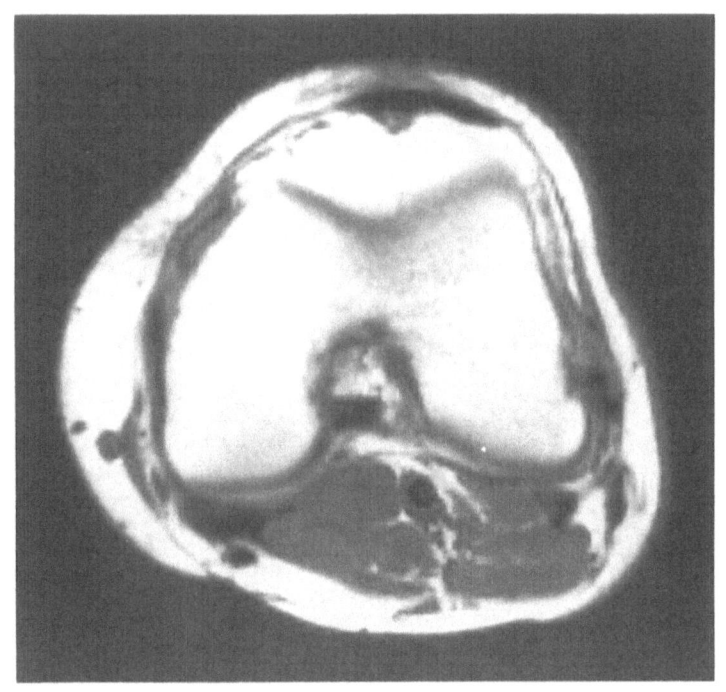

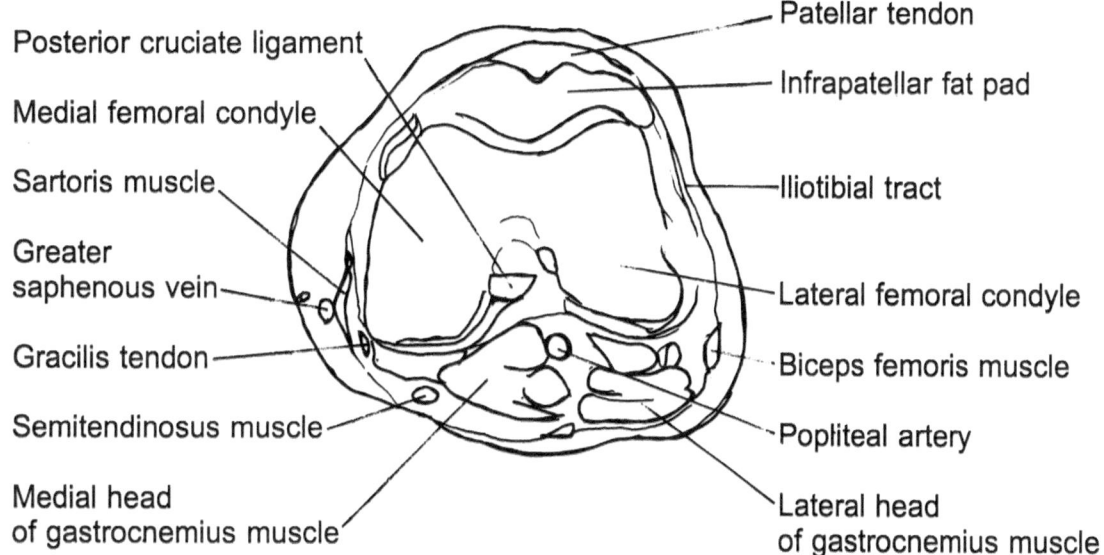

Posterior cruciate ligament

Medial femoral condyle

Sartoris muscle

Greater
saphenous vein

Gracilis tendon

Semitendinosus muscle

Medial head
of gastrocnemius muscle

Patellar tendon

Infrapatellar fat pad

Iliotibial tract

Lateral femoral condyle

Biceps femoris muscle

Popliteal artery

Lateral head
of gastrocnemius muscle

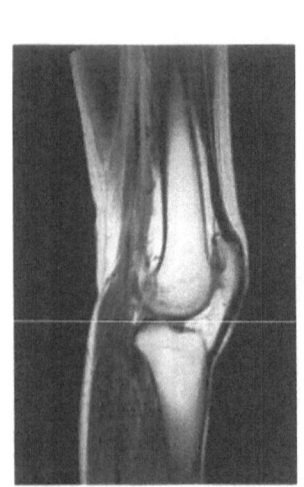

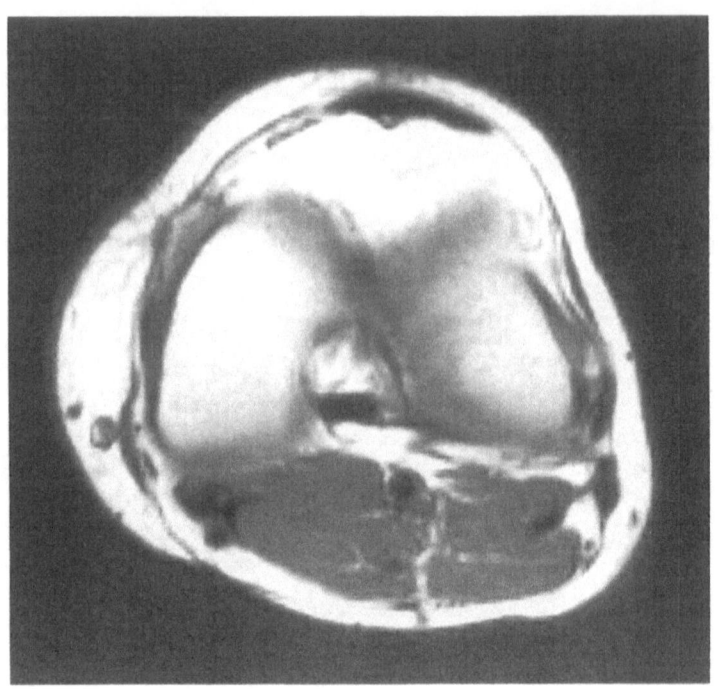

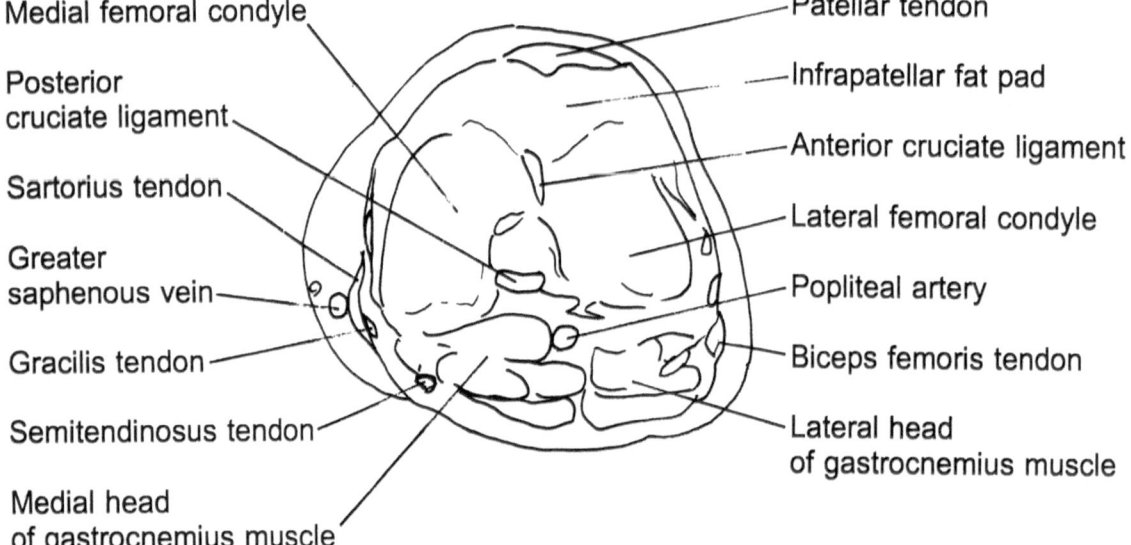

Medial femoral condyle

Posterior
cruciate ligament

Sartorius tendon

Greater
saphenous vein

Gracilis tendon

Semitendinosus tendon

Medial head
of gastrocnemius muscle

Patellar tendon

Infrapatellar fat pad

Anterior cruciate ligament

Lateral femoral condyle

Popliteal artery

Biceps femoris tendon

Lateral head
of gastrocnemius muscle

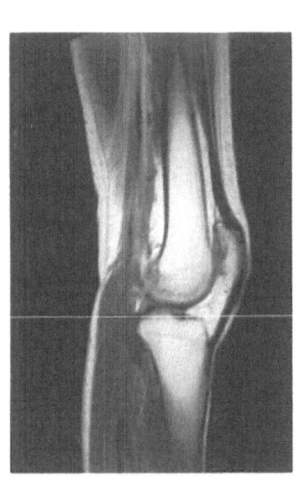
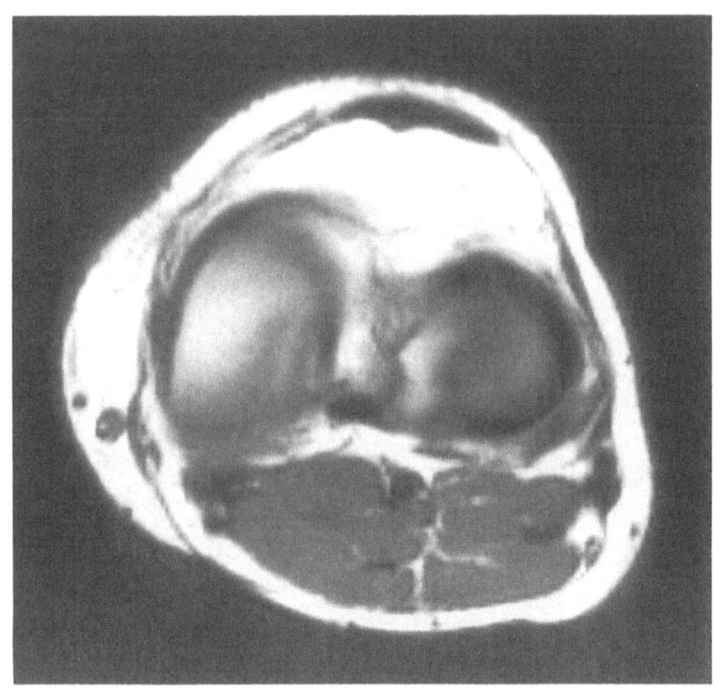

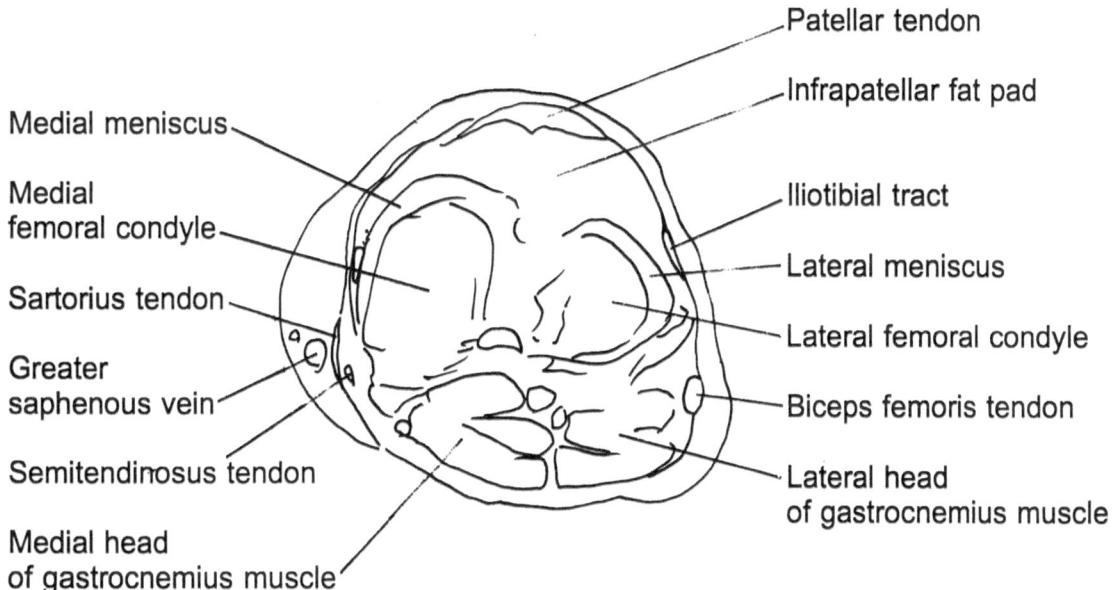

Medial meniscus

Medial
femoral condyle

Sartorius tendon

Greater
saphenous vein

Semitendinosus tendon

Medial head
of gastrocnemius muscle

Patellar tendon

Infrapatellar fat pad

Iliotibial tract

Lateral meniscus

Lateral femoral condyle

Biceps femoris tendon

Lateral head
of gastrocnemius muscle

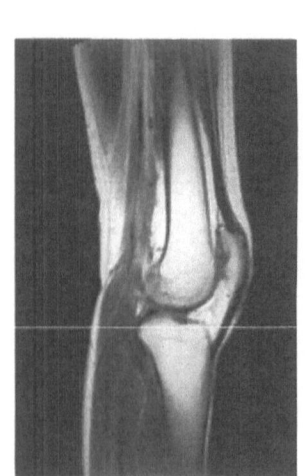

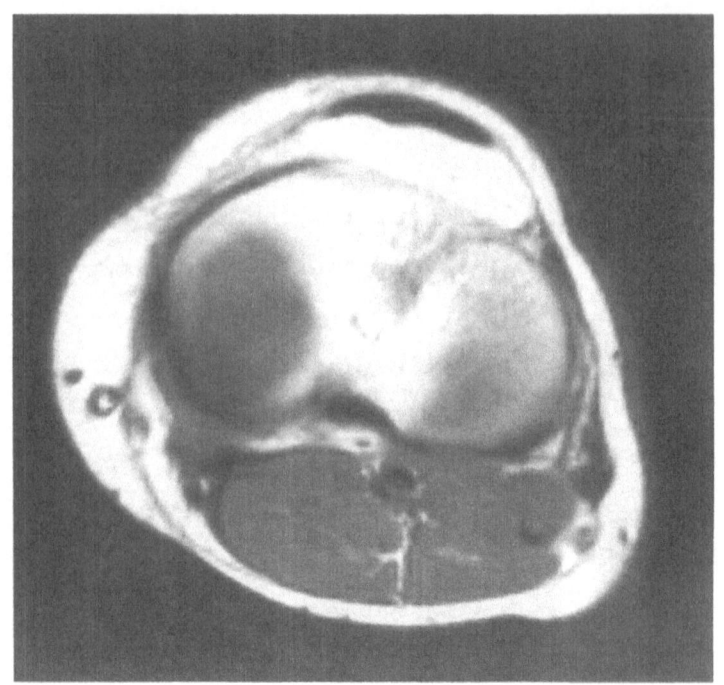

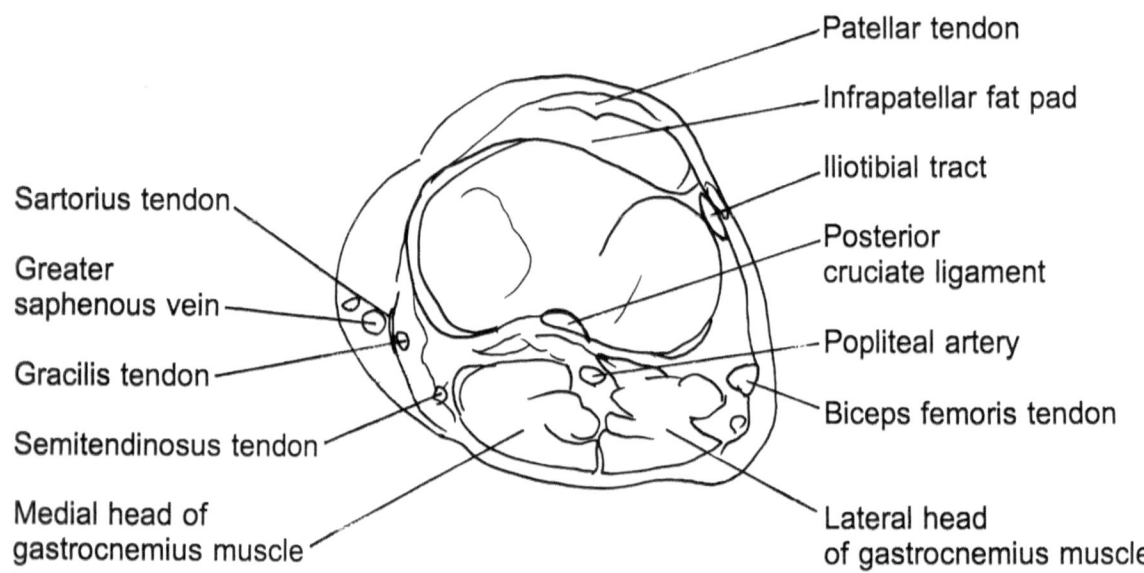

Patellar tendon

Infrapatellar fat pad

Iliotibial tract

Posterior
cruciate ligament

Popliteal artery

Biceps femoris tendon

Lateral head
of gastrocnemius muscle

Sartorius tendon

Greater
saphenous vein

Gracilis tendon

Semitendinosus tendon

Medial head of
gastrocnemius muscle

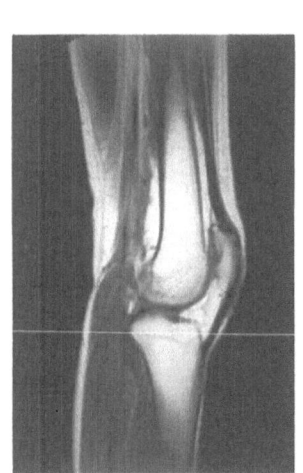

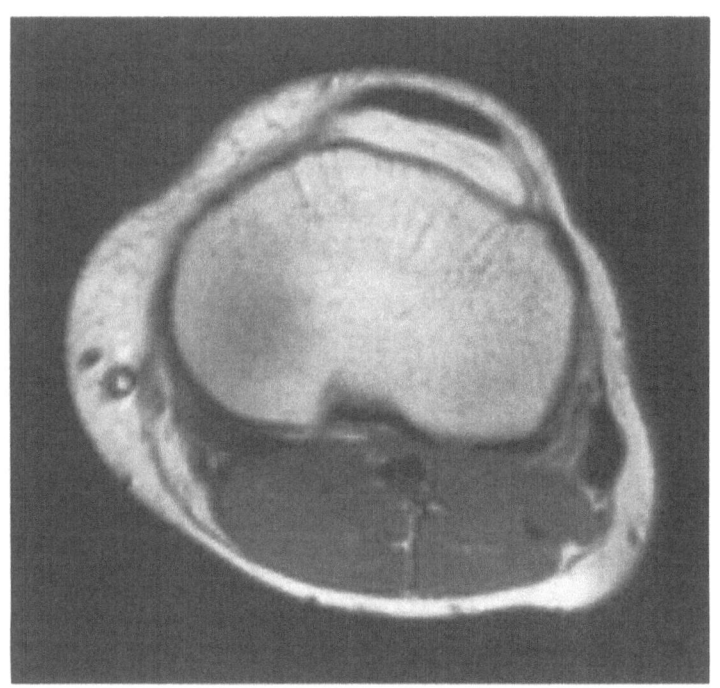

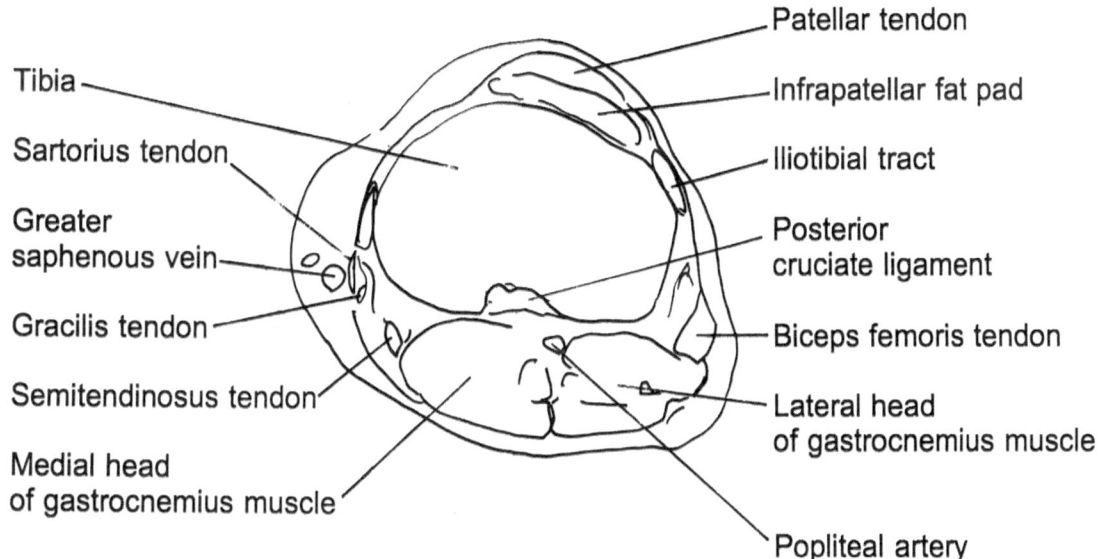

Tibia

Sartorius tendon

Greater
saphenous vein

Gracilis tendon

Semitendinosus tendon

Medial head
of gastrocnemius muscle

Patellar tendon

Infrapatellar fat pad

Iliotibial tract

Posterior
cruciate ligament

Biceps femoris tendon

Lateral head
of gastrocnemius muscle

Popliteal artery

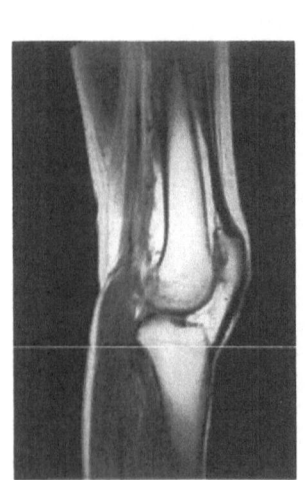

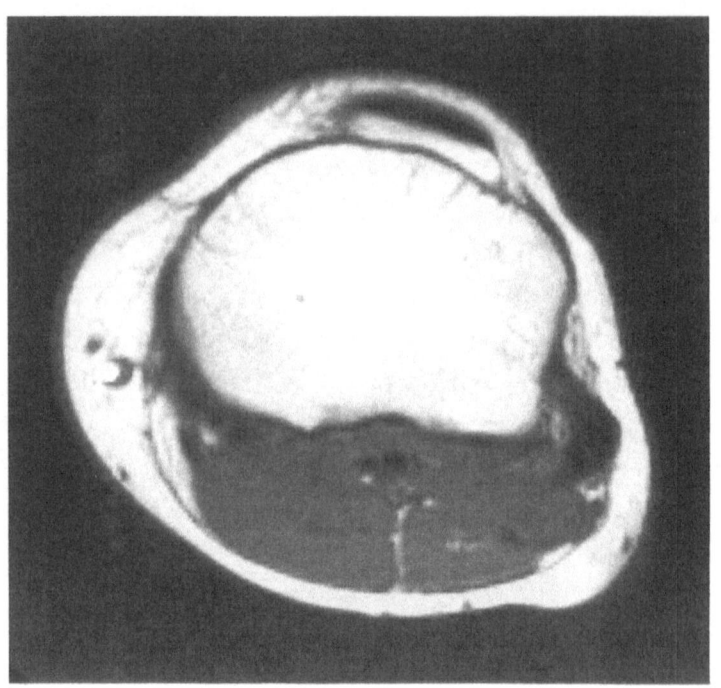

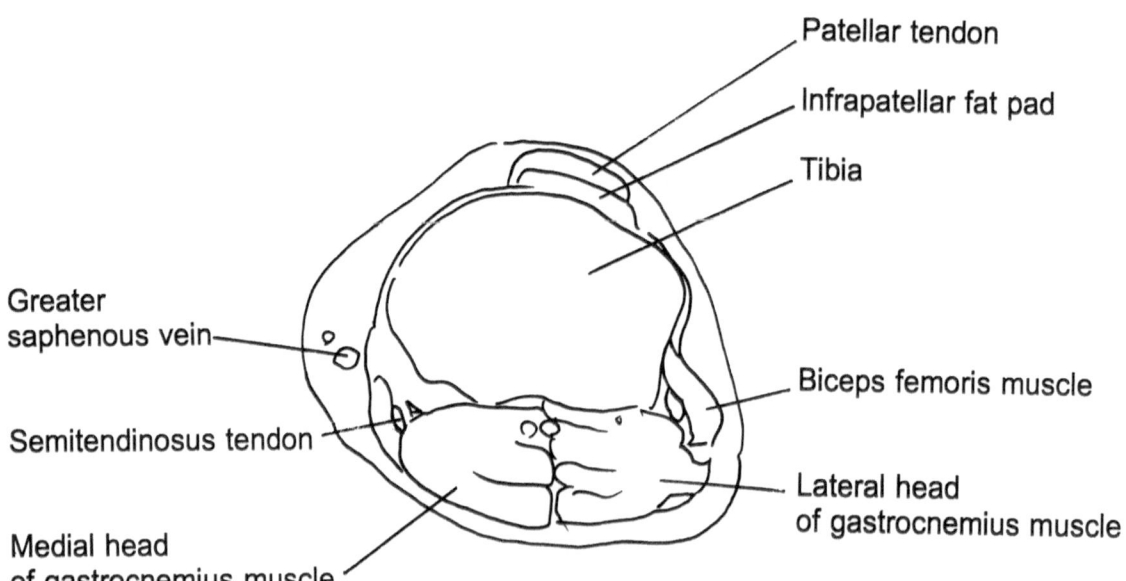

Patellar tendon

Infrapatellar fat pad

Tibia

Greater
saphenous vein

Biceps femoris muscle

Semitendinosus tendon

Lateral head
of gastrocnemius muscle

Medial head
of gastrocnemius muscle

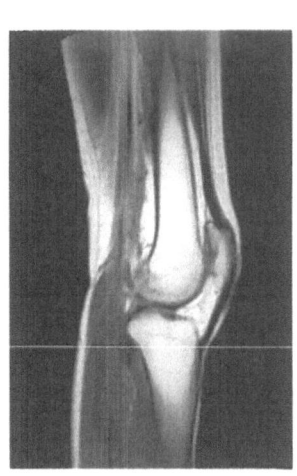

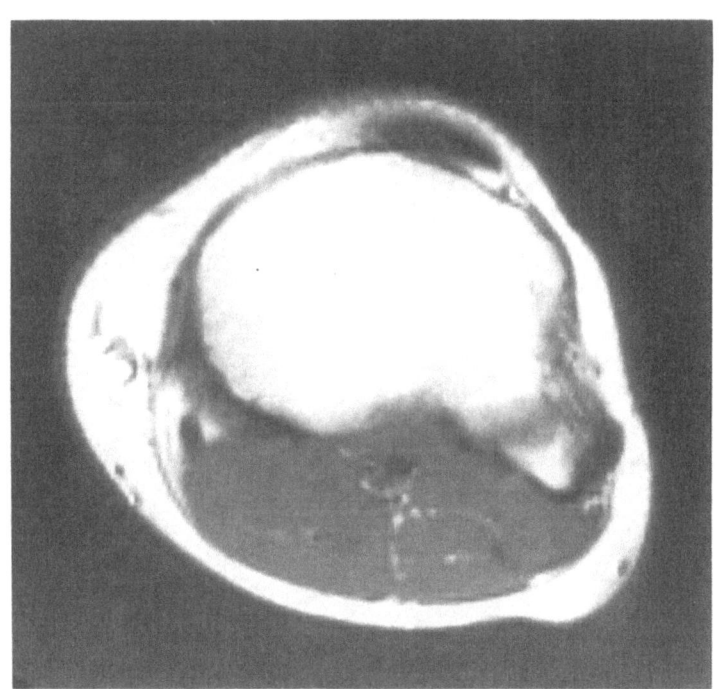

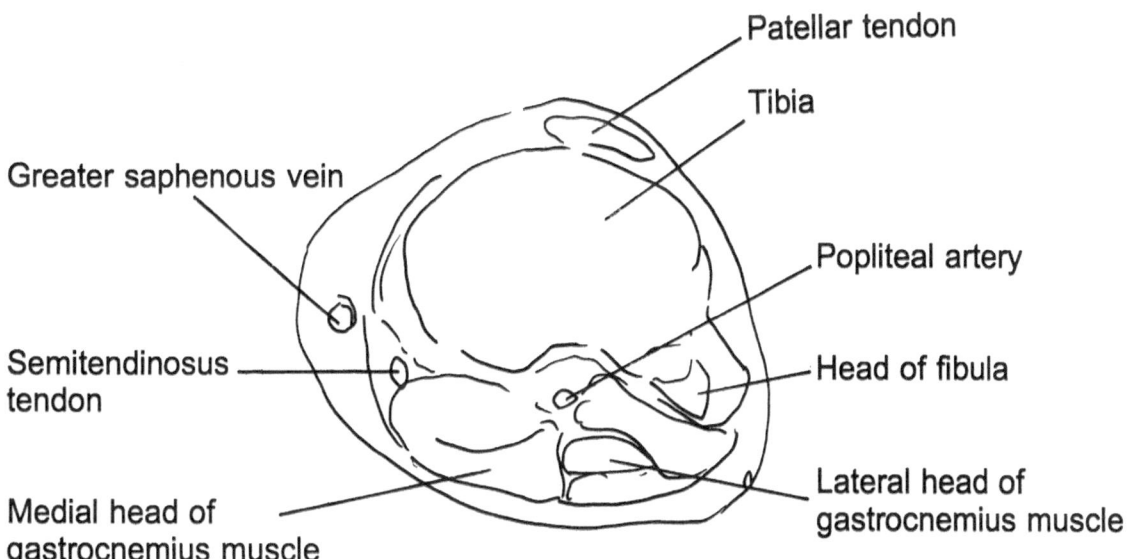

Patellar tendon

Tibia

Greater saphenous vein

Popliteal artery

Semitendinosus tendon

Head of fibula

Medial head of gastrocnemius muscle

Lateral head of gastrocnemius muscle

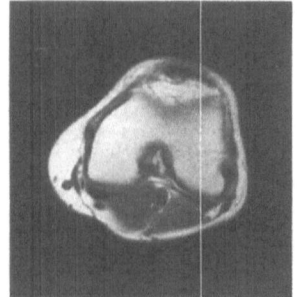

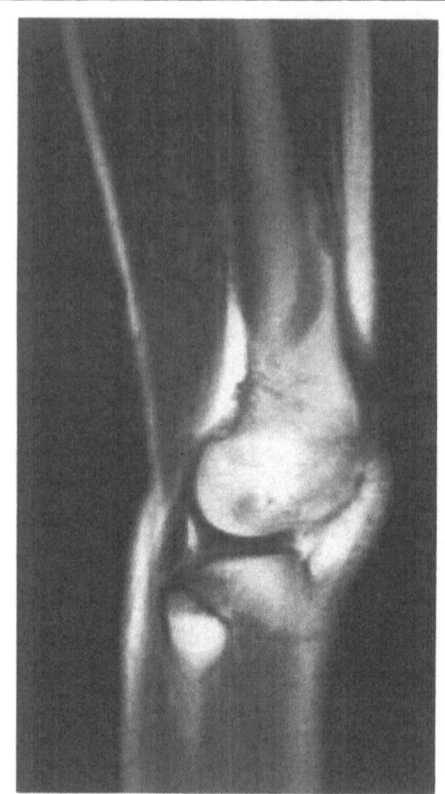

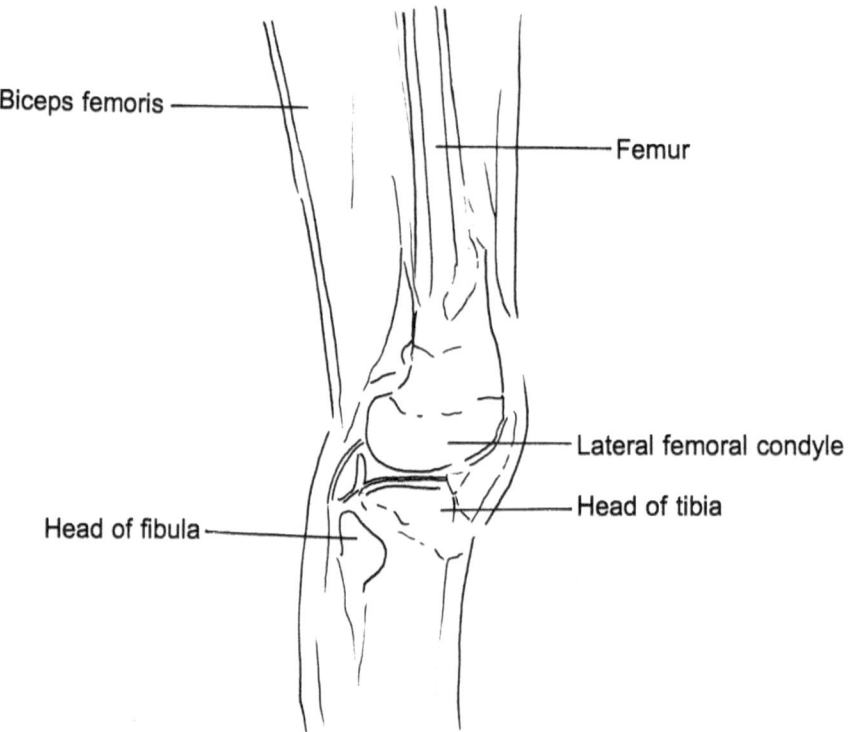

Biceps femoris

Femur

Lateral femoral condyle

Head of tibia

Head of fibula

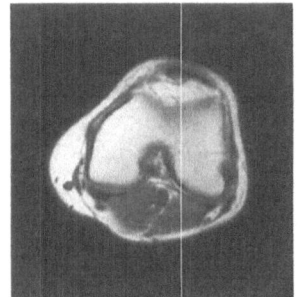

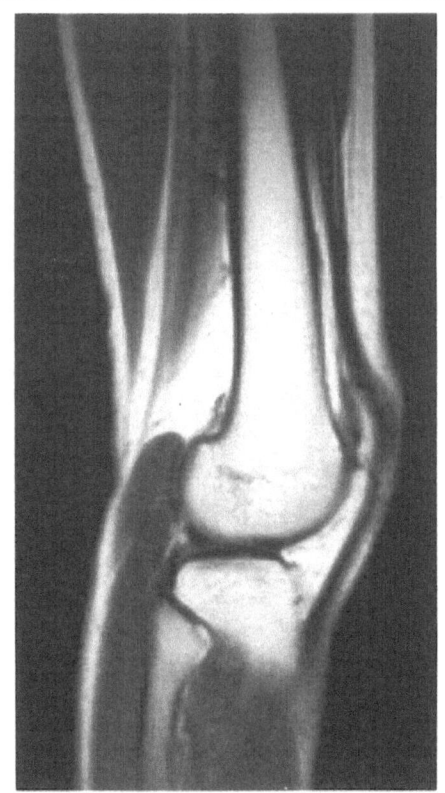

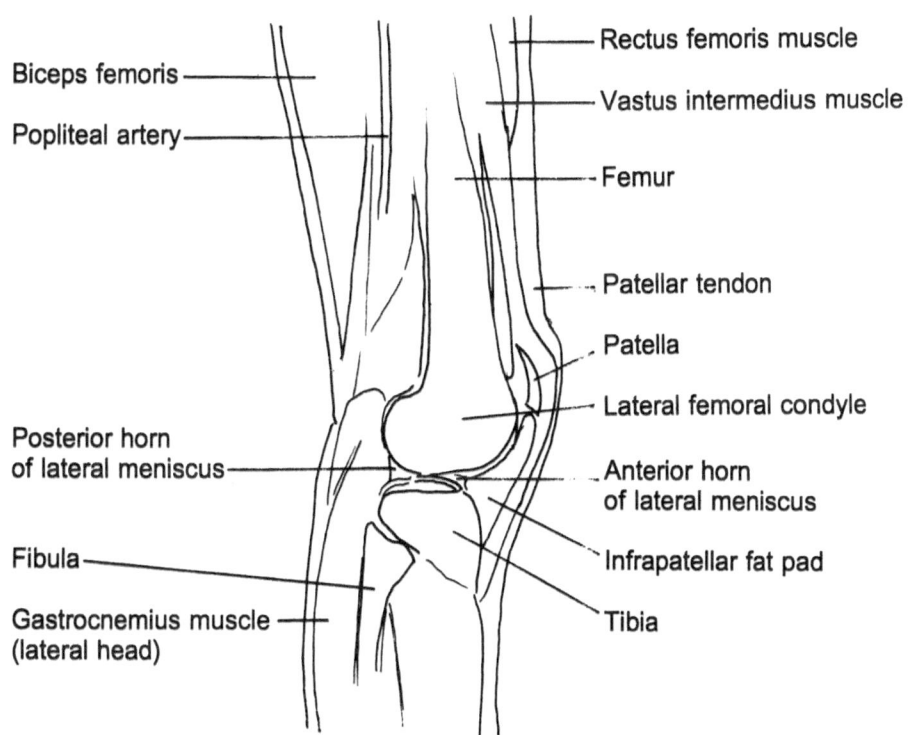

Biceps femoris

Popliteal artery

Posterior horn
of lateral meniscus

Fibula

Gastrocnemius muscle
(lateral head)

Rectus femoris muscle

Vastus intermedius muscle

Femur

Patellar tendon

Patella

Lateral femoral condyle

Anterior horn
of lateral meniscus

Infrapatellar fat pad

Tibia

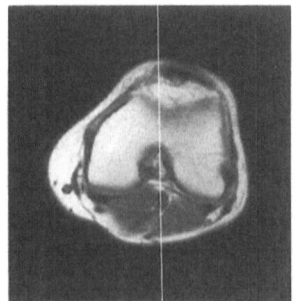

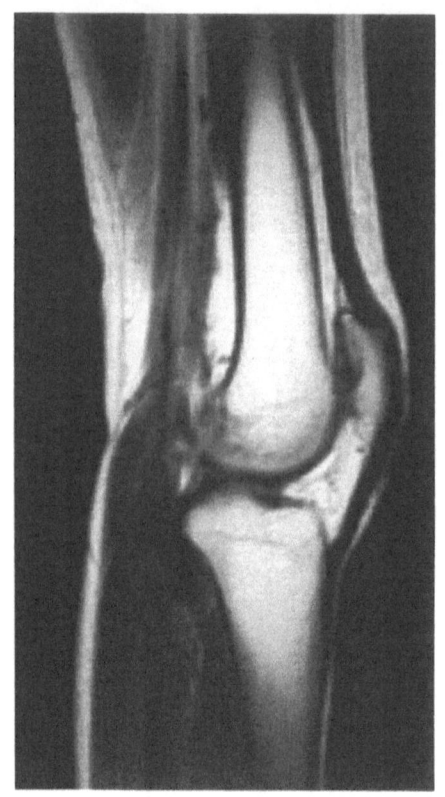

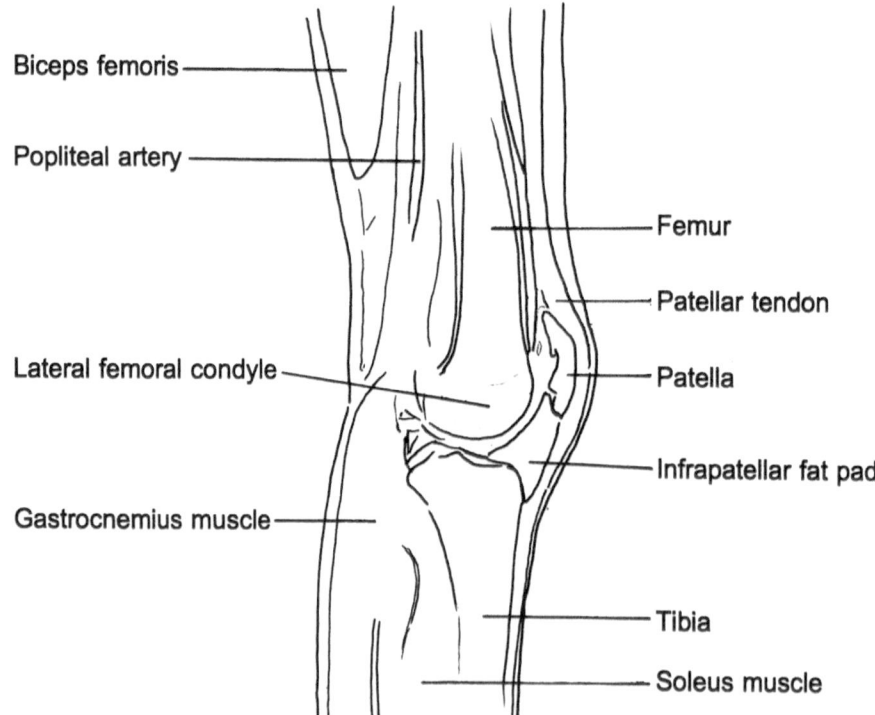

Biceps femoris

Popliteal artery

Lateral femoral condyle

Gastrocnemius muscle

Femur

Patellar tendon

Patella

Infrapatellar fat pad

Tibia

Soleus muscle

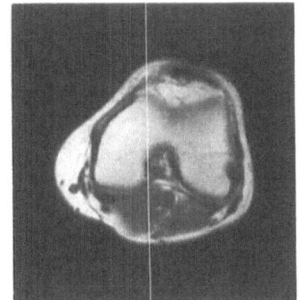

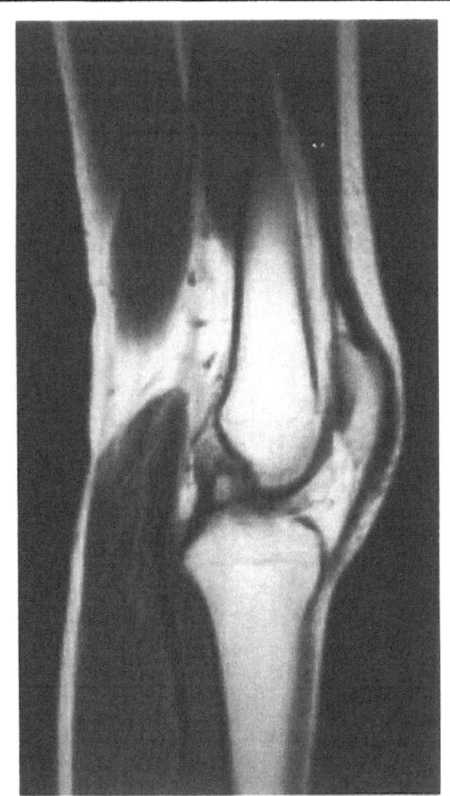

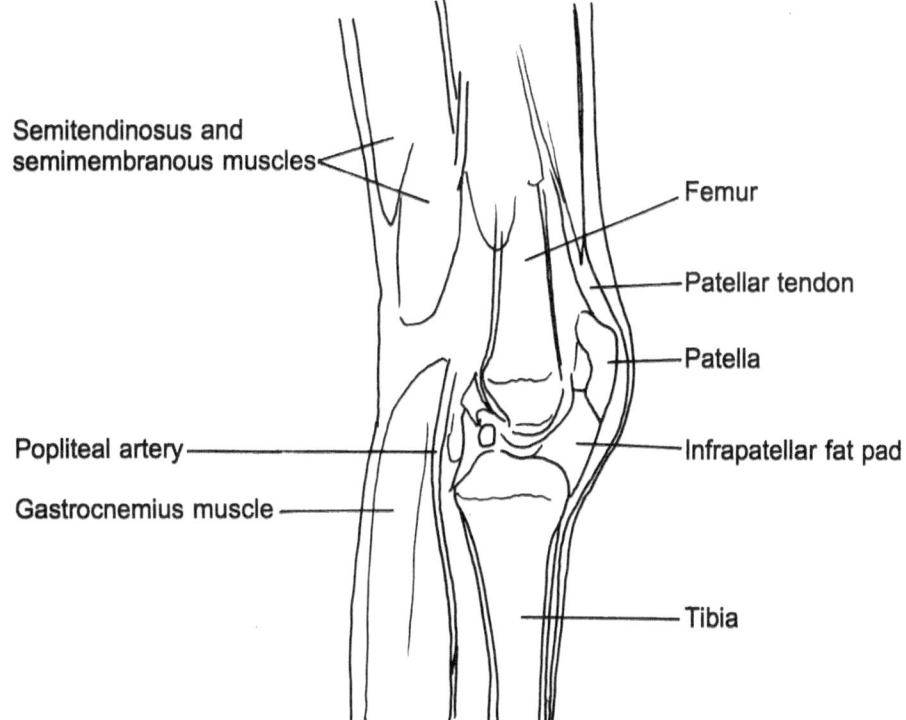

Semitendinosus and
semimembranous muscles

Femur

Patellar tendon

Patella

Popliteal artery

Infrapatellar fat pad

Gastrocnemius muscle

Tibia

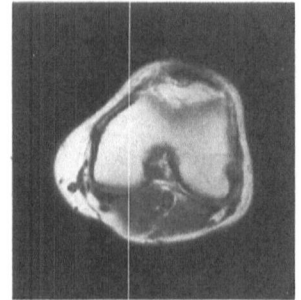

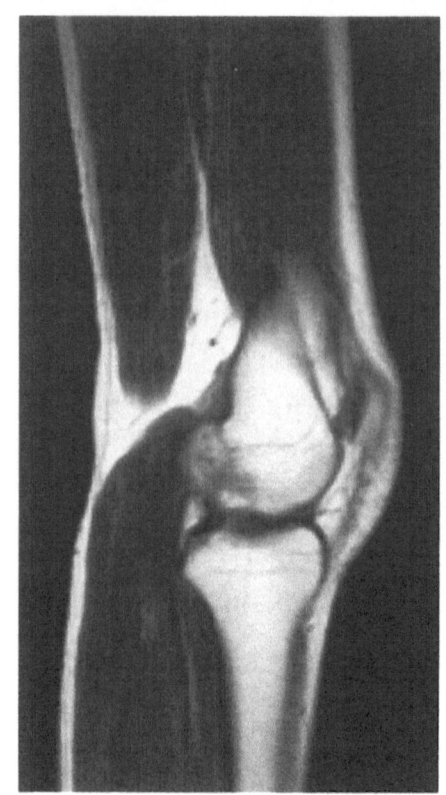

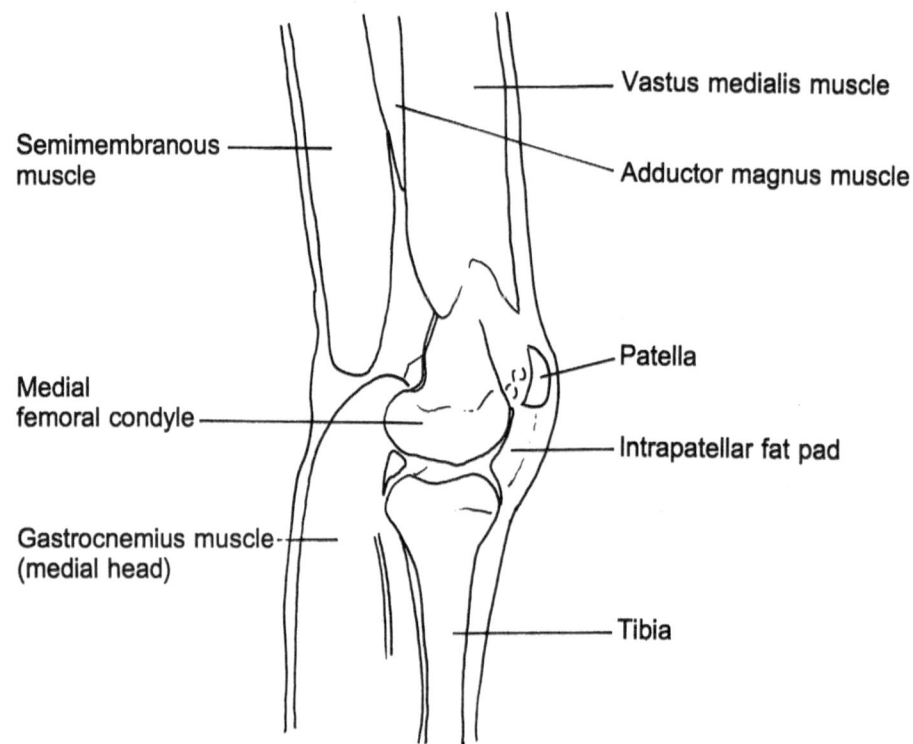

Semimembranous muscle

Vastus medialis muscle

Adductor magnus muscle

Medial femoral condyle

Patella

Intrapatellar fat pad

Gastrocnemius muscle (medial head)

Tibia

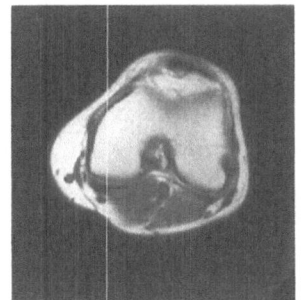

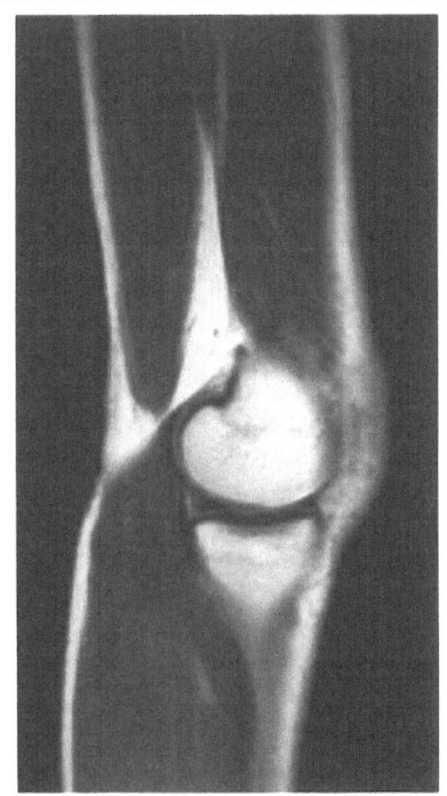

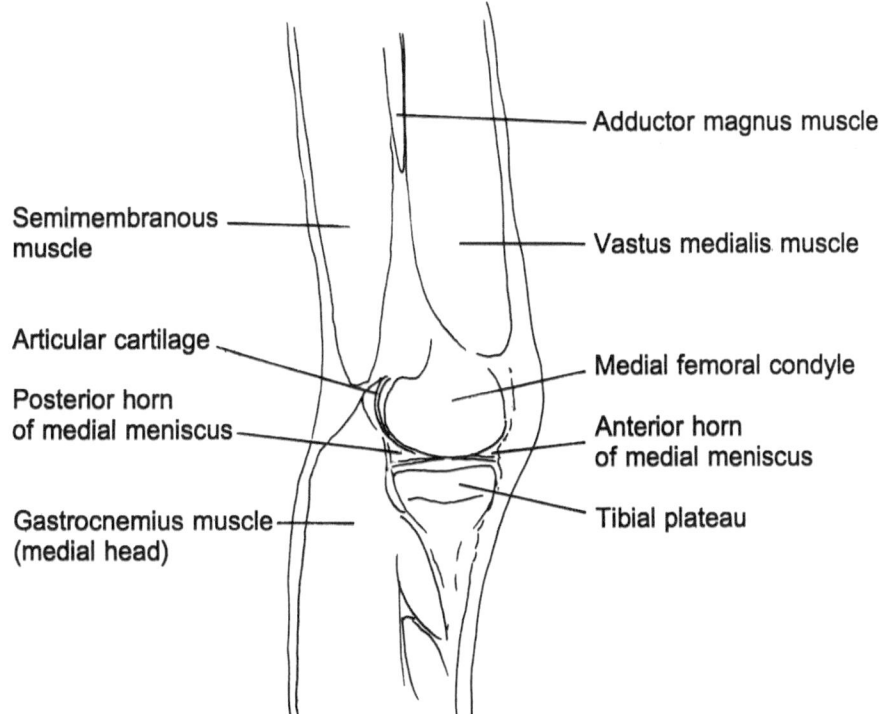

Adductor magnus muscle

Semimembranous muscle

Vastus medialis muscle

Articular cartilage

Medial femoral condyle

Posterior horn of medial meniscus

Anterior horn of medial meniscus

Gastrocnemius muscle (medial head)

Tibial plateau

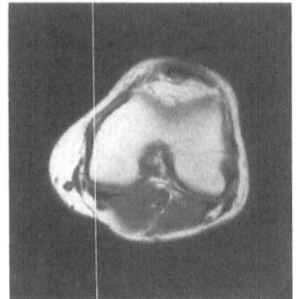

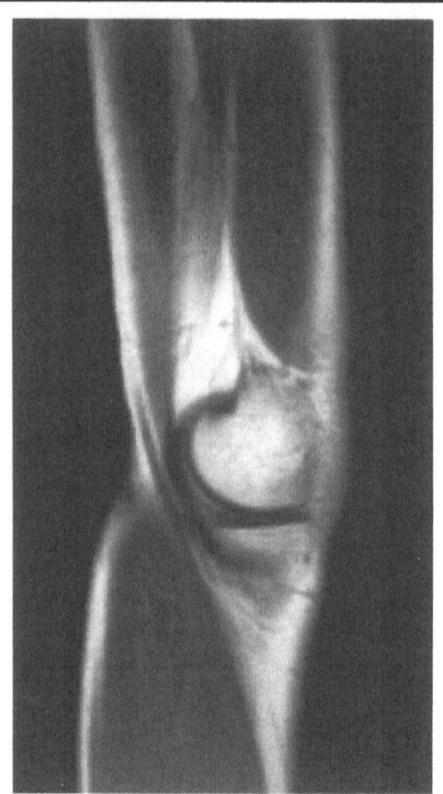

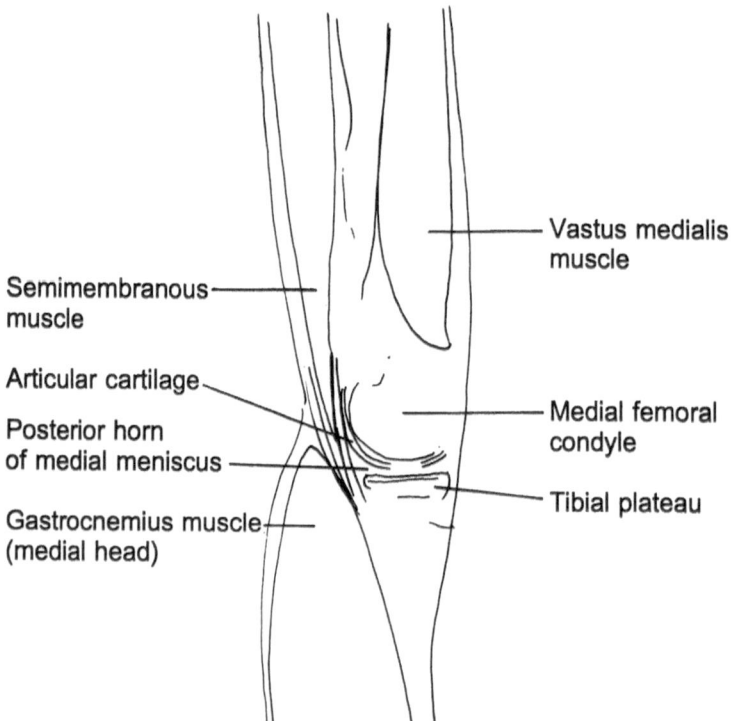

Vastus medialis muscle

Semimembranous muscle

Articular cartilage

Posterior horn of medial meniscus

Gastrocnemius muscle (medial head)

Medial femoral condyle

Tibial plateau

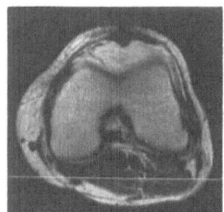

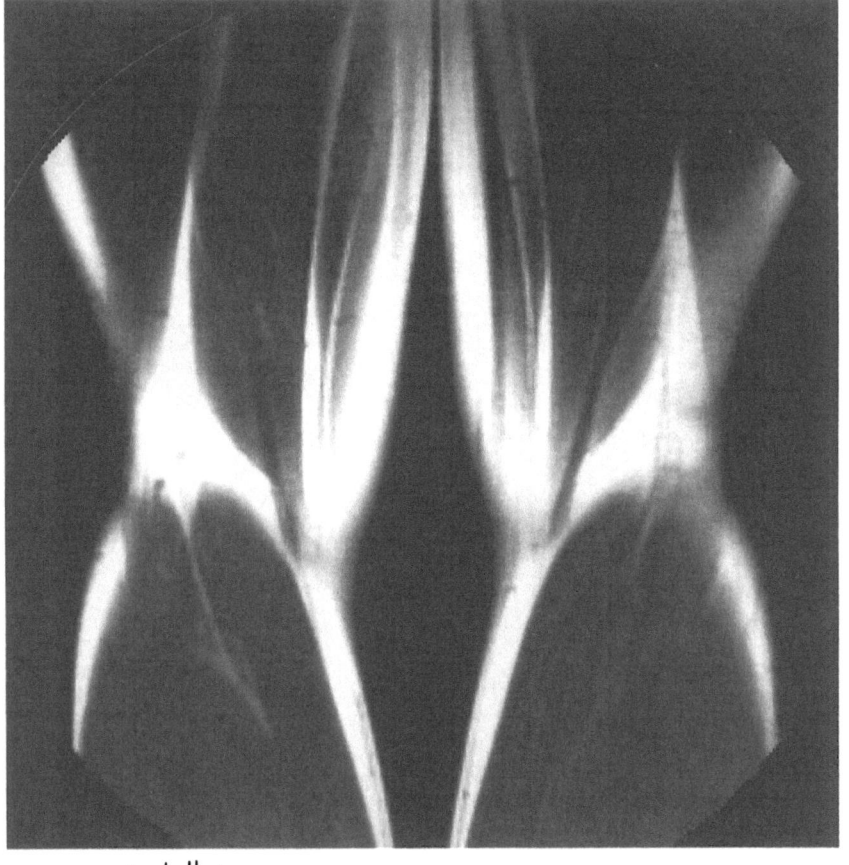

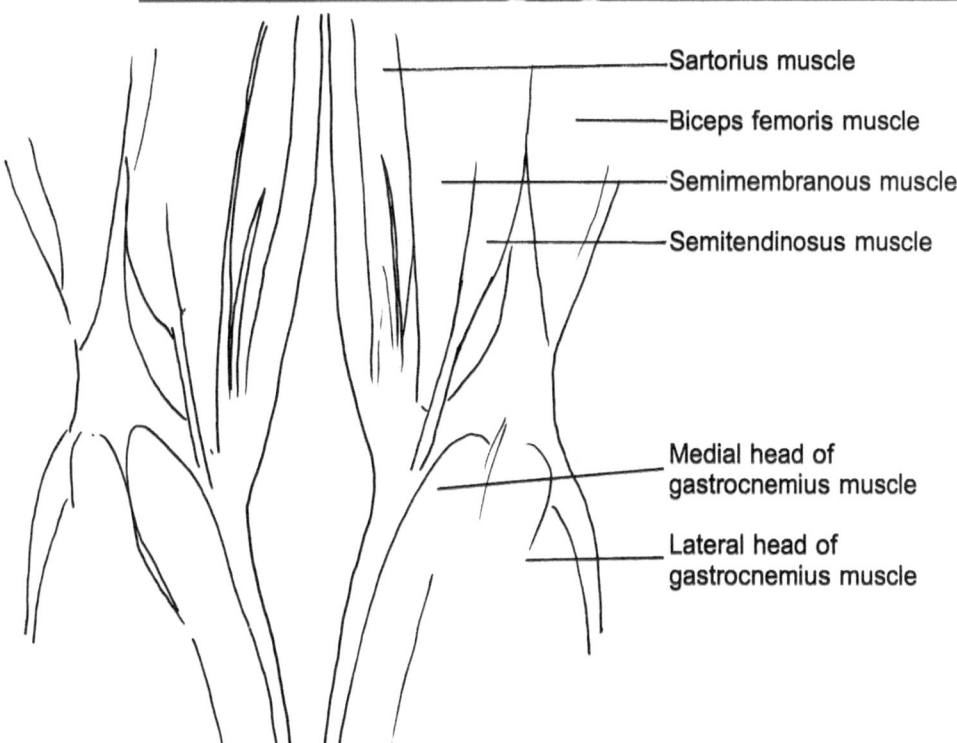

Sartorius muscle

Biceps femoris muscle

Semimembranous muscle

Semitendinosus muscle

Medial head of
gastrocnemius muscle

Lateral head of
gastrocnemius muscle

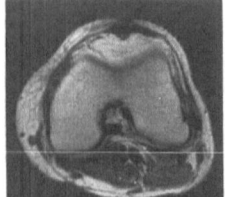

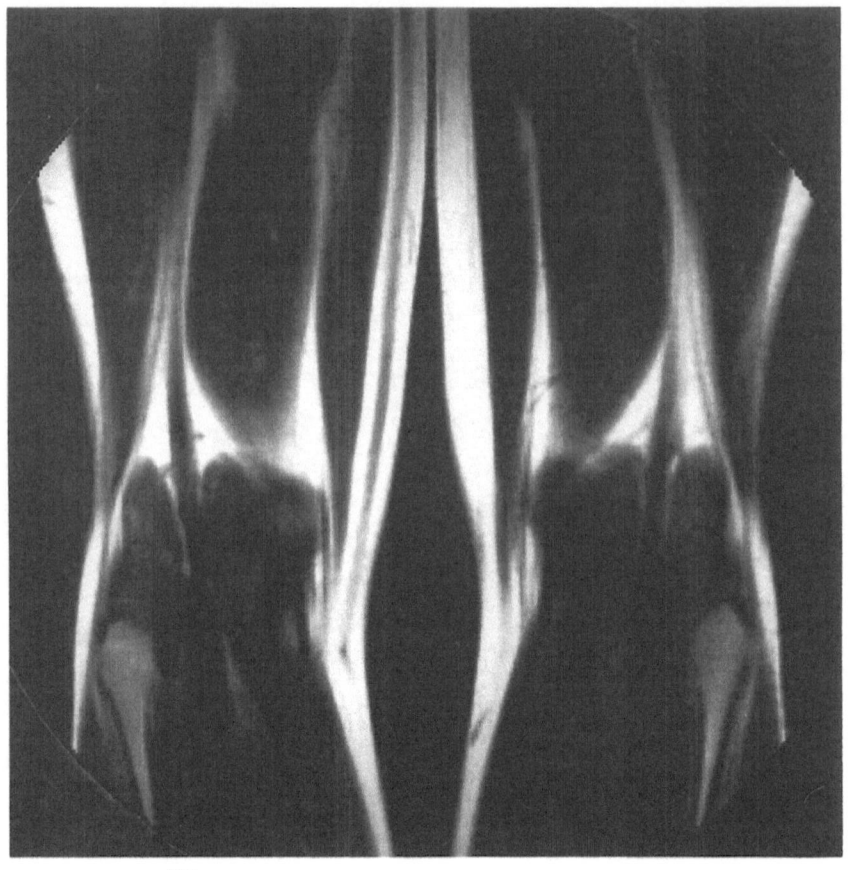

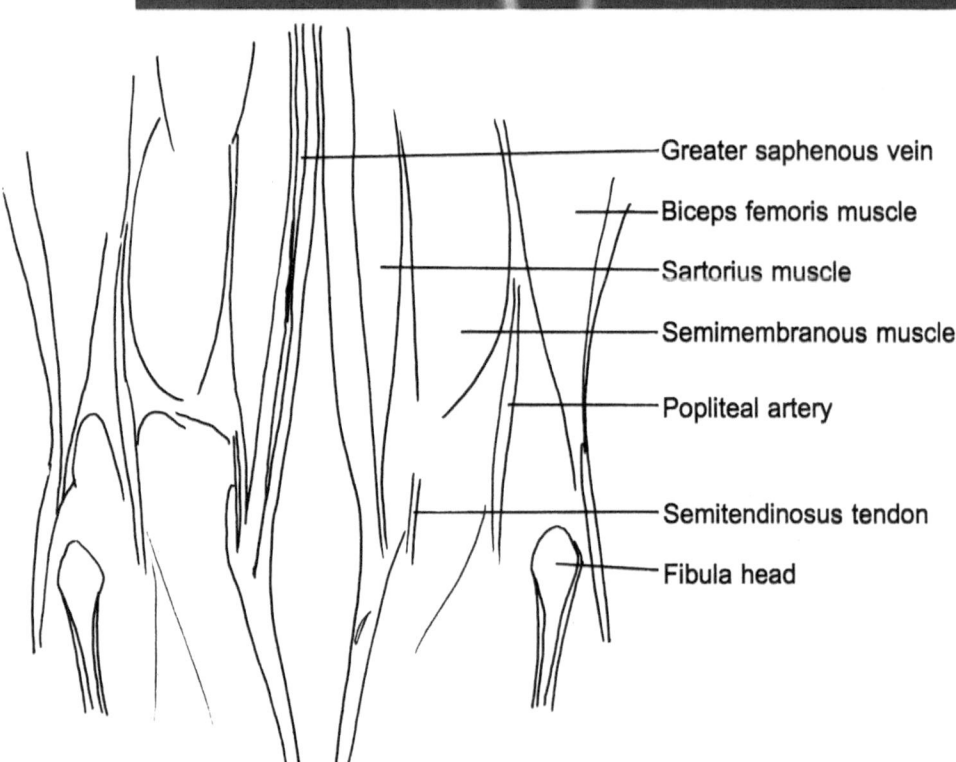

Greater saphenous vein

Biceps femoris muscle

Sartorius muscle

Semimembranous muscle

Popliteal artery

Semitendinosus tendon

Fibula head

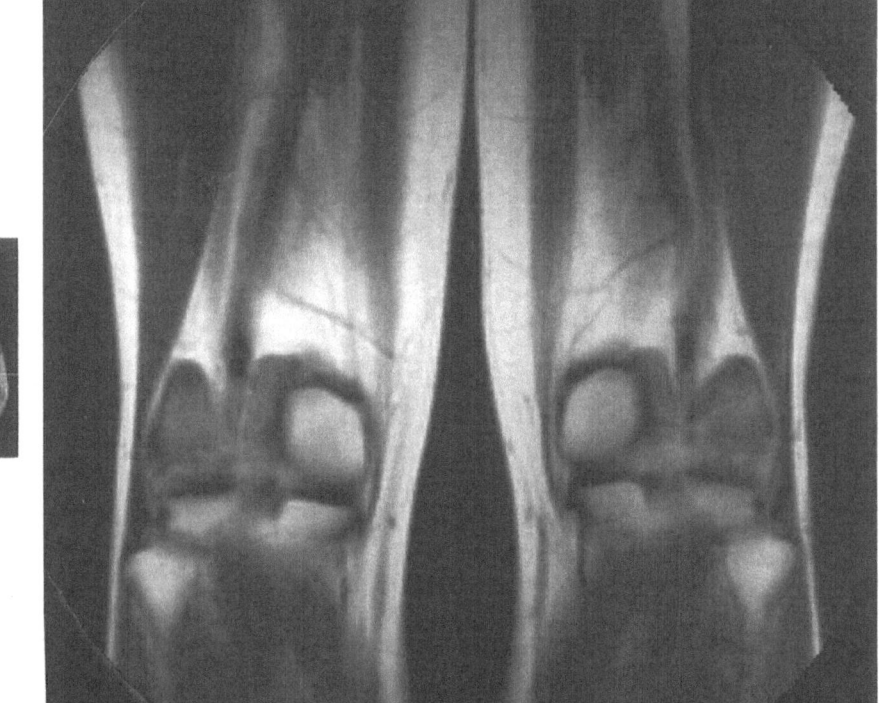

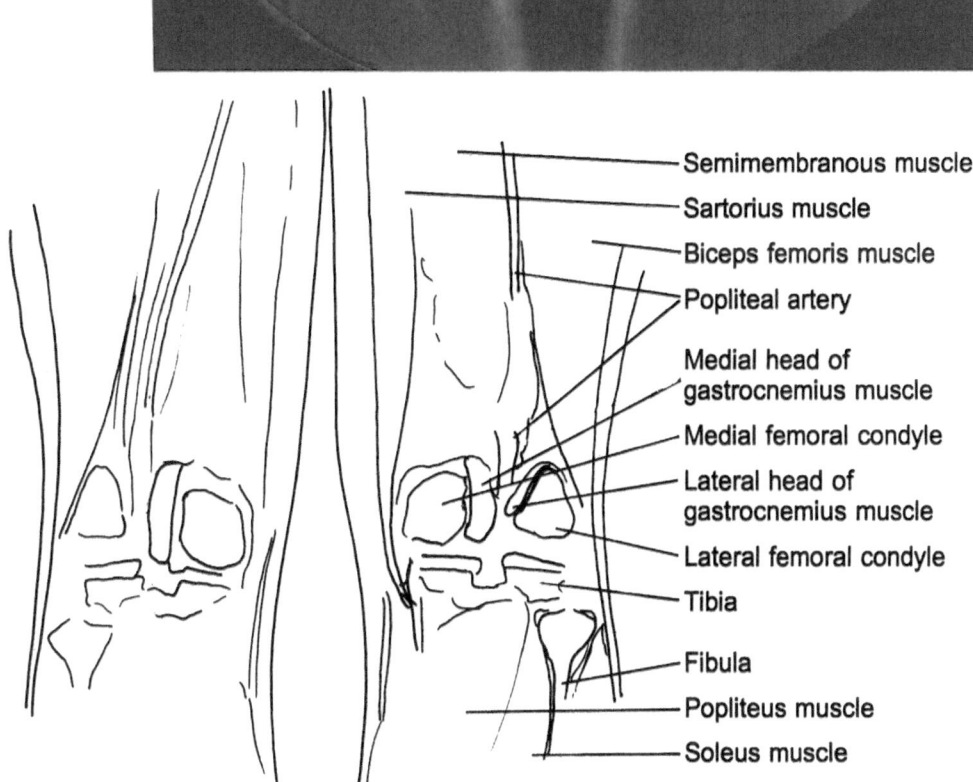

Semimembranous muscle

Sartorius muscle

Biceps femoris muscle

Popliteal artery

Medial head of
gastrocnemius muscle

Medial femoral condyle

Lateral head of
gastrocnemius muscle

Lateral femoral condyle

Tibia

Fibula

Popliteus muscle

Soleus muscle

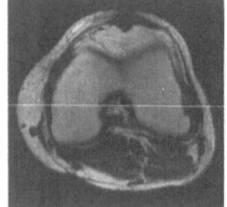

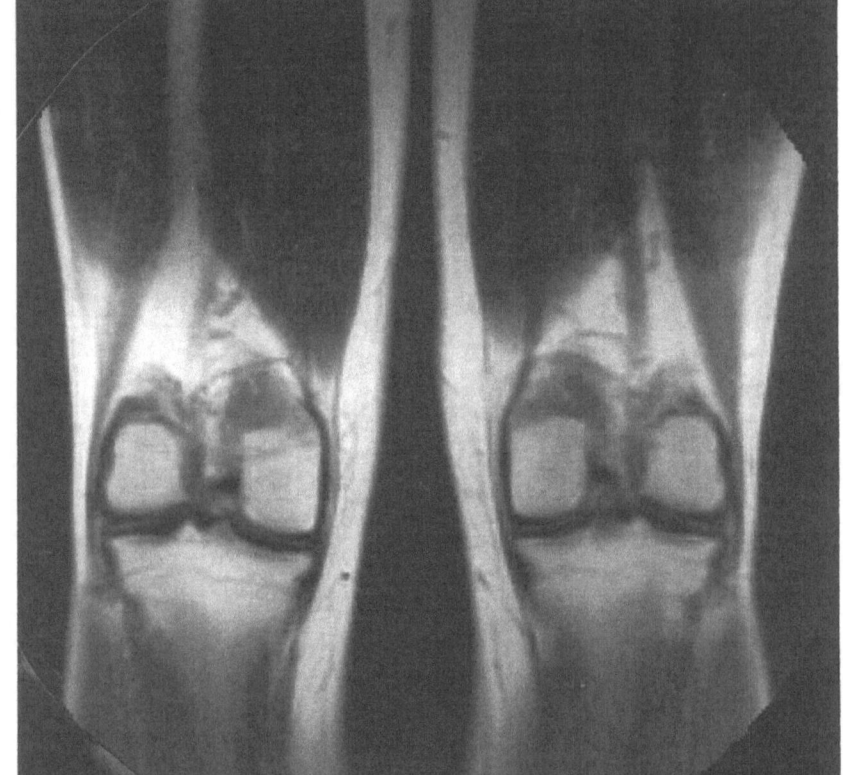

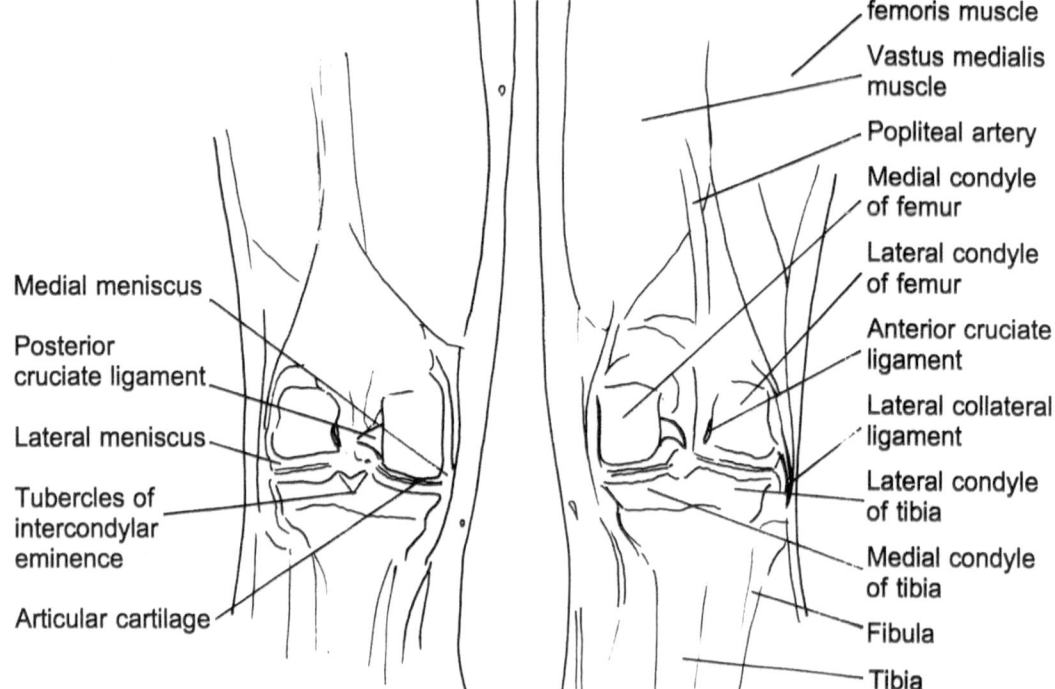

Biceps
femoris muscle

Vastus medialis
muscle

Popliteal artery

Medial condyle
of femur

Lateral condyle
of femur

Anterior cruciate
ligament

Lateral collateral
ligament

Lateral condyle
of tibia

Medial condyle
of tibia

Fibula

Tibia

Medial meniscus

Posterior
cruciate ligament

Lateral meniscus

Tubercles of
intercondylar
eminence

Articular cartilage

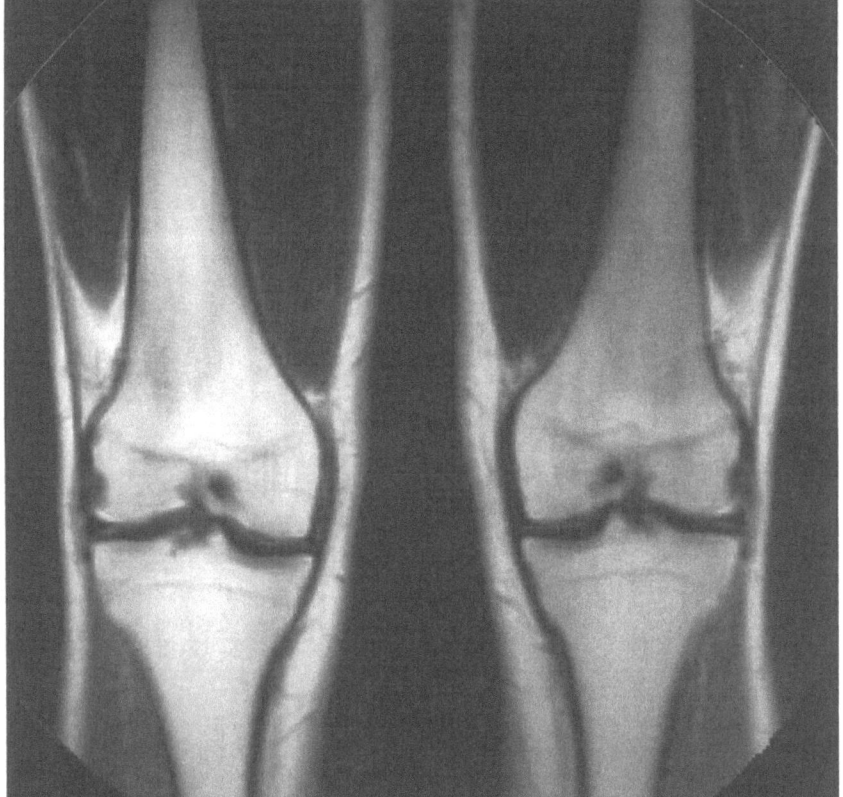

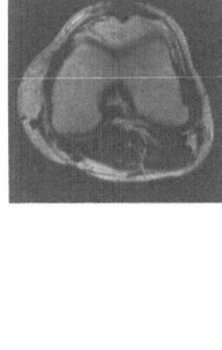

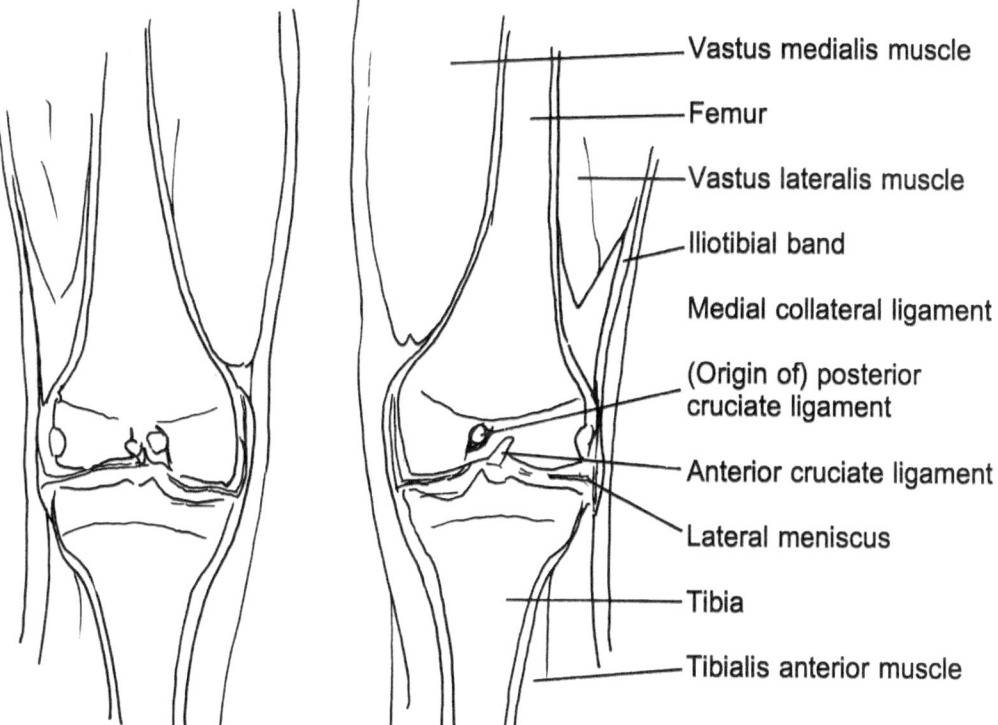

Vastus medialis muscle

Femur

Vastus lateralis muscle

Iliotibial band

Medial collateral ligament

(Origin of) posterior cruciate ligament

Anterior cruciate ligament

Lateral meniscus

Tibia

Tibialis anterior muscle

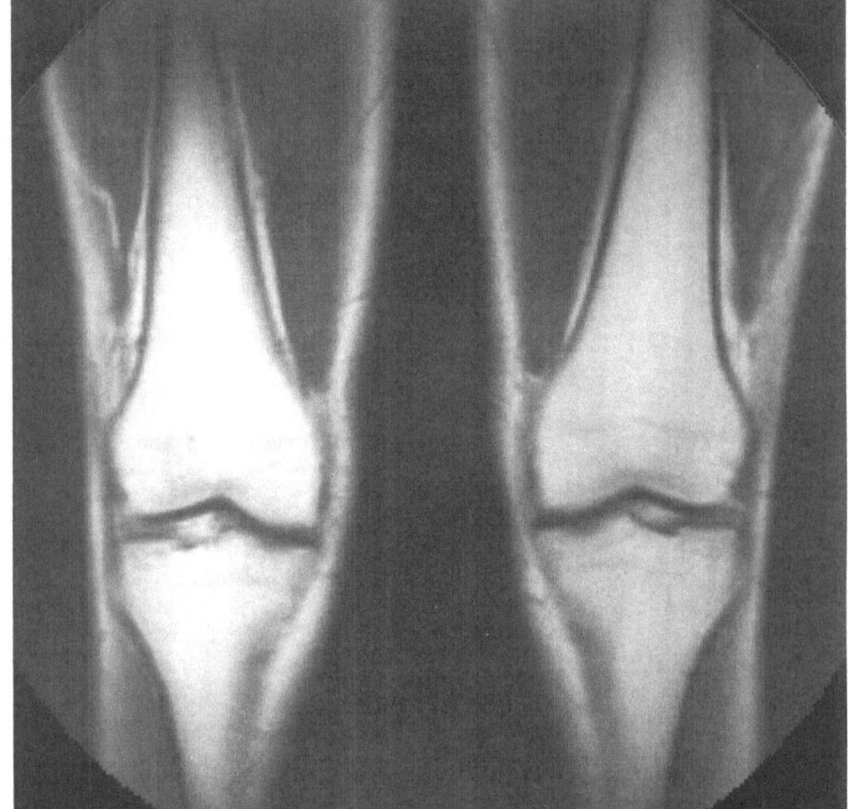
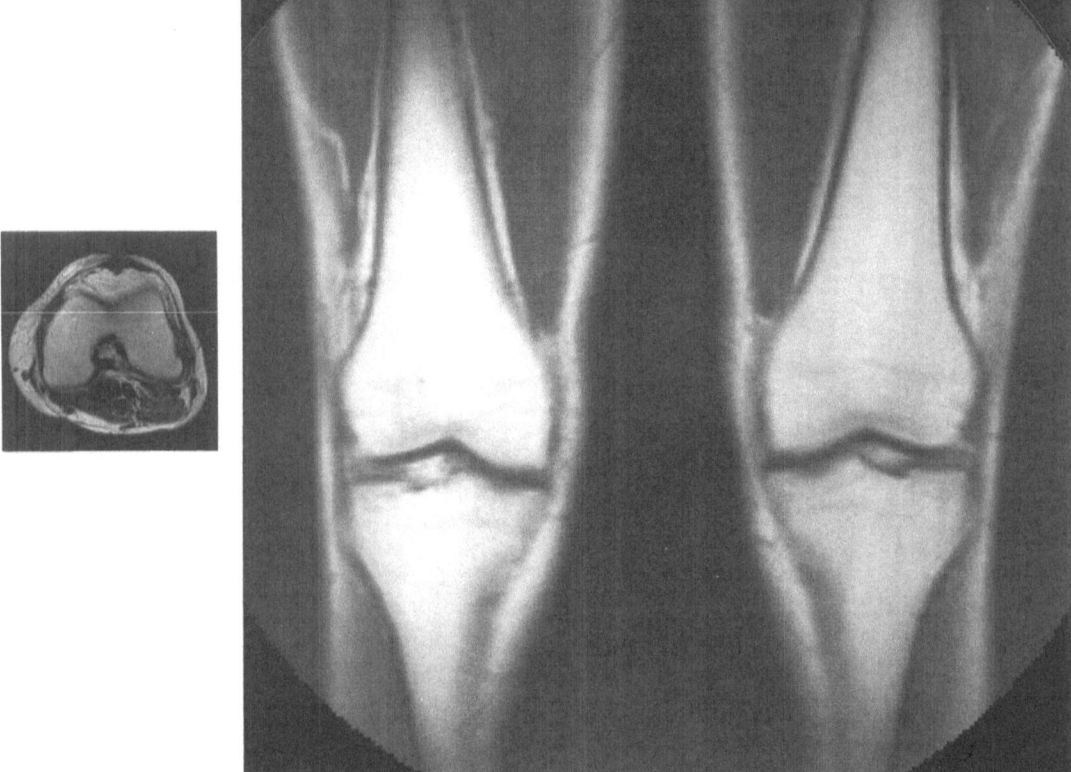
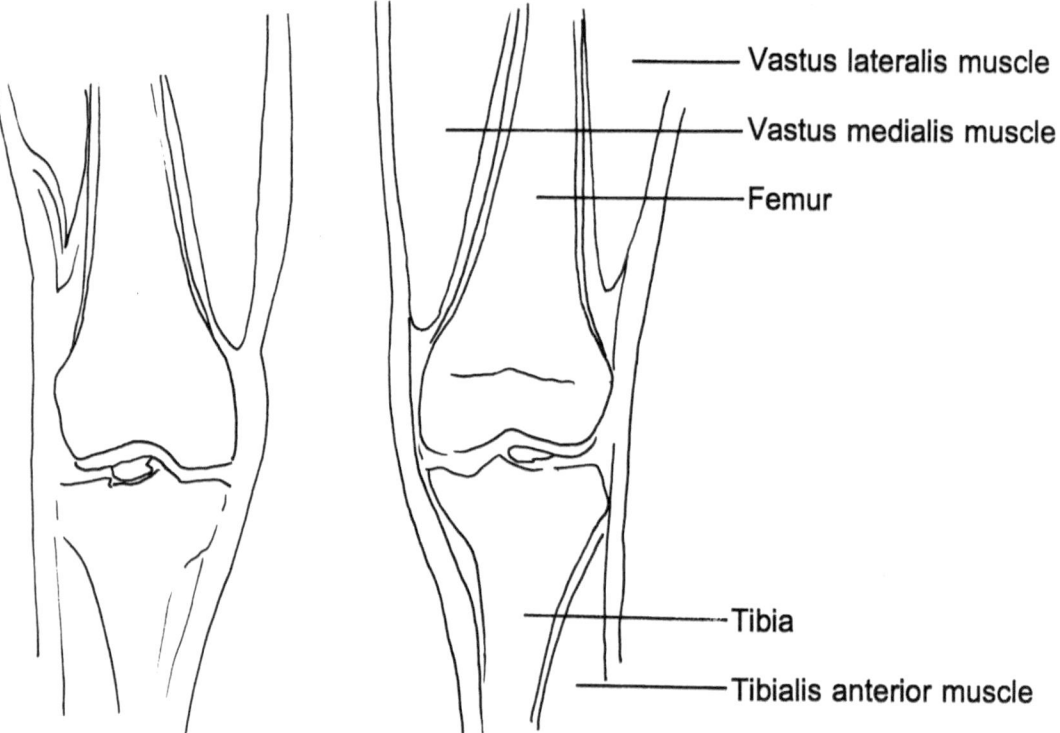

Vastus lateralis muscle

Vastus medialis muscle

Femur

Tibia

Tibialis anterior muscle

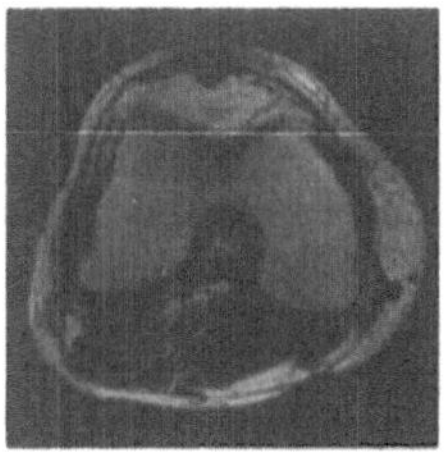

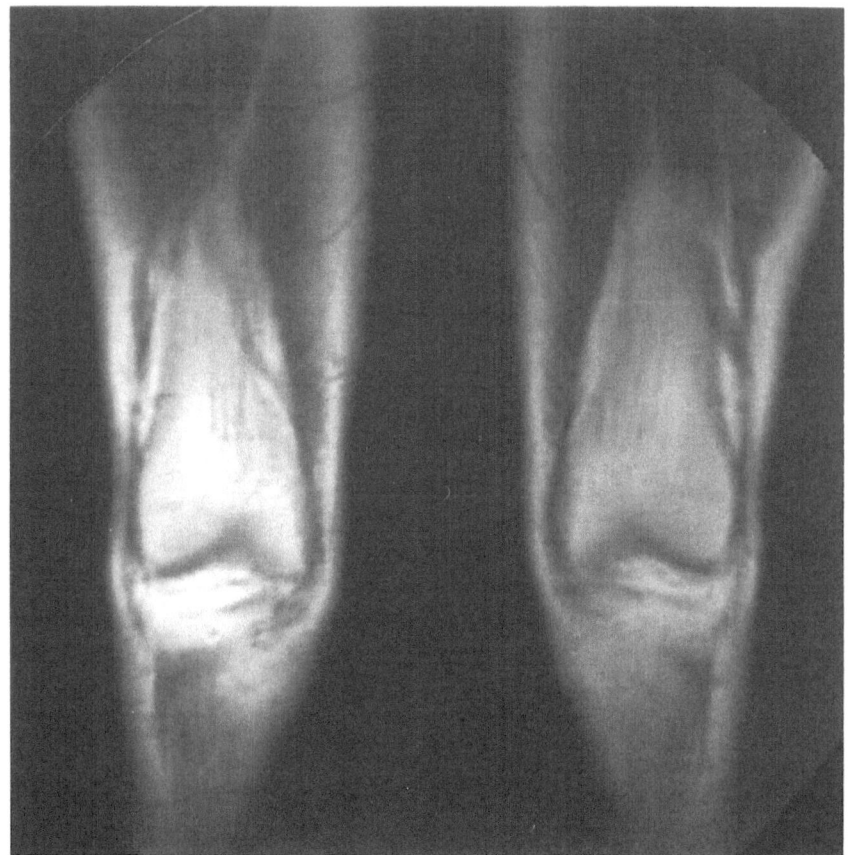

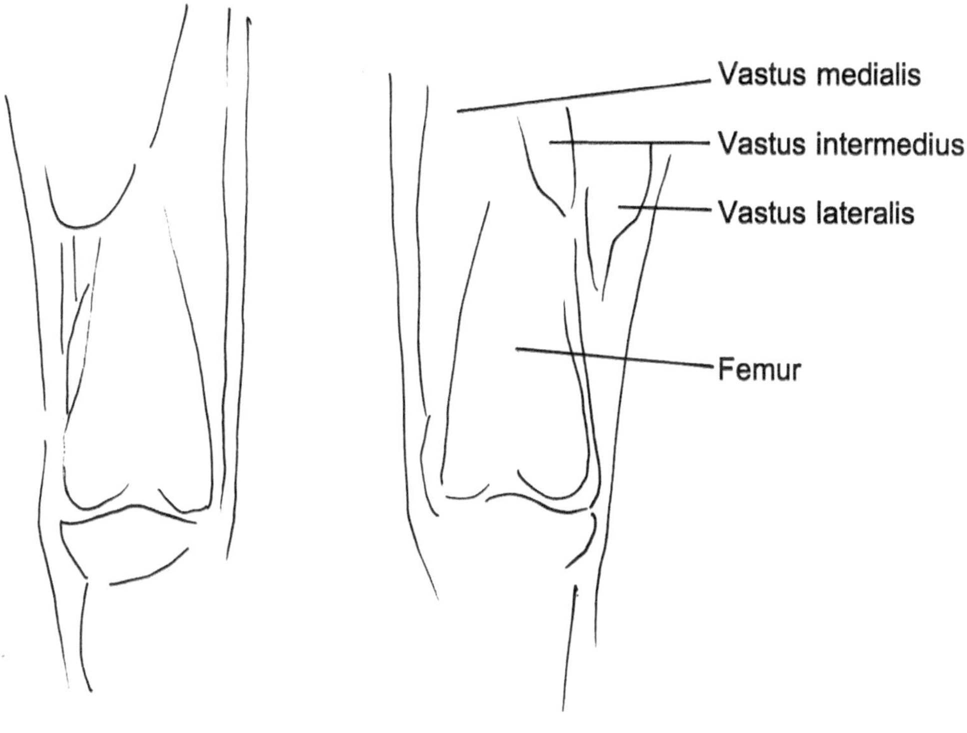

Vastus medialis

Vastus intermedius

Vastus lateralis

Femur

FOOT

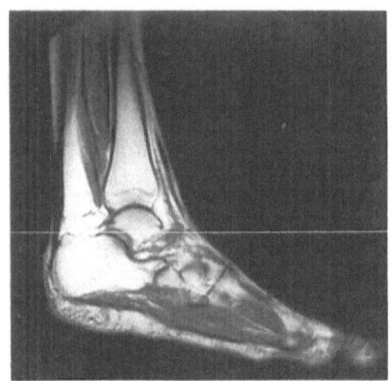
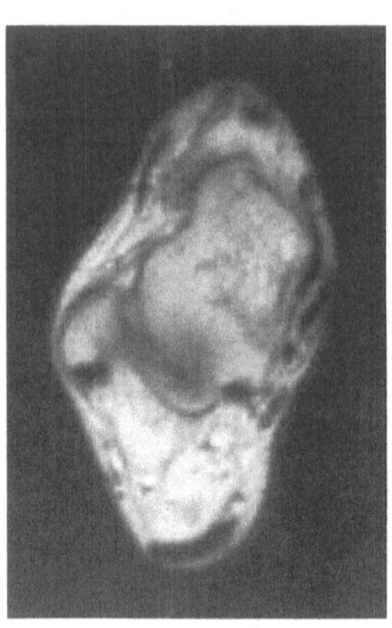

Extensor digitorum longus

Peroneus tertius

Extensor hallucis longus

Lateral malleolus

Talus

Deltoid ligament

Tibialis posterior

Peronei brevis and longus

Flexor digitorum longus

Flexor hallucis longus tendon

Calcaneus

Posterior talofibular ligament

Achilles tendon

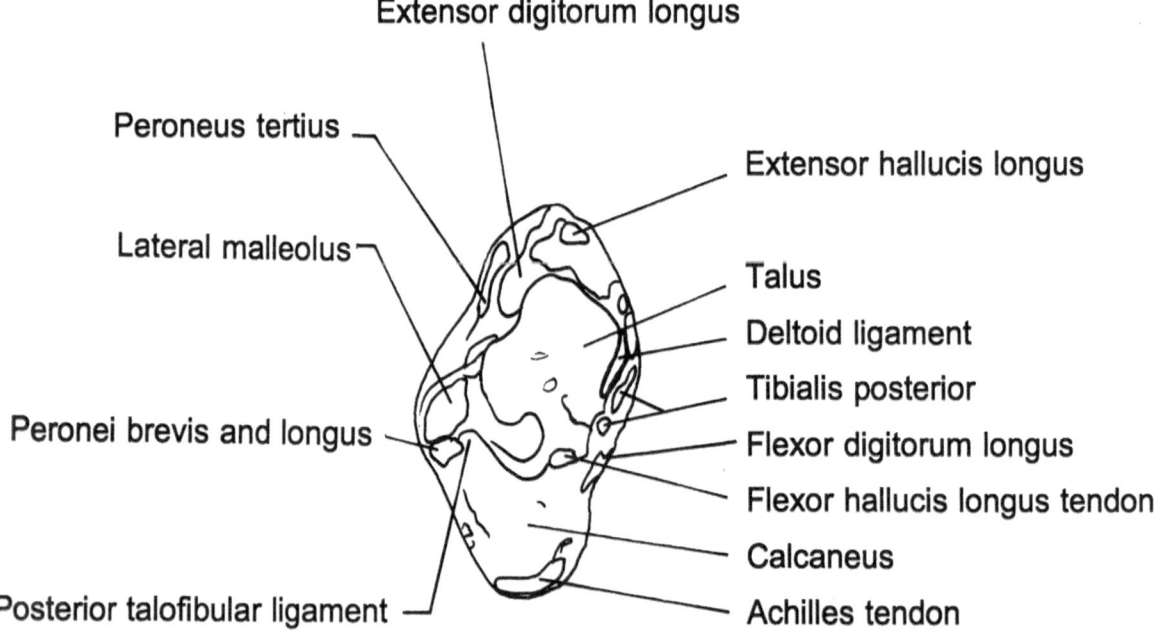

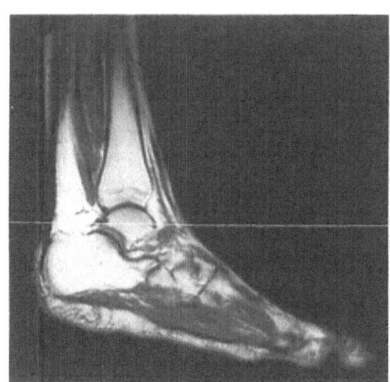
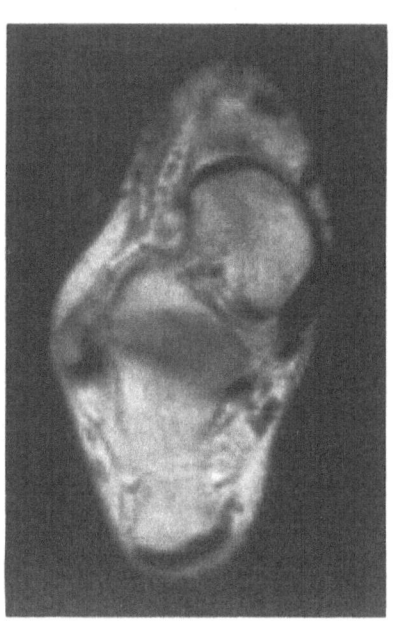

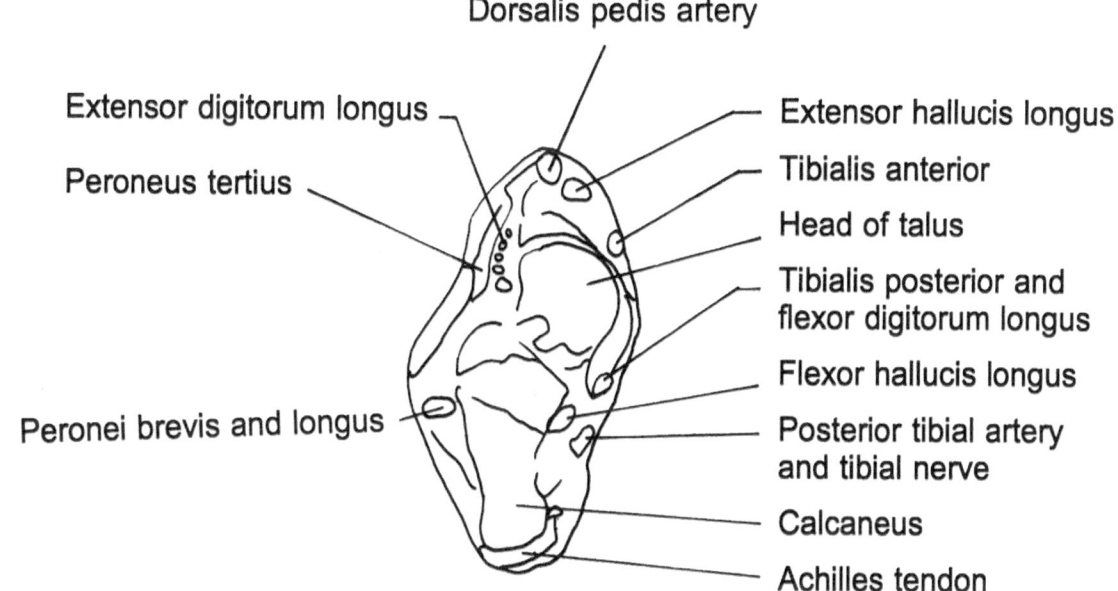

Dorsalis pedis artery

Extensor digitorum longus

Peroneus tertius

Peronei brevis and longus

Extensor hallucis longus

Tibialis anterior

Head of talus

Tibialis posterior and
flexor digitorum longus

Flexor hallucis longus

Posterior tibial artery
and tibial nerve

Calcaneus

Achilles tendon

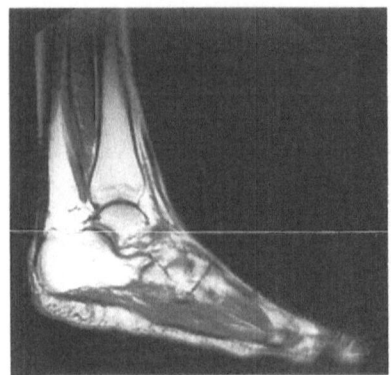
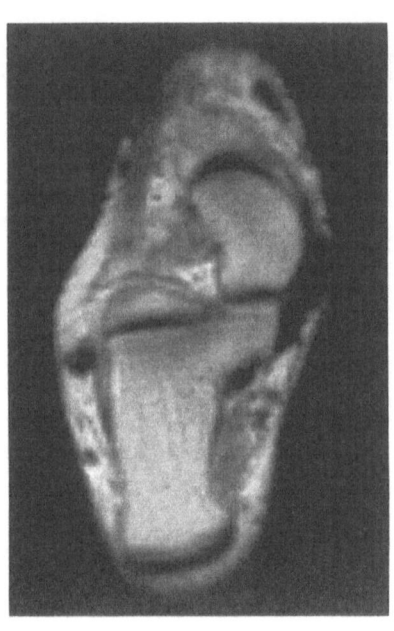

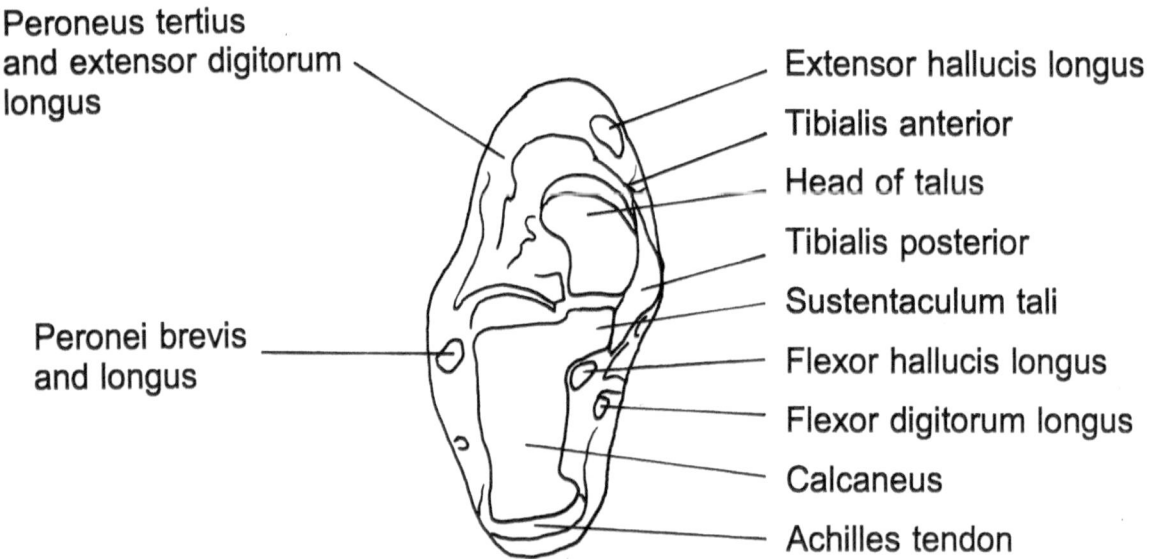

Peroneus tertius
and extensor digitorum
longus

Peronei brevis
and longus

Extensor hallucis longus

Tibialis anterior

Head of talus

Tibialis posterior

Sustentaculum tali

Flexor hallucis longus

Flexor digitorum longus

Calcaneus

Achilles tendon

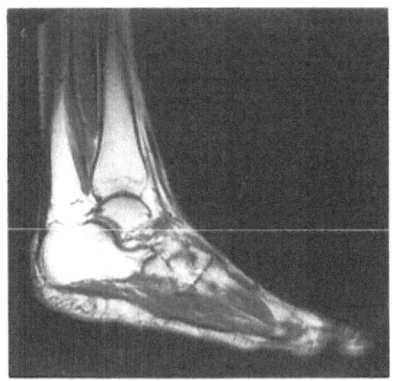

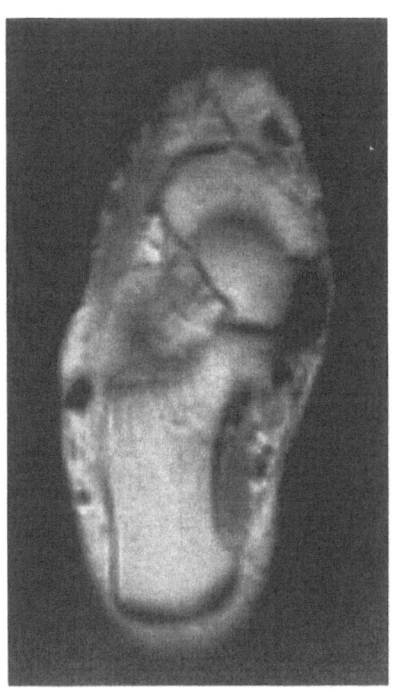

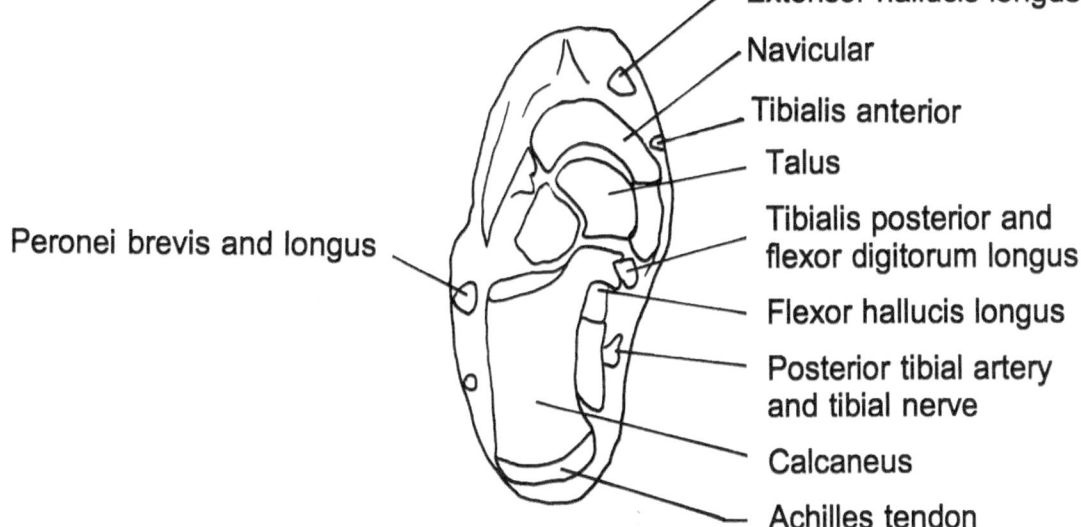

Peronei brevis and longus

Extensor hallucis longus

Navicular

Tibialis anterior

Talus

Tibialis posterior and flexor digitorum longus

Flexor hallucis longus

Posterior tibial artery and tibial nerve

Calcaneus

Achilles tendon

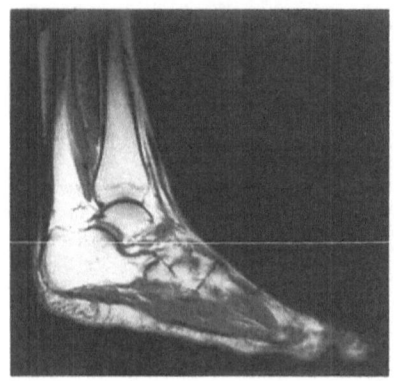

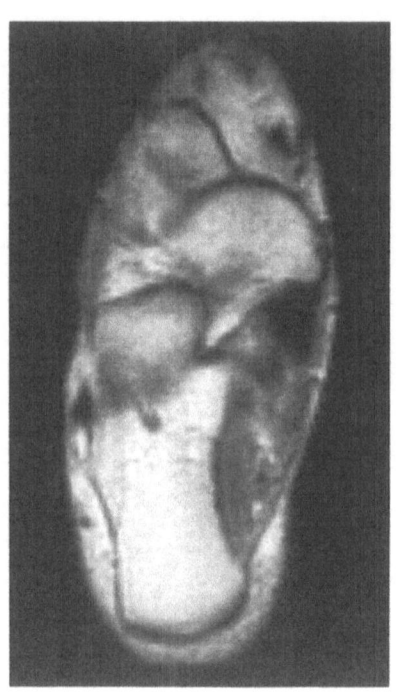

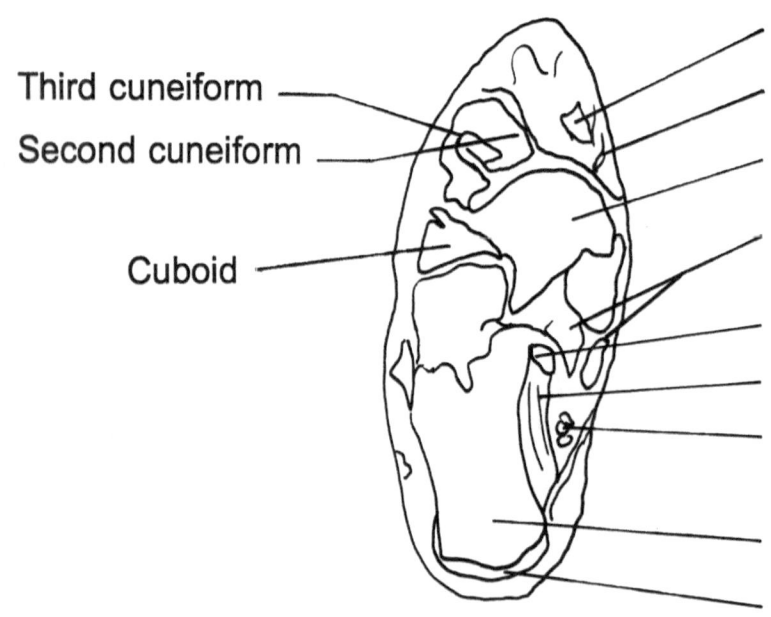

Third cuneiform

Second cuneiform

Cuboid

Extensor hallucis longus

Tibialis anterior

Talus

Tibialis posterior and
flexor digitorum longus

Flexor hallucis longus

Quadratus plantae

Posterior tibial artery
and tibial nerve

Calcaneus

Achilles tendon

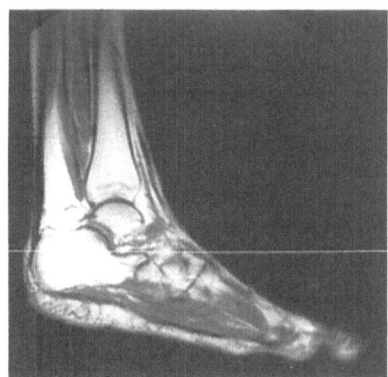

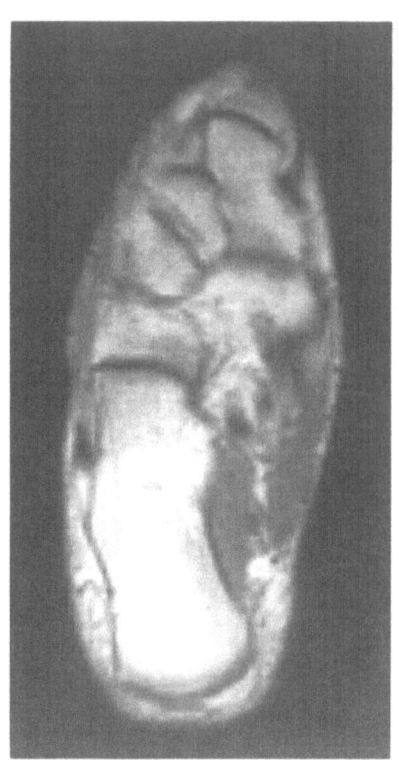

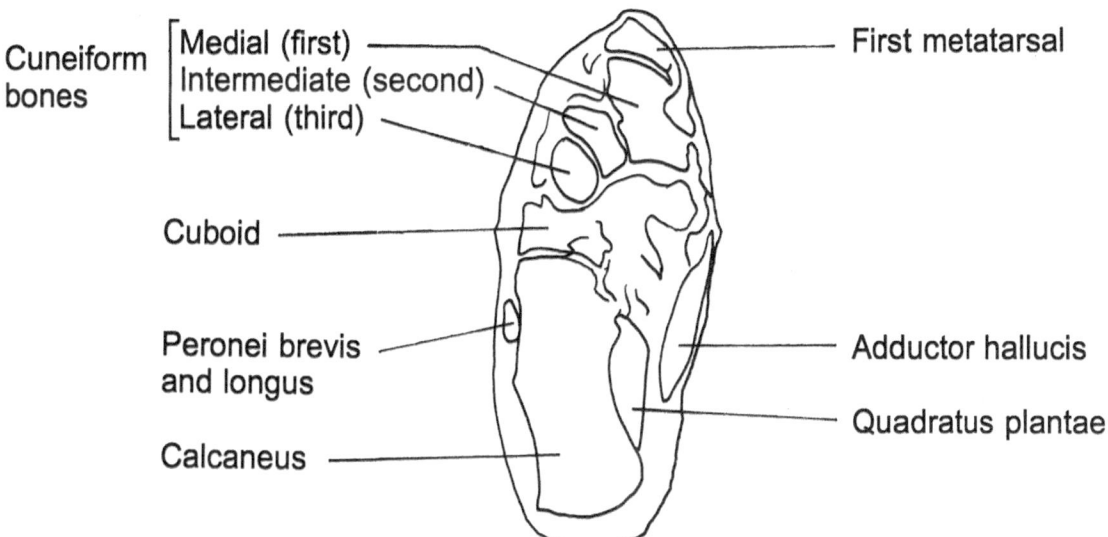

Cuneiform bones
- Medial (first) —————— First metatarsal
- Intermediate (second) —
- Lateral (third) ——————

Cuboid ——————

Peronei brevis and longus ——————

Calcaneus ——————

Adductor hallucis

Quadratus plantae

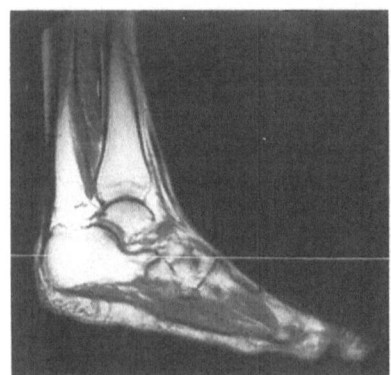

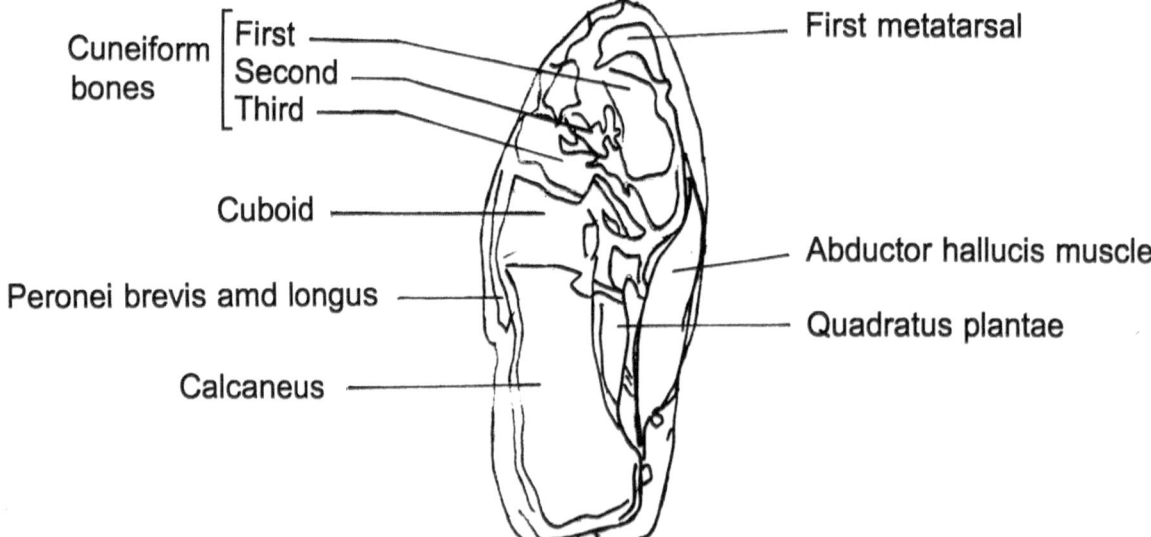

Cuneiform bones — First, Second, Third · First metatarsal · Cuboid · Peronei brevis amd longus · Calcaneus · Abductor hallucis muscle · Quadratus plantae

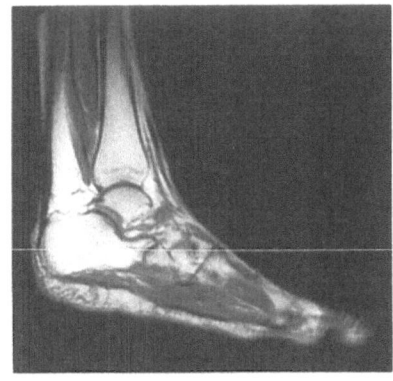
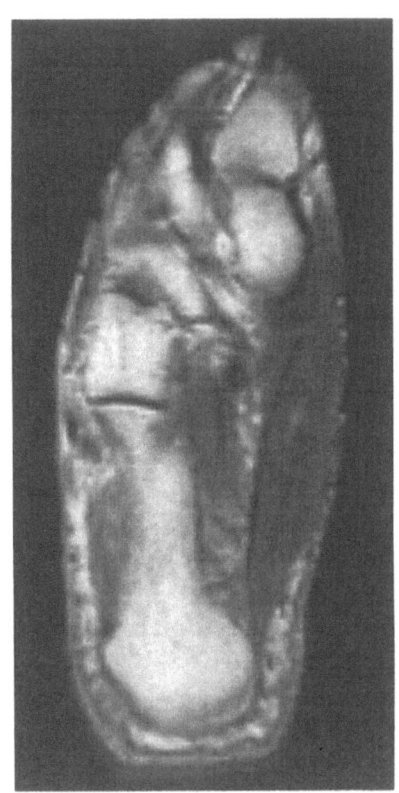

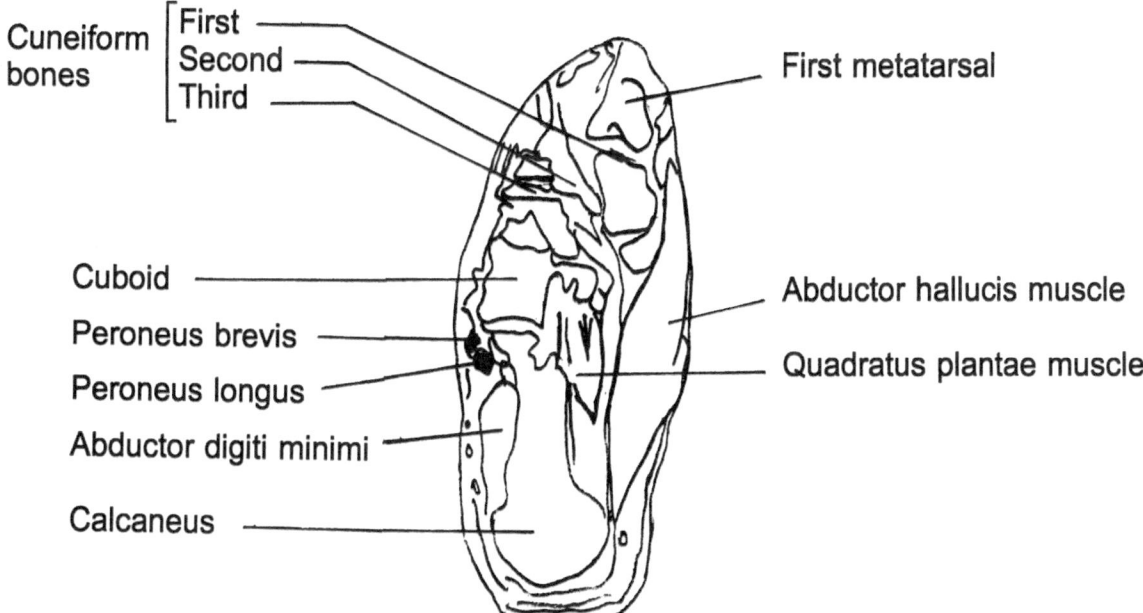

Cuneiform bones
- First
- Second
- Third

First metatarsal

Cuboid

Peroneus brevis

Peroneus longus

Abductor digiti minimi

Calcaneus

Abductor hallucis muscle

Quadratus plantae muscle

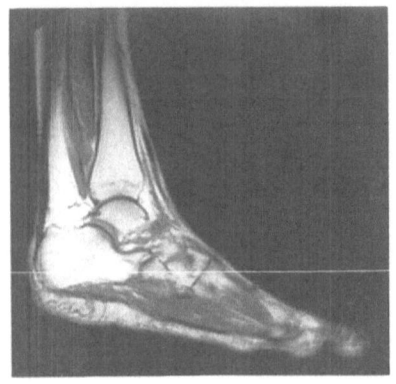

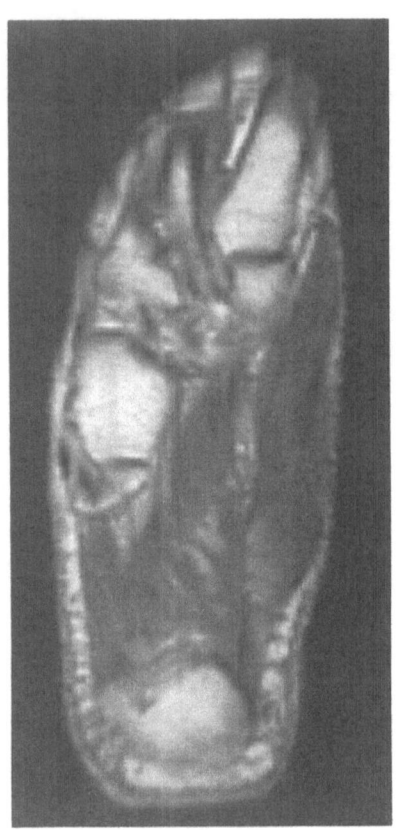

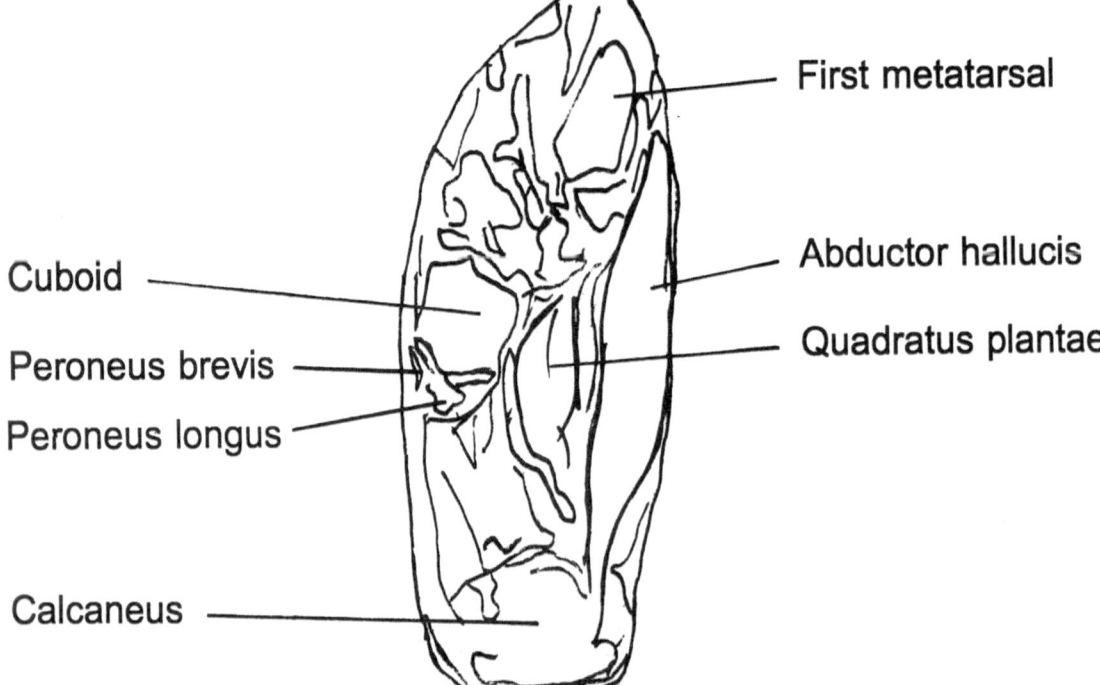

First metatarsal

Abductor hallucis

Quadratus plantae

Cuboid

Peroneus brevis

Peroneus longus

Calcaneus

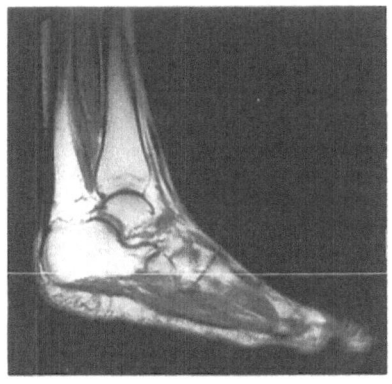

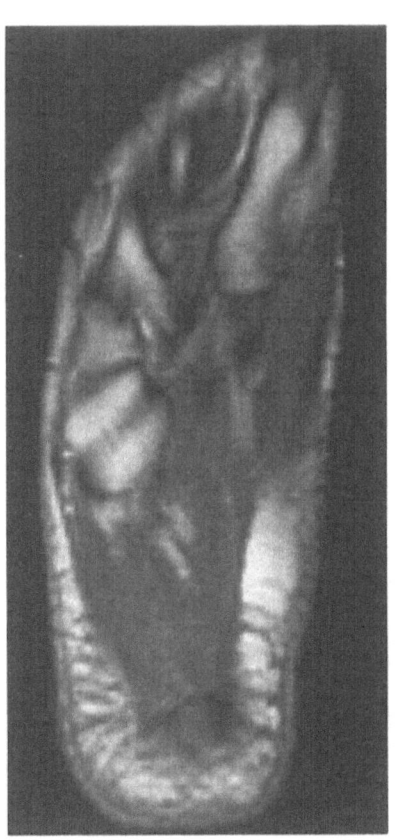

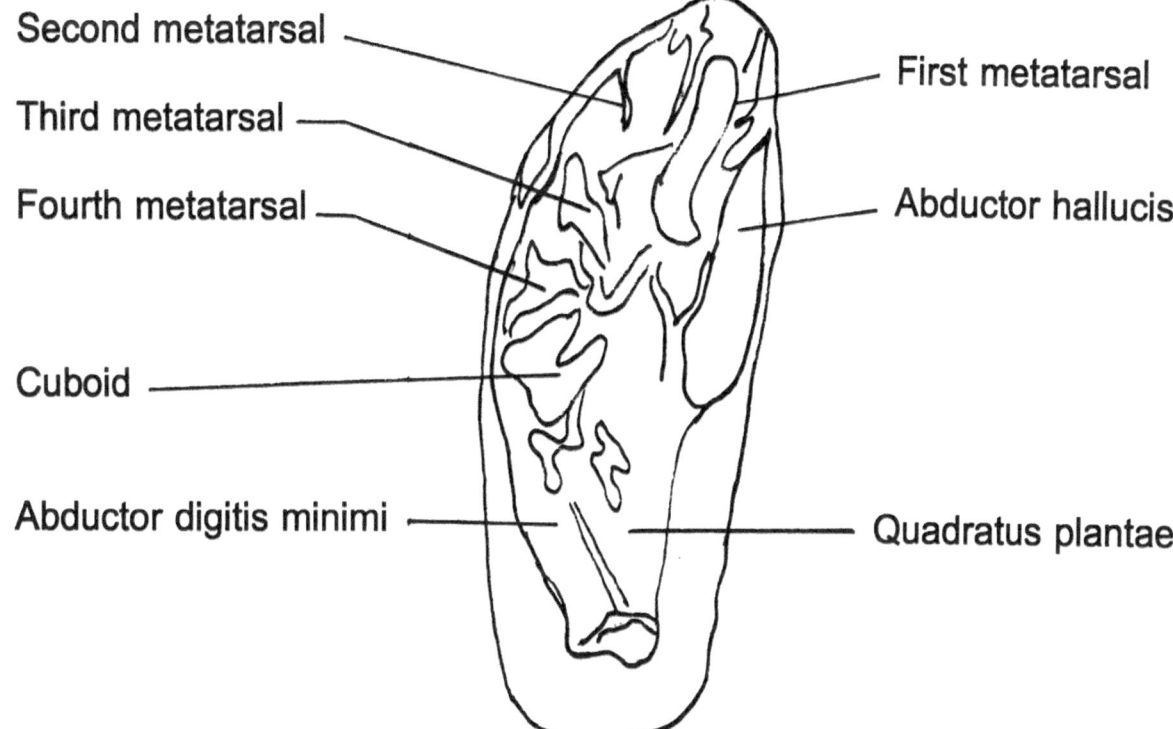

Second metatarsal

Third metatarsal

Fourth metatarsal

Cuboid

Abductor digitis minimi

First metatarsal

Abductor hallucis

Quadratus plantae

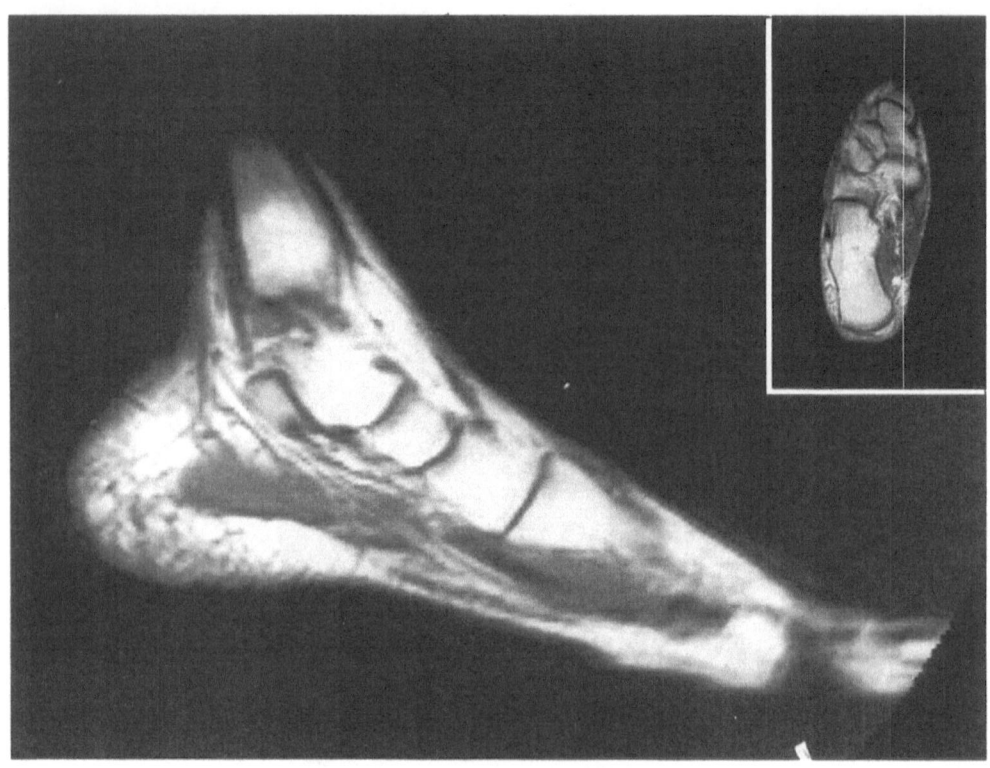

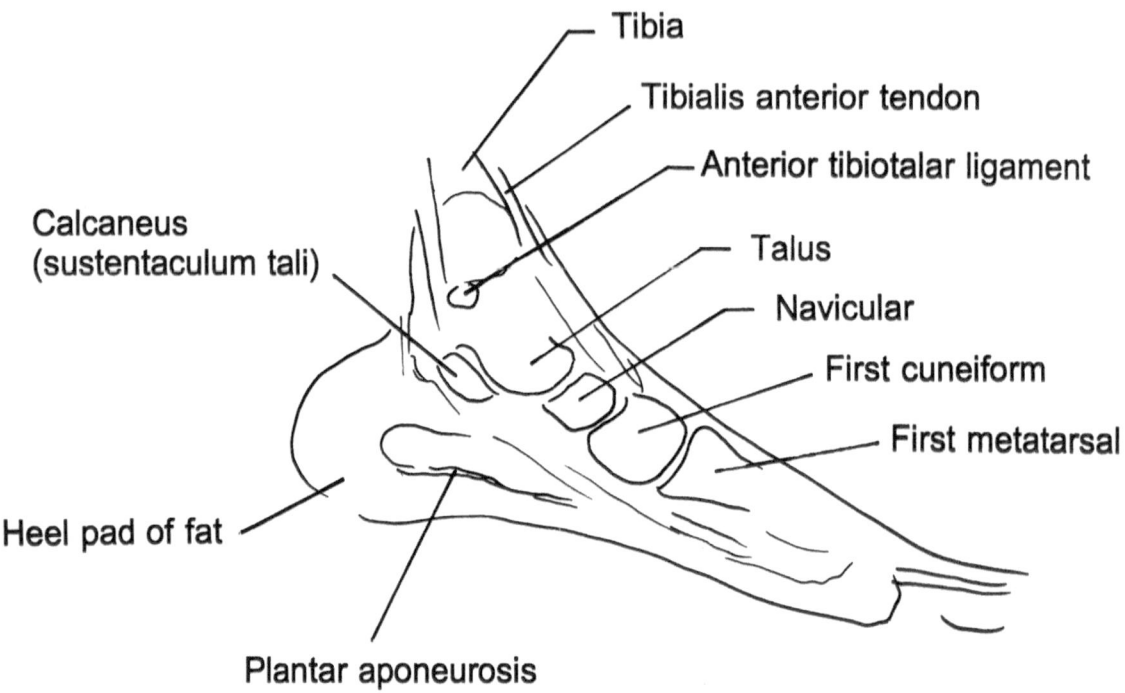

Tibia

Tibialis anterior tendon

Anterior tibiotalar ligament

Calcaneus
(sustentaculum tali)

Talus

Navicular

First cuneiform

First metatarsal

Heel pad of fat

Plantar aponeurosis

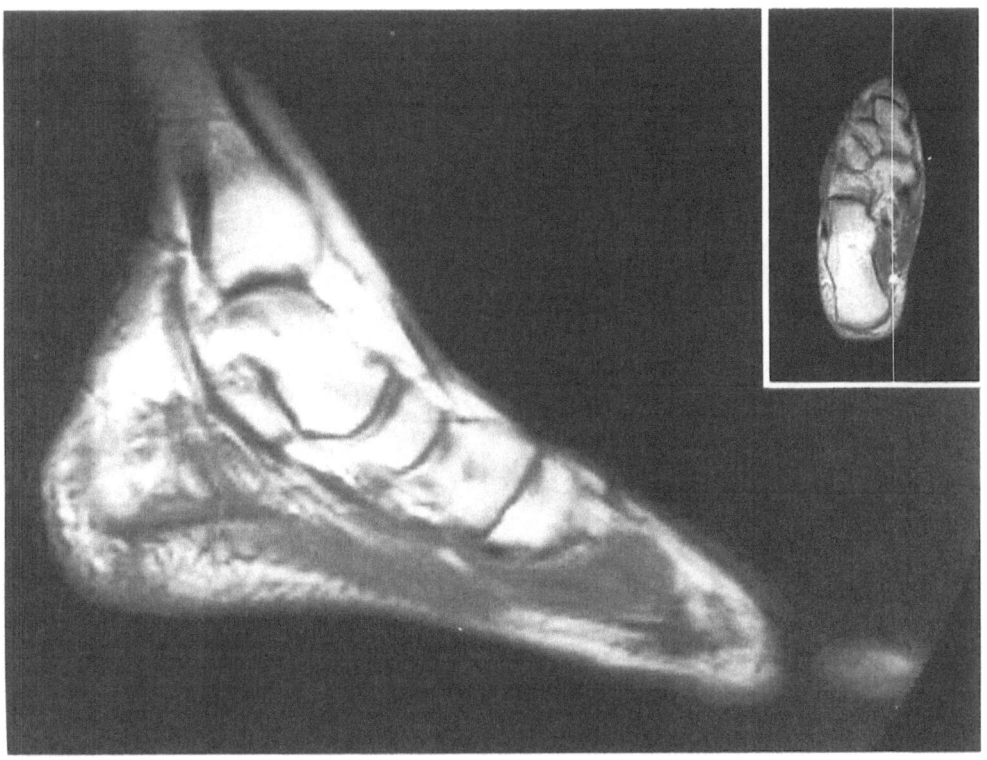

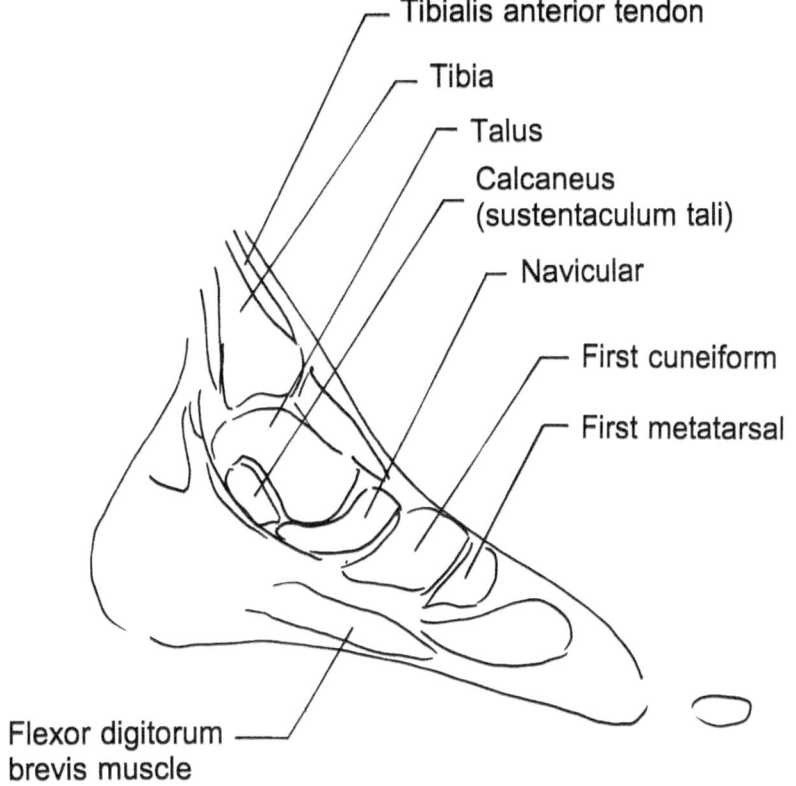

Tibialis anterior tendon

Tibia

Talus

Calcaneus
(sustentaculum tali)

Navicular

First cuneiform

First metatarsal

Flexor digitorum
brevis muscle

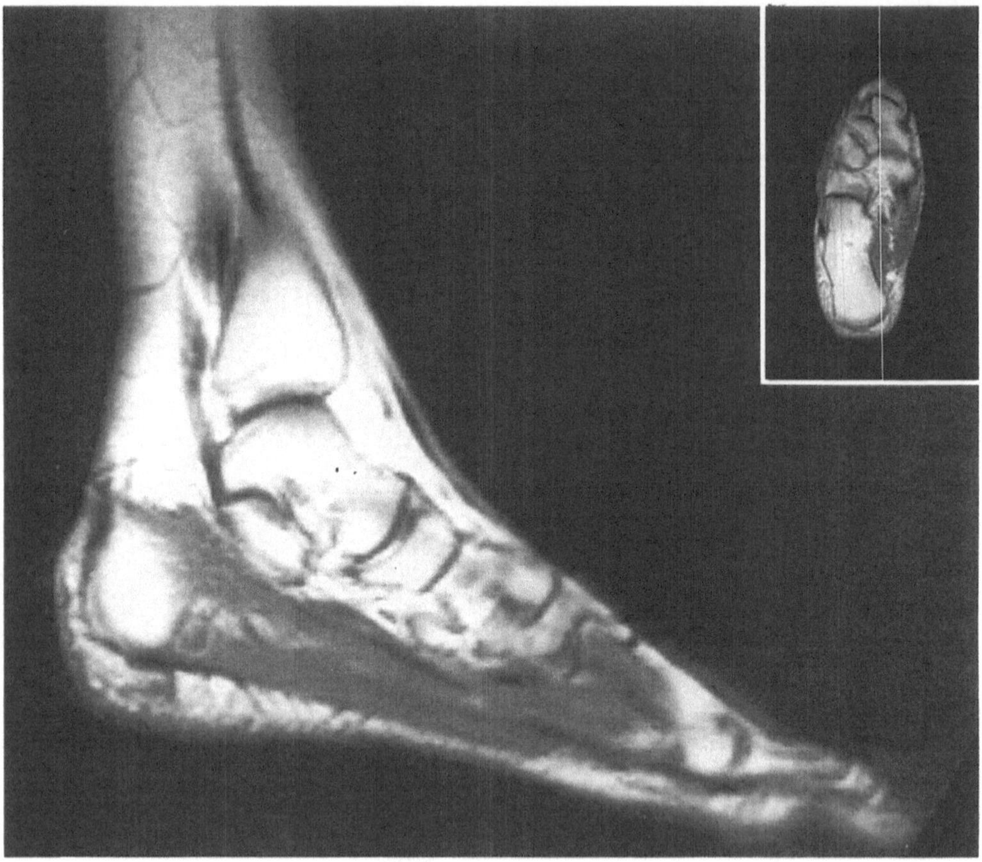

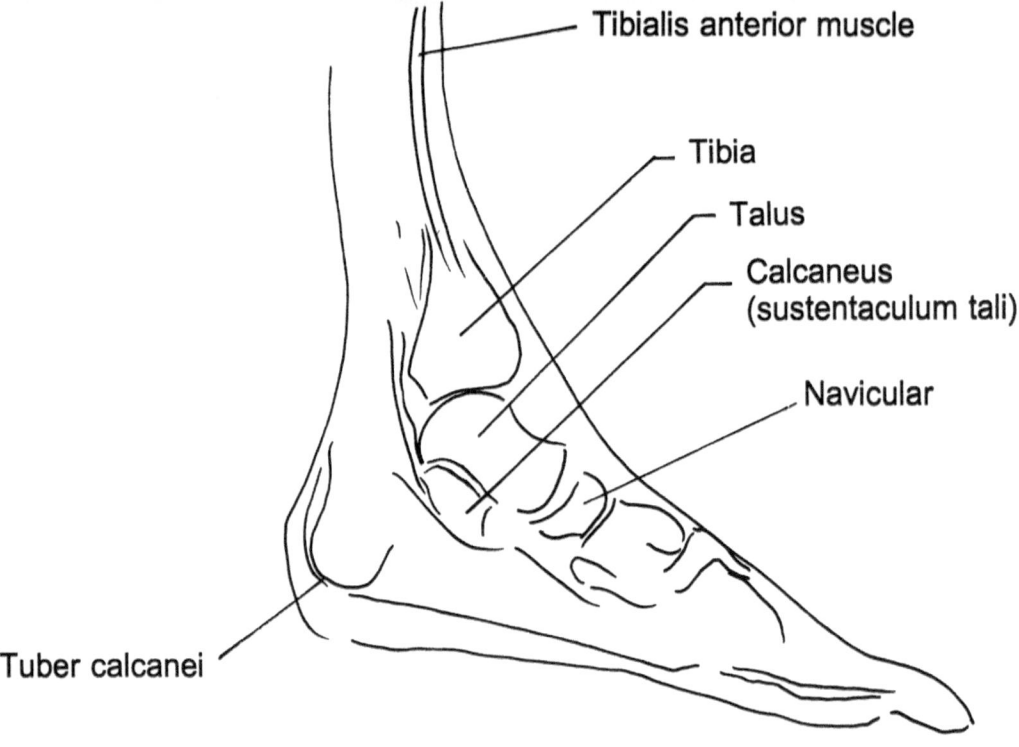

Tibialis anterior muscle

Tibia

Talus

Calcaneus
(sustentaculum tali)

Navicular

Tuber calcanei

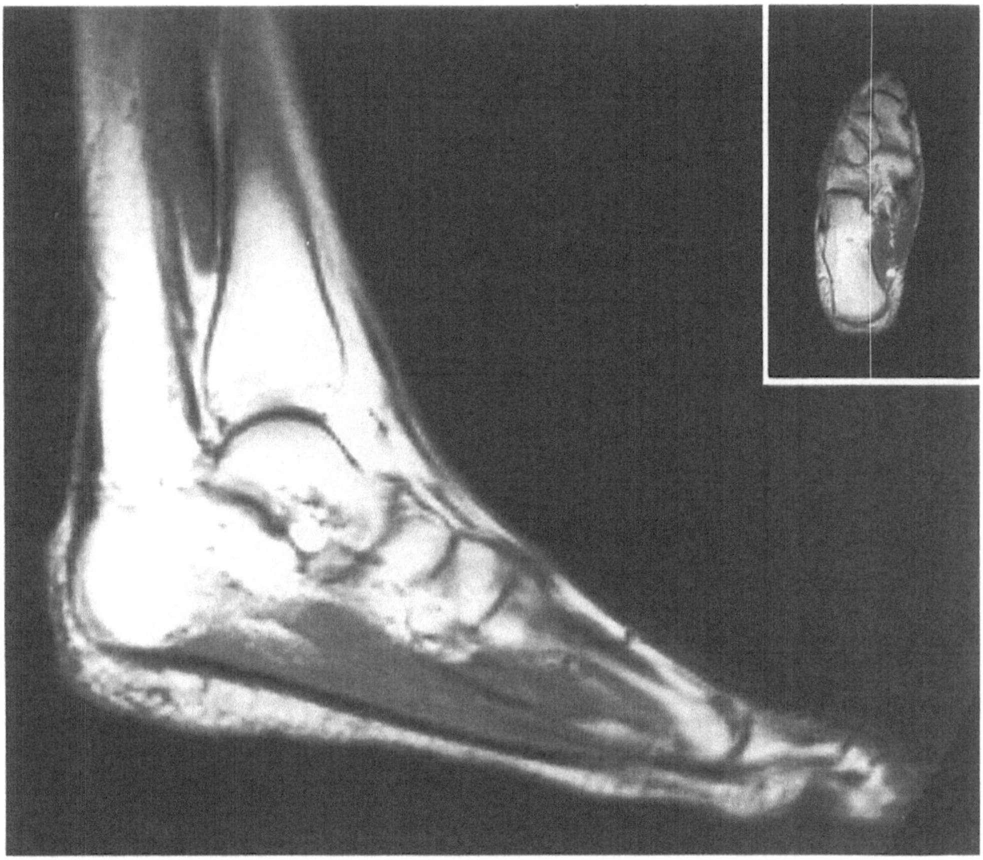

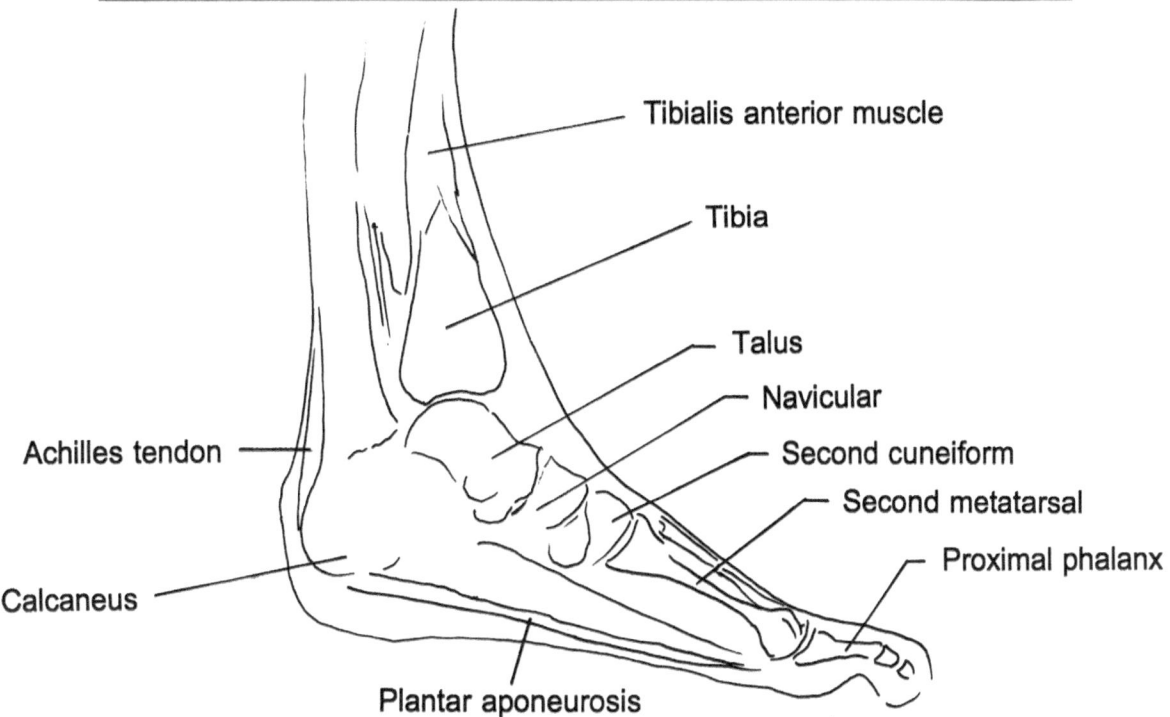

Tibialis anterior muscle

Tibia

Talus

Navicular

Second cuneiform

Second metatarsal

Proximal phalanx

Achilles tendon

Calcaneus

Plantar aponeurosis

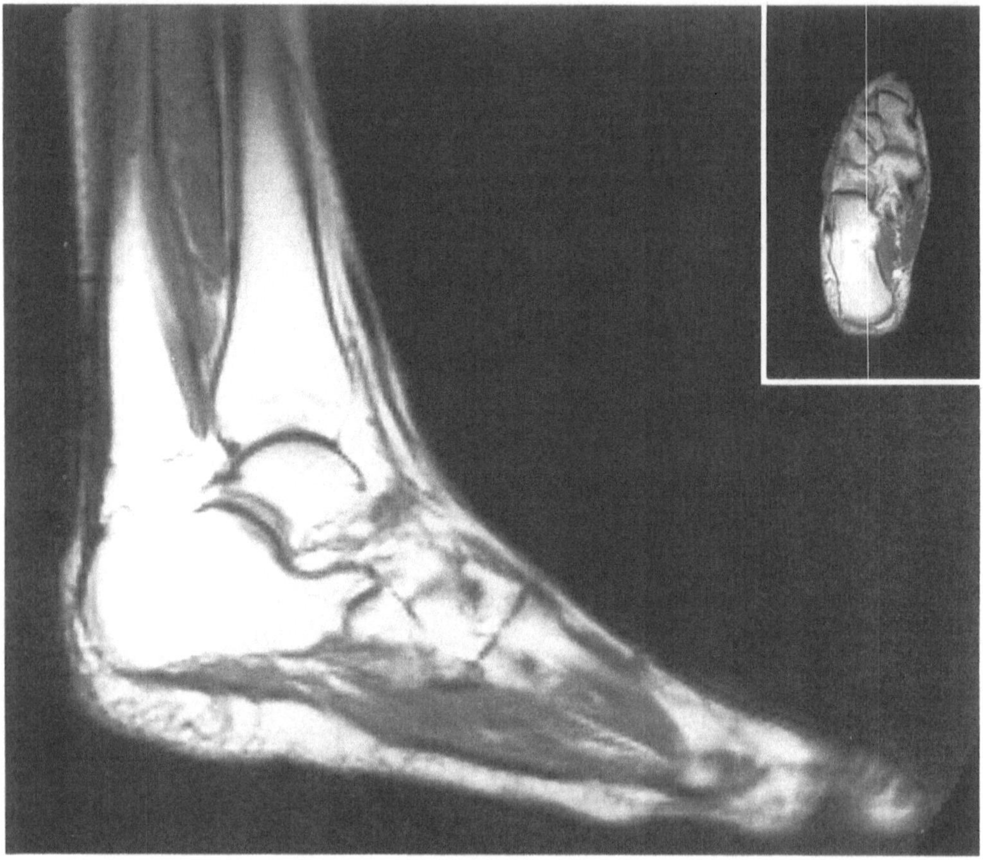

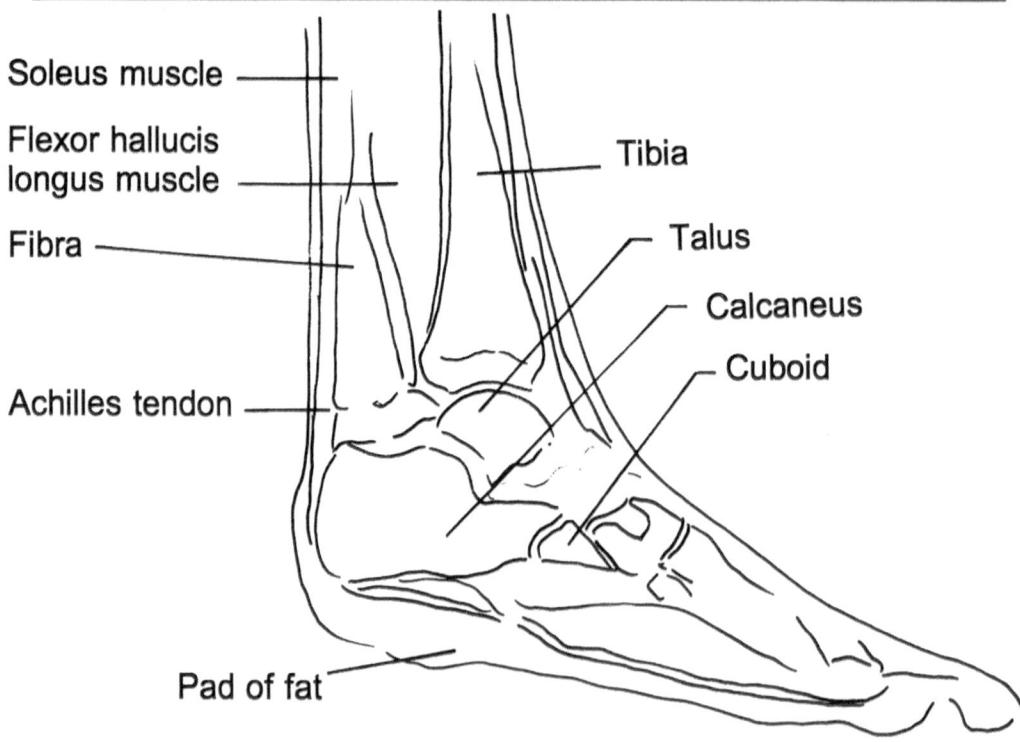

Soleus muscle

Flexor hallucis
longus muscle

Fibra

Achilles tendon

Tibia

Talus

Calcaneus

Cuboid

Pad of fat

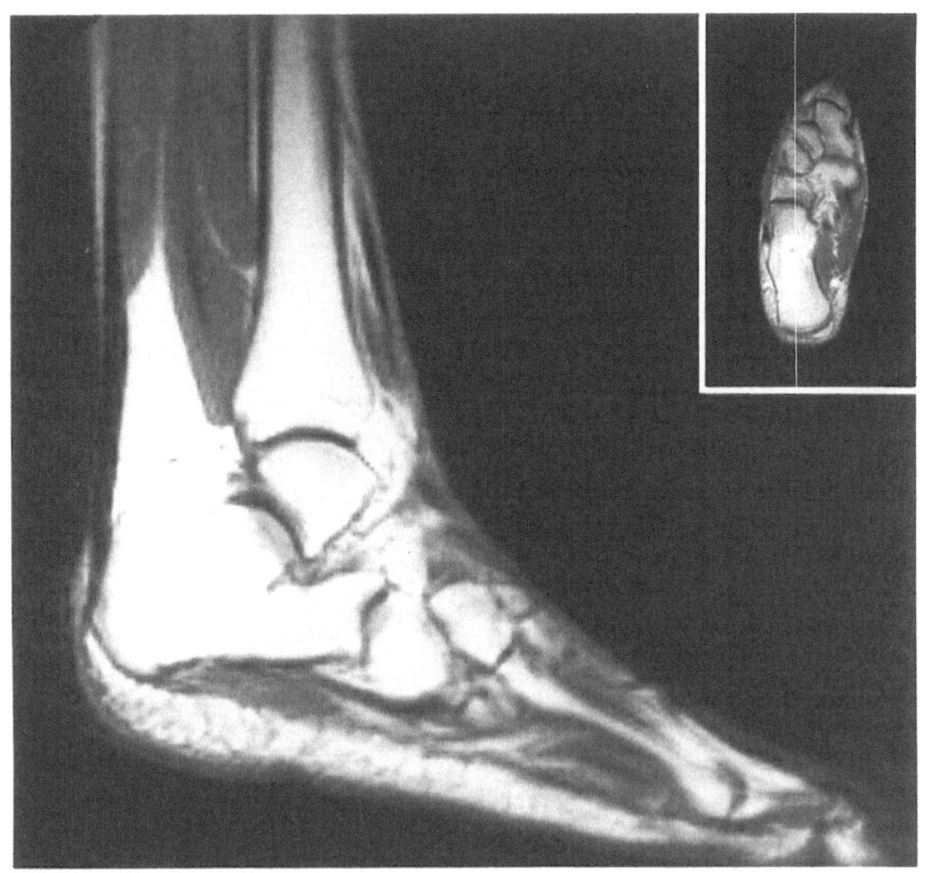

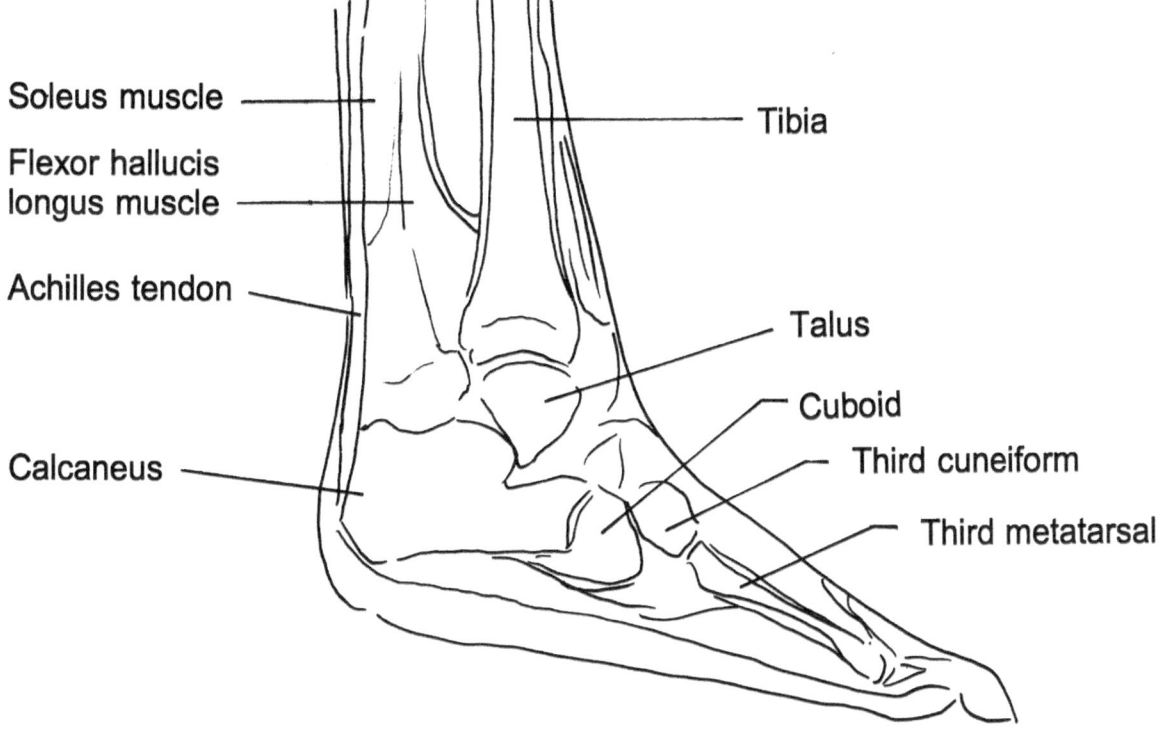

Soleus muscle

Flexor hallucis longus muscle

Achilles tendon

Calcaneus

Tibia

Talus

Cuboid

Third cuneiform

Third metatarsal

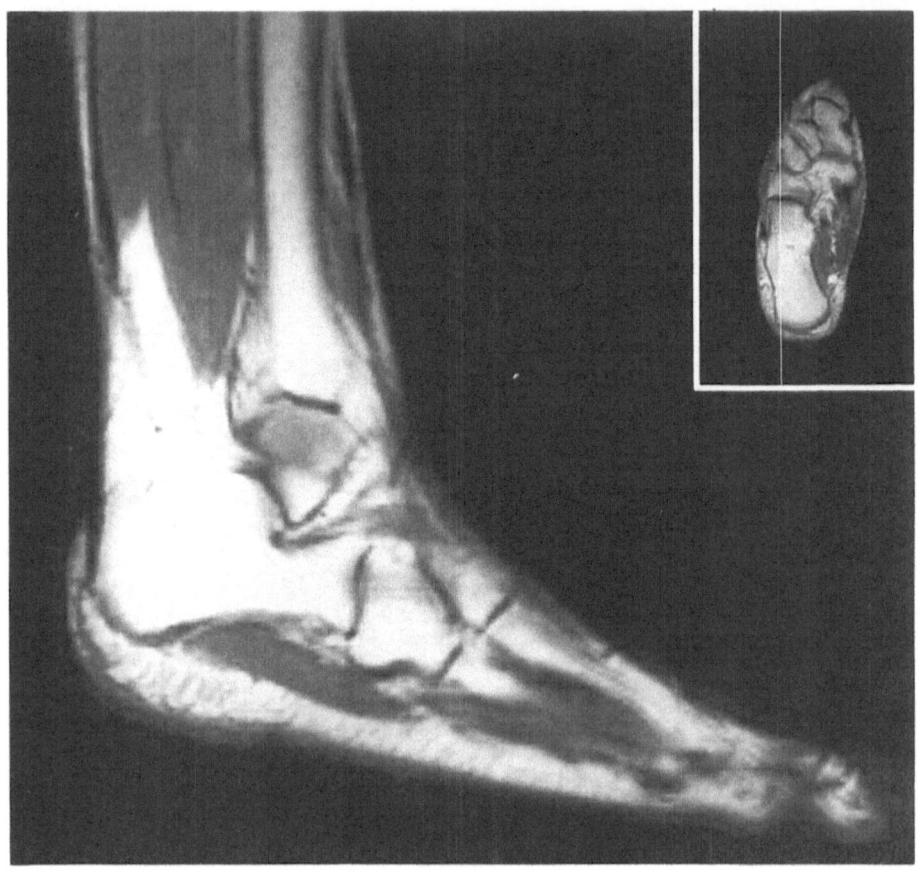

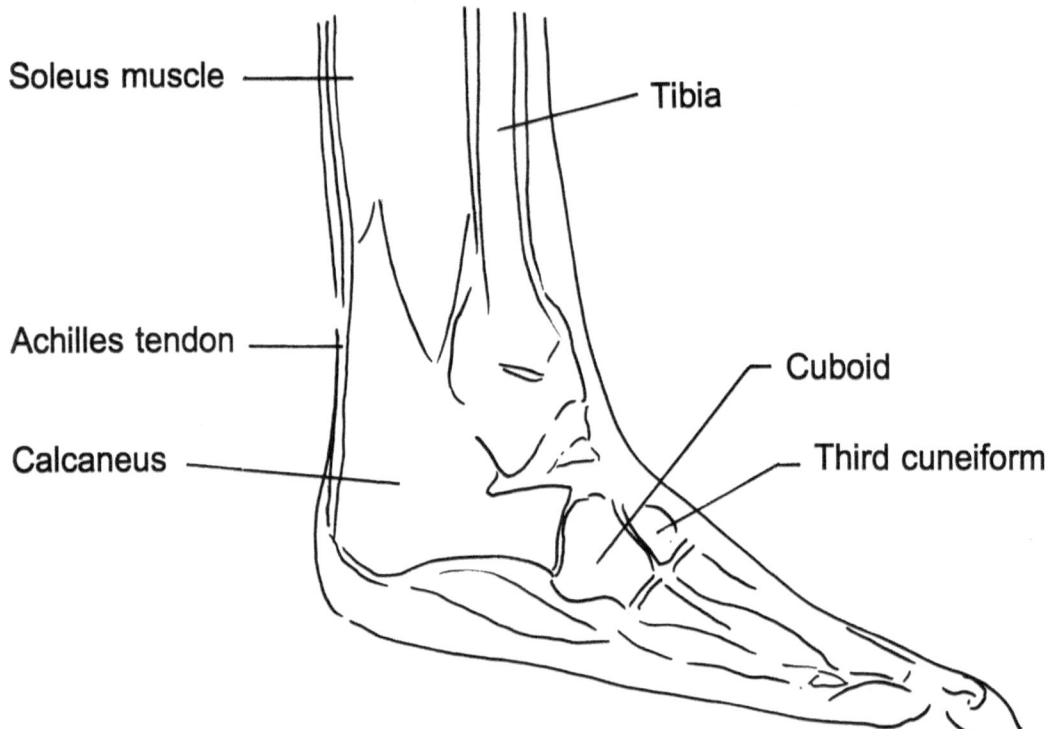

Soleus muscle

Tibia

Achilles tendon

Cuboid

Calcaneus

Third cuneiform

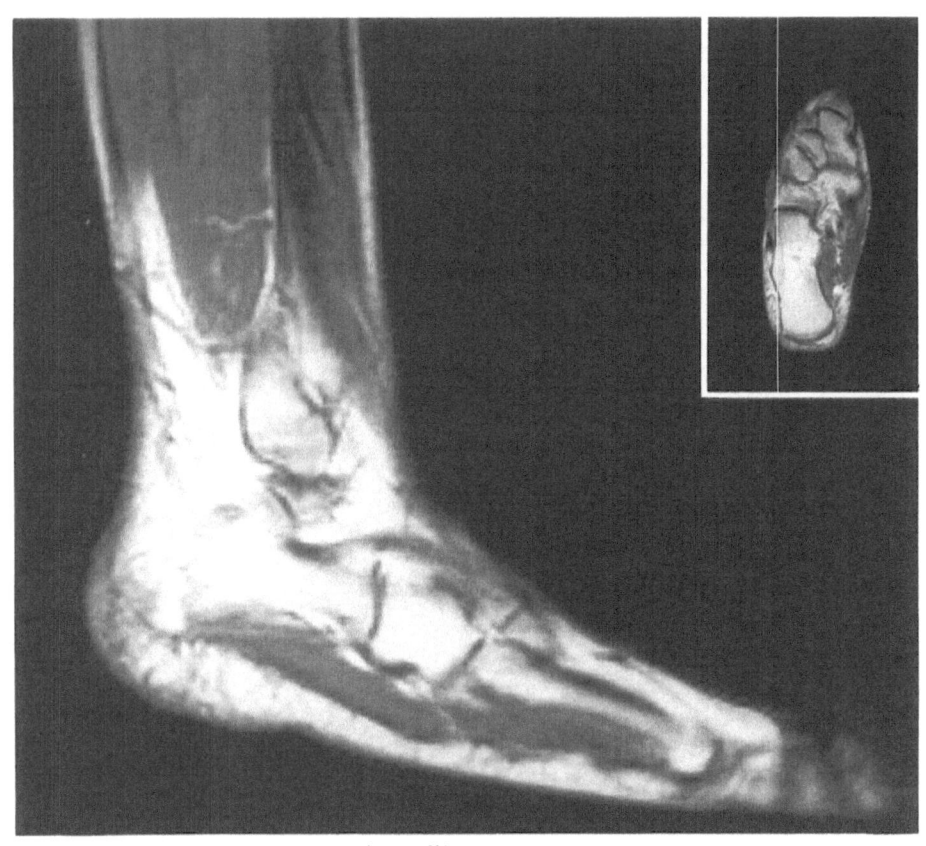

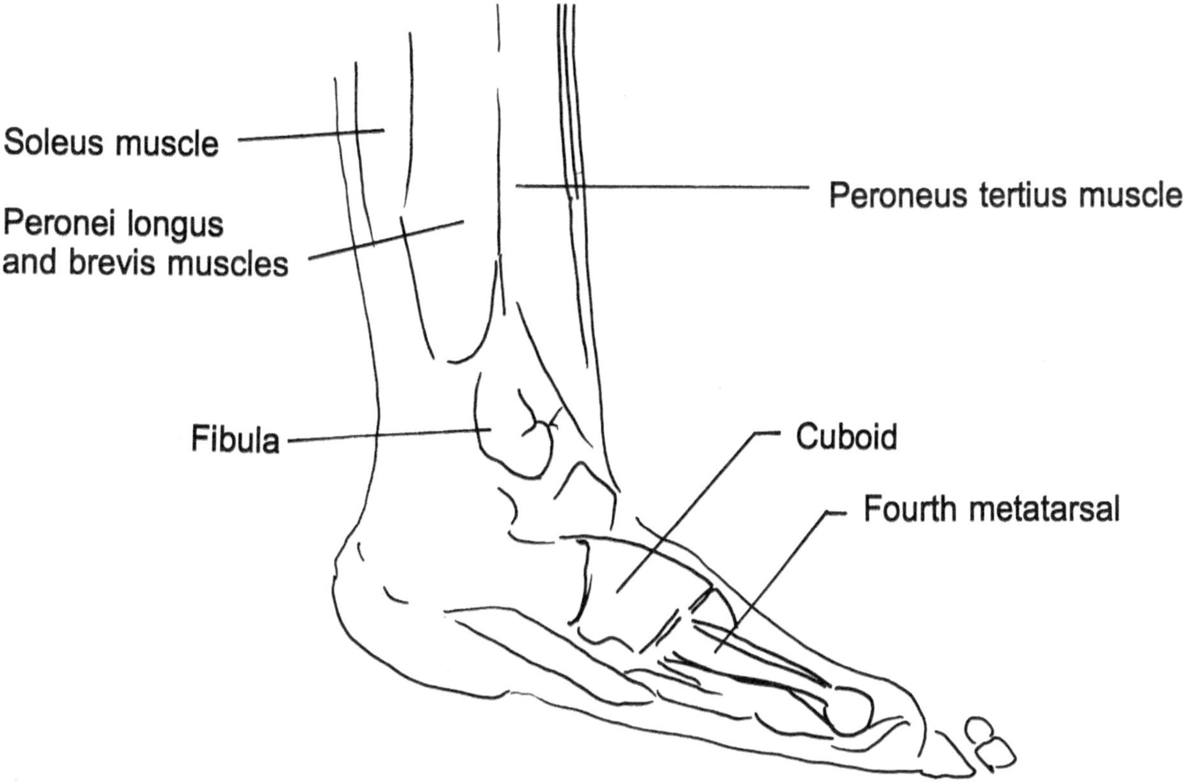

Soleus muscle

Peronei longus
and brevis muscles

Fibula

Peroneus tertius muscle

Cuboid

Fourth metatarsal

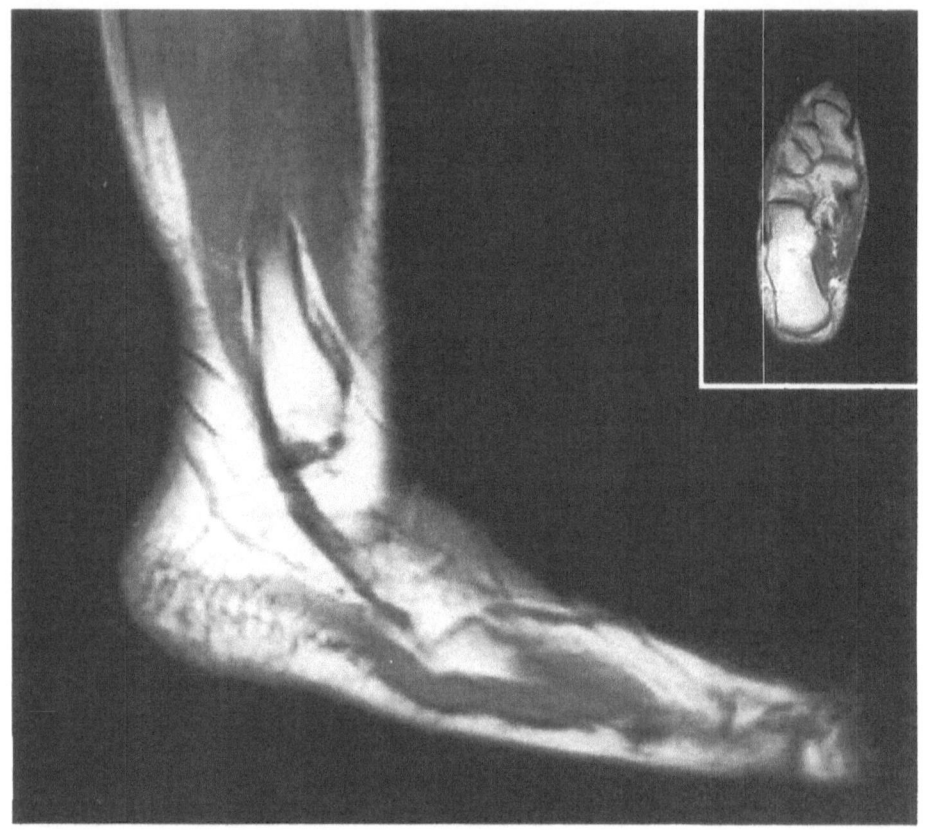

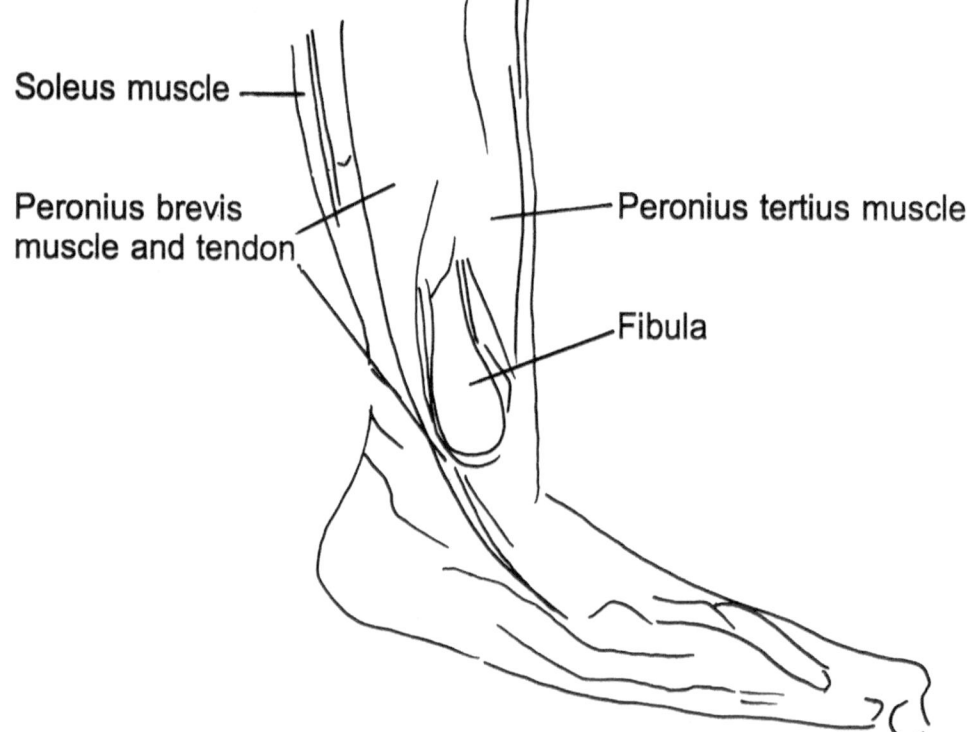

Soleus muscle

Peronius brevis
muscle and tendon

Peronius tertius muscle

Fibula